Stefan Reichl

Inverse Dynamics and Trajectory Tracking

Stefan Reichl

Inverse Dynamics and Trajectory Tracking

of Underactuated Multibody Systems

Südwestdeutscher Verlag für Hochschulschriften

Impressum/Imprint (nur für Deutschland/only for Germany)
Bibliografische Information der Deutschen Nationalbibliothek: Die Deutsche Nationalbibliothek verzeichnet diese Publikation in der Deutschen Nationalbibliografie; detaillierte bibliografische Daten sind im Internet über http://dnb.d-nb.de abrufbar.
Alle in diesem Buch genannten Marken und Produktnamen unterliegen warenzeichen-, marken- oder patentrechtlichem Schutz bzw. sind Warenzeichen oder eingetragene Warenzeichen der jeweiligen Inhaber. Die Wiedergabe von Marken, Produktnamen, Gebrauchsnamen, Handelsnamen, Warenbezeichnungen u.s.w. in diesem Werk berechtigt auch ohne besondere Kennzeichnung nicht zu der Annahme, dass solche Namen im Sinne der Warenzeichen- und Markenschutzgesetzgebung als frei zu betrachten wären und daher von jedermann benutzt werden dürften.

Verlag: Südwestdeutscher Verlag für Hochschulschriften GmbH & Co. KG
Dudweiler Landstr. 99, 66123 Saarbrücken, Deutschland
Telefon +49 681 37 20 271-1, Telefax +49 681 37 20 271-0
Email: info@svh-verlag.de

Approved by: Wien, TU, Diss., 2011

Herstellung in Deutschland:
Schaltungsdienst Lange o.H.G., Berlin
Books on Demand GmbH, Norderstedt
Reha GmbH, Saarbrücken
Amazon Distribution GmbH, Leipzig
ISBN: 978-3-8381-2169-7

Imprint (only for USA, GB)
Bibliographic information published by the Deutsche Nationalbibliothek: The Deutsche Nationalbibliothek lists this publication in the Deutsche Nationalbibliografie; detailed bibliographic data are available in the Internet at http://dnb.d-nb.de.
Any brand names and product names mentioned in this book are subject to trademark, brand or patent protection and are trademarks or registered trademarks of their respective holders. The use of brand names, product names, common names, trade names, product descriptions etc. even without a particular marking in this works is in no way to be construed to mean that such names may be regarded as unrestricted in respect of trademark and brand protection legislation and could thus be used by anyone.

Publisher: Südwestdeutscher Verlag für Hochschulschriften GmbH & Co. KG
Dudweiler Landstr. 99, 66123 Saarbrücken, Germany
Phone +49 681 37 20 271-1, Fax +49 681 37 20 271-0
Email: info@svh-verlag.de

Printed in the U.S.A.
Printed in the U.K. by (see last page)
ISBN: 978-3-8381-2169-7

Copyright © 2011 by the author and Südwestdeutscher Verlag für Hochschulschriften GmbH & Co. KG and licensors
All rights reserved. Saarbrücken 2011

Preface

This dissertation originates from the doctoral program at the Institute of Mechanics and Mechatronics at the Vienna University of Technology and my time as a research associate at the Upper Austria University of Applied Sciences between 2007 and 2011. I would like to thank my supervisor Ao.Univ.Prof. Dipl.-Ing. Dr.techn. Alois Steindl for giving me the opportunity to this program and for supporting me the whole time during my research. I also express my gratitude to the second reviewer Univ.Prof. Dipl.-Ing. Dr.techn. Stefan Jakubek. Special thanks go to Priv.Doz. Dipl.-Ing. Dr.techn. Wolfgang Steiner. He inspired me to this work and helped me with his knowledge and several fruitful discussions. I would like to express my deep gratitude to Dr. Michael Steinbatz who supported me in modeling and simulation techniques of multibody systems and finite element models. Furthermore, my thanks go to Dr. Thomas Reiter, Priv.Doz. Dr. Martin Egger, Priv.Doz. Dr. Klaus Schiefermayr and Dr. Gernot Grabmair for discussions that advanced my work.

From the company Pöttinger I would like to thank Dr. Manfred Hofer, Wolfgang Schimpl, Daniela Weberndorfer, Robert Hehenberger and Gerald Mittergeber. The cooperation proceeded well and especially in the work concerning virtual iteration the communication between simulation and measurement team turned out to be very instructive and important. Dr. Gerald Wimmer from the company Siemens VAI receives my gratitude for discussions regarding to the AOD steel converter.

I thank all my office colleagues for their encouragement, even if it is not related to mechanics. Many thanks to Wolfgang Hauer, Günther Mayr, Rainer Widmann, Stefan Mairhofer, Stephan Hutterer, Christian Kneidinger, Josef Pühringer, Gerhard Hanis, Dr. Michael Olbrich, Dr. Christian Forsich and Dr. Roman Froschauer. We had a pleasant working atmosphere and I really enjoyed this time.
Most of all I would like to thank my family and my friends for their unconditional moral support. They always encouraged me in personal and occupational matters and helped me to focus on the real important issues in life. Without them I would not be where I am.

This project was supported by the program Regionale Wettbewerbsfähigkeit OÖ 2010-2013, which is financed by the European Regional Development Fund and the Government of Upper Austria. Additional financial support was acquired by the industrial project "AOD converter vibrations" with the company Siemens VAI. Further funding was received by the programs "SimSÜ" (Simulation geregelter Schlechtweg-Überfahrten von landtechnischen Gespannen) and "vLDE" (virtuelle Lastdatenermittlung) which were financed by the FFG (Forschungsförderungsgesellschaft).

Abstract

Inverse dynamics problems arise in several mechanical systems. The aim is to calculate the inputs of a system in order that the outputs are identical to predefined or measured target signals. The motivation for inverse methods is related to practical applications in robotics, cranes or test rigs in the automotive and agricultural industry. A multibody system is called underactuated if the number of control inputs is less than the number of degrees of freedom. The control of underactuated systems is much more challenging compared to fully actuated systems.

The thesis considers four mathematical methods regarding to inverse problems in underactuated multibody systems. The method of virtual iteration is based on a linearization of the nonlinear system and an inverse computation of the excitations in the frequency domain. The algorithm is suitable for large multibody systems and finite element models, which are nearly linear.

The second method formulates the equations of motion as differential-algebraic equations and introduces so called control or servo constraints. This results in a system of high index, which can be solved by appropriate numerical algorithms.

The inverse problem can also be formulated as an optimal control problem. The basis is a cost functional, which includes the system outputs and the targets. The goal is to minimize this performance measure. Here it is distinguished between indirect and direct methods. In indirect optimal control the necessary optimality conditions are derived and the resulting boundary value problem has to be solved. Direct methods discretize the system and reformulate the optimal control problem to static optimization problems.

The fourth method under consideration is a flatness-based trajectory tracking control. In specific systems the state and input variables can be parameterized by the outputs and their time derivatives up to a certain order. Such systems are called differentially flat and the outputs are known as flat outputs.

The considered methods are applied to academic and industrial examples. A nonlinear oscillator, an underactuated planar crane and an underactuated rotary crane are studied. Finite element models and hybrid multibody systems of a steel converter, a trailed cultivator and a plough are representative examples of industrial problems regarding to inverse dynamics. The different methods are compared with respect to their applicability and efficiency.

Keywords: underactuated multibody system, inverse dynamics, virtual iteration, optimal control, control constraints, differentially flat system

Kurzfassung

Inverse dynamische Probleme treten in zahlreichen mechanischen Systemen auf. Das Ziel ist die Berechnung von Eingangsvariablen des Systems, sodass die Ausgänge identisch zu vordefinierten oder gemessenen Targetsignalen sind. Die Motivation zu inversen Methoden stammt aus praktischen Anwendungen in der Robotik, bei Kränen oder Prüfständen im Automobilbereich und der Landmaschinenindustrie. Ein Mehrkörpersystem wird als unteraktuiert bezeichnet, wenn die Anzahl der Steuereingänge geringer ist als die Anzahl an Freiheitsgraden. Die Regelung von unteraktuierten Systemen ist um Größenordnungen schwieriger als von voll aktuierten Systemen.
Diese Dissertation untersucht vier mathematische Methoden im Bezug auf inverse Probleme bei unteraktuierten Mehrkörpersystemen. Die Methode der virtuellen Iteration basiert auf einer Linearisierung des nichtlinearen Systems und einer inversen Berechnung der Anregungen im Frequenzbereich. Der Algorithmus ist für große Mehrkörpersysteme und Finite Elemente Modelle, welche beinahe linear sind, geeignet.
Die zweite Methode formuliert die Bewegungsgleichungen als differential-algebraische Gleichungen und führt sogenannte Steuerungszwangsbedingungen ein. Dies führt zu einem System von hohem Index, welches durch geeignete numerische Algorithmen gelöst wird.
Das inverse Problem kann auch als Optimalsteuerungsproblem formuliert werden. Die Basis ist ein Kostenfunktional, welches Systemausgänge und Targets inkludiert. Das Ziel liegt in der Minimierung dieses Funktionals. Hierbei wird zwischen indirekten und direkten Methoden unterschieden. In einer indirekten Optimalsteuerung werden die notwendigen Optimalitätsbedingungen hergeleitet und das daraus resultierende Randwertproblem gelöst. Direkte Methoden diskretisieren das System und formen das Optimalsteuerungs-problem in statische Optimierungsprobleme um.
Die vierte untersuchte Methode ist eine flachheitsbasierte Trajektorienfolgeregelung. In bestimmten Systemen können die Zustands- und Eingangsvariablen durch die Ausgänge und ihre zeitlichen Ableitungen bis zu einem bestimmten Grad parametriert werden. Diese Systeme werden als differentiell flach und die Ausgänge als flache Ausgänge bezeichnet.

Die jeweiligen Methoden werden auf akademische und industrielle Problemstellungen angewandt. Ein nichtlinearer Massenschwinger, ein unteraktuierter ebener Kran und ein rotierender Kran werden untersucht. Finite Elemente Modelle sowie hybride Mehrkörper-systeme eines Stahlkonverters, eines Grubbers und eines Pfluges stellen repräsentative Beispiele von industriellen Problemen im Bezug auf inverse Dynamik dar. Die verschiedenen Methoden werden hinsichtlich Anwendbarkeit und Effizienz verglichen.

Schlagwörter: unteraktuiertes Mehrkörpersystem, inverse Dynamik, virtuelle Iteration, optimale Steuerung, Steuerungszwangsbedingung, differenziell flaches System

Contents

1 **Introduction**   1
   1.1 Overview 1
       1.1.1 Multibody Systems 2
       1.1.2 Brief Historical Review 3
       1.1.3 State of the Art 4
       1.1.4 Types of Problems 8
   1.2 Aim of the Study 10
   1.3 Outline of the Present Work 14
   1.4 Scope of the Present Work 16

2 **Dynamics of Multibody Systems**   17
   2.1 Kinematics of Rigid Bodies 17
       2.1.1 Parametrization of the Rotation Matrix 18
       2.1.2 Velocity of a Rigid Body 21
       2.1.3 Formulation of the Angular Velocity 22
   2.2 Newton-Euler equations 23
   2.3 Types of Constraints 24
   2.4 Variational Principles 25
       2.4.1 Principles of d'Alembert, Jourdain and Gauß 25
       2.4.2 Hamilton's Principle 27
   2.5 Lagrange's Equations of the First Kind, Descriptor Form . . . 28
   2.6 Lagrange's Equations of the Second Kind 30
   2.7 Kinetic and Potential Energy of Rigid Bodies 30
   2.8 Numerical Solution of DAEs 32
       2.8.1 Index of the Descriptor Form 32
       2.8.2 Index Reduction 33
       2.8.3 Drift Problem of the Index 1 System 34
       2.8.4 Stabilization Methods 34
       2.8.5 Index 3 Solver 37
   2.9 Flexible Multibody Systems 38
       2.9.1 Overview 38
       2.9.2 Determination of Shape Functions 41
       2.9.3 Guyan Reduction (Static Reduction) 41
       2.9.4 Modal Reduction 42
       2.9.5 Craig-Bampton Reduction 43

| | | 2.9.6 | Component Mode Synthesis (CMS) . . . . | 45 |

3 Virtual Iteration 47
- 3.1 System Identification 47
  - 3.1.1 System Identification via Noise Excitations 48
  - 3.1.2 Linearization of the Nonlinear Equations of Motion 51
  - 3.1.3 Calculation of the Transfer Matrix 54
- 3.2 Target Simulation 56
  - 3.2.1 Virtual Iteration Algorithm 58
  - 3.2.2 The Inverse of the Transfer Matrix 59

4 DAE Approach with Control Constraints 63
- 4.1 Formulation of the DAEs 64
- 4.2 Dependent versus Independent Coordinates 65
- 4.3 Index Reduction Procedure 66
- 4.4 Numerical Solution 68

5 Optimization and Optimal Control 71
- 5.1 Static Optimization Problems 72
  - 5.1.1 Formulation of the Problem 72
  - 5.1.2 Optimality Conditions of Unconstrained Optimization Problems 72
- 5.2 Numerical Methods for Unconstrained Static Problems 73
  - 5.2.1 Steepest Descent Method 73
  - 5.2.2 Conjugate Gradient Method 74
  - 5.2.3 Newton's Method 75
  - 5.2.4 Quasi-Newton Methods 75
  - 5.2.5 Levenberg-Marquardt Algorithm 76
- 5.3 Numerical Methods for Constrained Static Problems 77
  - 5.3.1 Stationarity Condition for a Single Equality Constraint 77
  - 5.3.2 Stationarity Condition for a Single Inequality Constraint . . . 78
  - 5.3.3 Karush-Kuhn-Tucker Conditions 78
  - 5.3.4 Sequential Quadratic Programming 79
- 5.4 Dynamic Optimization Problems 79
  - 5.4.1 Formulation of the Optimal Control Problem 79
  - 5.4.2 Unconstrained Problems 81
  - 5.4.3 Optimality Conditions 82
  - 5.4.4 Solution Strategies 85
  - 5.4.5 Singular Case 86
  - 5.4.6 Constrained Problems 87
  - 5.4.7 Pontryagin's Maximum Principle 87
- 5.5 Numerical Methods for Dynamic Optimization Problems 88
  - 5.5.1 Indirect Methods: Solving the Optimality Conditions 88
  - 5.5.2 Direct Methods: Reduction to Static Optimization Problems . 98
  - 5.5.3 Comparison of Direct and Indirect Methods 105
  - 5.5.4 Software 105

6 Flatness-Based Trajectory Tracking — 108
- 6.1 Nonlinear Feedback for SISO-systems 108
  - 6.1.1 Exact Input-Output Linearization 108
  - 6.1.2 Transformation to the Byrnes-Isidori Normal Form 109
  - 6.1.3 Zero Dynamics 111
  - 6.1.4 Exact Input-State Linearization 112
  - 6.1.5 Trajectory Tracking Control 114
- 6.2 Nonlinear Feedback for MIMO-systems 116
  - 6.2.1 Exact Input-Output Linearization 116
  - 6.2.2 Transformation to the Byrnes-Isidori Normal Form 119
  - 6.2.3 Zero Dynamics 120
  - 6.2.4 Exact Input-State Linearization 120
  - 6.2.5 Trajectory Tracking Control 121
- 6.3 Flatness Based Trajectory Tracking 121

7 Academic Examples — 124
- 7.1 Nonlinear Oscillator 124
  - 7.1.1 Problem Description 124
  - 7.1.2 Equations of Motion 125
  - 7.1.3 DAE Approach with Control Constraints 125
  - 7.1.4 Flatness-Based Trajectory Tracking 126
  - 7.1.5 Optimal Control 127
  - 7.1.6 Results 130
  - 7.1.7 Discussion 132
- 7.2 Planar Overhead Crane 133
  - 7.2.1 Problem Description 133
  - 7.2.2 Equations of Motion 133
  - 7.2.3 DAE Approach with Control Constraints 136
  - 7.2.4 Flatness-Based Trajectory Tracking 137
  - 7.2.5 Optimal Control 139
  - 7.2.6 Results 143
  - 7.2.7 Discussion 146
- 7.3 3D Rotary Crane 148
  - 7.3.1 Problem Description 148
  - 7.3.2 Equations of Motion 149
  - 7.3.3 DAE Approach with Control Constraints 153
  - 7.3.4 Flatness-Based Trajectory Tracking 154
  - 7.3.5 Optimal Control 156
  - 7.3.6 Results 160
  - 7.3.7 Discussion 162

8 Examples from Industrial Applications — 163
- 8.1 AOD Converter 163
  - 8.1.1 Model Description 164
  - 8.1.2 Measuring Setup 165

		8.1.3	Modal Analysis 166

        8.1.3   Modal Analysis 166
        8.1.4   Transfer Functions 167
        8.1.5   Inverse Computation of the Excitations 169
        8.1.6   Verification 171
        8.1.7   Sensitivity Analysis 172
        8.1.8   Calculation of Foundation Forces 174
        8.1.9   Movement of the Converter 174
        8.1.10 Discussion 177
  8.2   Trailed Cultivator: Synkro 6003T 178
        8.2.1   Model Description 178
        8.2.2   Tire Modeling 179
        8.2.3   Measuring Setup 180
        8.2.4   Modal Analysis, Verification 181
        8.2.5   Transfer Functions 182
        8.2.6   Virtual Iteration 186
        8.2.7   Discussion 187
  8.3   Plough: Servo 6.50 191
        8.3.1   Model Description 191
        8.3.2   Measuring Setup 192
        8.3.3   Modal Analysis, Verification 193
        8.3.4   Transfer Functions 196
        8.3.5   Virtual Iteration 196
        8.3.6   Discussion 201

9 Conclusion   203

A The Gradient of a Functional   207

List of Symbols

Chapter 1:

$\mathbf{u}(t) \in \mathbb{R}^{m_c}$	vector of input variables
$\mathbf{x}(t) \in \mathbb{R}^{2n}$	vector of state variables (positions, velocities)
$\mathbf{y}(t) \in \mathbb{R}^{k}$	vector of output variables
$\tilde{\mathbf{y}}(t) \in \mathbb{R}^{k}$	vector of target signals
$\mathbf{q}(t) \in \mathbb{R}^{n}$	vector of (generalized or redundant) coordinates
$\mathbf{q}^t, \mathbf{q}^r, \mathbf{q}^f$	translational, rotational and flexible degrees of freedom
$\mathbf{v}(t) \in \mathbb{R}^{n}$	velocity vector

Chapter 2:

$\mathbf{R}$	position regarding to a floating reference frame
$\mathbf{r}$	position regarding to the inertial frame
$\mathbf{u}$	coordinates of a body-fixed coordinate system
$\mathbf{A}$	rotation matrix
$\mathbf{A}_x(\phi), \mathbf{A}_y(\phi), \mathbf{A}_z(\phi)$	elementary rotation matrices
α, β, γ	Tait-Bryan angles
ϕ, ψ, θ	Euler angles
$\mathbf{a}$	unit vector to define rotation axis
θ	rotation angle about rotation axis $\mathbf{a}$
e_0, e_1, e_2, e_3	Euler parameters
$\gamma_1, \gamma_2, \gamma_3$	Rodriguez parameters
$\mathbf{\Omega}$	angular velocity vector in the body-fixed coordinate system
$\tilde{\mathbf{\Omega}}$	skew-symmetric matrix of angular velocities
$\mathbf{H}$	matrix which maps $\dot{\mathbf{q}}^r$ to $\mathbf{\Omega}$
$\mathbf{v}_M$	velocity of the center of mass of a rigid body
$\mathbf{I}_M$	tensor of inertia
$\mathbf{M}(\mathbf{q}) \in \mathbb{R}^{n \times n}$	mass matrix
$\mathbf{f}(t, \mathbf{q}, \dot{\mathbf{q}}) \in \mathbb{R}^{n}$	vector of applied, conservative and gyroscopic forces
$\mathbf{G}(\mathbf{q}) \in \mathbb{R}^{m \times n}$	constraint Jacobian
$\boldsymbol{\lambda}, \boldsymbol{\nu}, \boldsymbol{\mu}, \boldsymbol{\xi}$	Lagrange multipliers

LIST OF SYMBOLS

$\mathbf{f}_i^o, \mathbf{f}_i^i$	outer / inner forces which act on body i
$\mathbf{f}_i^a, \mathbf{f}_i^r$	applied / reaction forces which act on body i
$\mathbf{M}_i^o, \mathbf{M}_i^i$	outer / inner torques which act on body i
$\mathbf{M}_i^a, \mathbf{M}_i^r$	applied / reaction torques which act on body i
m_i	mass of body i
$\mathbf{I}_i$	moment of inertia of body i
$\mathbf{a}_i(t)$	acceleration of body i with respect to the inertial frame
$\tilde{\mathbf{\Omega}}_i(t)$	skew-symmetric matrix of angular velocities of body i
$\mathbf{\Omega}_i(t)$	vector of angular velocities of body i
$\mathbf{g}(\mathbf{q}) = \mathbf{0}$	scleronomous holonomic constraints
$\mathbf{g}(\mathbf{q}, t) = \mathbf{0}$	rheonomous holonomic constraints
$\mathbf{g}(\mathbf{q}, \dot{\mathbf{q}}, t) = \mathbf{0}$	nonholonomic constraints
S	action integral
$L = L(\mathbf{q}, \dot{\mathbf{q}}, \boldsymbol{\lambda})$	Lagrangian
$T = T(\mathbf{q}, \dot{\mathbf{q}})$	kinetic energy
$V = V(\mathbf{q})$	potential energy
$E = E(\mathbf{q}, \dot{\mathbf{q}})$	total energy
W_{nc}	work done by non-conservative forces
$\mathbf{Q} = \mathbf{Q}(\mathbf{q}, \dot{\mathbf{q}}, t)$	non-conservative forces
$\delta \mathbf{r}$	virtual displacements
$\delta \boldsymbol{\varphi}$	virtual rotations
$\delta \mathbf{v}$	virtual velocities
$\delta \mathbf{a}$	virtual accelerations
δW	virtual work
δP	virtual power
δC	least constraint
ζ	viscous damping factor
ω_0	undamped eigenfrequency of a single degree-of-freedom system
$\mathbf{W}(X, Y, Z, t)$	displacement field of a flexible body
$\mathbf{N}(X, Y, Z)$	matrix of shape functions
$\mathbf{C} \in \mathbb{R}^{n \times n}$	damping matrix
$\mathbf{K} \in \mathbb{R}^{n \times n}$	stiffness matrix
$\mathbf{u} \in \mathbb{R}^{n_f}$	nodal DOFs in a FEM
$\mathbf{u}_M$	DOFs of master nodes in a FEM
$\mathbf{u}_S$	DOFs of slave nodes in a FEM
$\mathbf{q}_N$	modal coordinates corresponding to normal modes in a FEM
ω_k, λ_i	eigenfrequency, (complex) eigenvalue
$\boldsymbol{\Phi}_k$	eigenvector

LIST OF SYMBOLS

$\mathbf{\Phi}$	transformation matrix between physical and modal coordinates
$\mathbf{\Phi}_M$	matrix of constraint (static) modes in a Craig-Bampton reduction
$\mathbf{\Phi}_N$	matrix of normal (dynamic) modes in a Craig-Bampton reduction
$\bar{\mathbf{K}}$	generalized stiffness matrix corresponding to Craig-Bampton modes
$\bar{\mathbf{M}}$	generalized mass matrix corresponding to Craig-Bampton modes
$\bar{\mathbf{q}}^f$	modal coordinates regarding to the CMS
$\bar{\mathbf{\Phi}}$	transformation matrix regarding to the CMS

Chapter 3:

$F(i\omega)$	Fourier transform of $f(t)$
$S_{xx}(\omega)$	power (auto) spectral density
$R_{xx}(\tau)$	auto correlation function
$S_{xy}(i\omega)$	cross spectral density
$R_{xy}(\tau)$	cross correlation function
$\gamma_{xy}^2(\omega)$	coherence function
$G(i\omega)$	transfer function
$\mathbf{G}(i\omega)$	transfer matrix
$\mathbf{x}_s$	equilibrium point
$\mathbf{A}, \mathbf{B}, \mathbf{C}, \mathbf{D}$	state-space matrices
$\hat{f}(s)$	Laplace transform of $f(t)$
m_l, k_l, c_l	modal mass, stiffness, damping
e_{RMS}, e_{MAX}	root mean square error, maximum deviation
$\mathbf{G}^{-1}(s)$	inverse of the transfer matrix
$\mathbf{G}^+(s)$	Moore-Penrose pseudoinverse
$\mathbf{G}^H(s)$	adjoint matrix
$\mathbf{\Gamma}, \mathbf{\Pi}, \mathbf{\Sigma}$	unitary matrices, matrix with singular values

Chapter 4:

$\mathbf{c}(\mathbf{q},t) = \mathbf{\Phi}(\mathbf{q}) - \boldsymbol{\gamma}(t)$	control constraints
$\mathbf{\Phi}(\mathbf{q})$	system outputs
$\boldsymbol{\gamma}(t)$	desired targets
$\mathbf{B}$	input transformation matrix
$\mathbf{C} = D\mathbf{\Phi}(\mathbf{q})$	projection matrix, projection to constrained subspace
$\mathbf{D}$	projection matrix, projection to unconstrained subspace
$\mathcal{C}, \mathcal{D}$	constrained / unconstrained subspace
$\mathcal{N}$	original configuration manifold
p	measure of control constraint realization
$R_\mathbf{q}, R_\mathbf{v}, R_\boldsymbol{\lambda}, R_\mathbf{u}$	residual vectors

LIST OF SYMBOLS

Chapter 5:

$\mathbf{q}(\mathbf{x}) = \mathbf{0}$	equality constraints
$\mathbf{h}(\mathbf{x}) \leq \mathbf{0}$	inequality constraints
$\mathbf{d}^k$	descent direction
α^k	step size
β^k	scalar factor for Fletcher-Reeves / Polak-Ribiere formula
$\mathbf{Q}^k$	approximation of the Hessian matrix
$J(\mathbf{u})$	cost functional, performance measure
$\Phi(\mathbf{x}(t_f), t_f)$	Mayer term in the cost functional
$\mathcal{L}(\mathbf{x}(t), \mathbf{u}(t), t)$	integral term (Lagrange term) in the cost functional
$\mathbf{p}(t)$	costates, adjoint variables
$H(\mathbf{x}, \mathbf{u}\mathbf{p}, t)$	Hamiltonian
$\mathbf{G}(\mathbf{x}(t_f)) = \mathbf{0}$	final conditions
$\mathbf{A}(t), \mathbf{B}(t), \mathbf{C}(t), \mathbf{D}(t)$	time-variant matrices of linearized differential equation
κ	step size in Kelley-Bryson method
α	weighting factor of error at final time t_f
$\mathbf{p}, \mathbf{w}$	adjoint variables corresponding to $\mathbf{q}, \mathbf{v}$
ε	weighting factor for Tikhonov regularization term

Chapter 6:

$\mathbf{f}(\mathbf{x}), \mathbf{g}(\mathbf{x})$	sufficiently smooth vector fields
$h(\mathbf{x})$	sufficiently smooth function
$L_\mathbf{f} h(\mathbf{x}), L_\mathbf{g} h(\mathbf{x})$	Lie-derivatives of scalar function $h(\mathbf{x})$ along $\mathbf{f}(\mathbf{x}), \mathbf{g}(\mathbf{x})$
r	relative degree of the system
$\mathbf{z} = \mathbf{\Phi}(\mathbf{x})$	state transformation to Byrnes-Isidori normal form
$\boldsymbol{\xi} = [z_1, ..., z_r]$	states of the controllable, observable system
$\boldsymbol{\eta} = [z_{r+1}, ..., z_{2n}]$	states of the non-observable system
$\dot{\boldsymbol{\eta}} = \mathbf{q}(\mathbf{0}, \boldsymbol{\eta})$	zero dynamics
$\mathbf{A}_r, \mathbf{A}_e$	dynamic matrix, error dynamic matrix
$[\mathbf{f}, \mathbf{g}](\mathbf{x})$	Lie-bracket
$ad_\mathbf{f}^k \mathbf{g}(\mathbf{x})$	k^{th} Lie-bracket
D	distribution
$\mathbf{w} = l(\mathbf{x})$	measurable variables
$\mathbf{A}(\mathbf{x})$	decoupling matrix
$\mathbf{v}$	vector of output-derivatives
$\mathbf{b}(\mathbf{x})$	vector of Lie-derivatives
β	number of highest time derivation
i	index of a DAE

List of Abbreviations

AI	affine input system
AOD	argon oxygen decarburization
ASD	auto spectral density
BDF	backward differentiation formula
BFGS	Broyden, Fletcher, Goldfarb, Shanon-update
BVP	boundary value problem
CFD	computational fluid dynamics
CMS	component mode synthesis
CSD	cross spectral density
DAE	differential-algebraic equation
DFP	Davidon, Fletcher, Powell-update
DOF	degree of freedom
FEM	finite element method / finite element model
FFT	fast Fourier transform
FRF	frequency response function
GGL	Gear-Gupta-Leimkuhler stabilization
HHT	Hilber-Hughes-Taylor integrator
IFFT	inverse fast Fourier transform
ILC	iterative learning control
IVP	initial value problem
KKT	Karush-Kuhn-Tucker conditions
LTI	linear time invariant

LIST OF ABBREVIATIONS

MAST	multi-axis shaker table
MBS	multibody system
MIMO	multiple input multiple output
MPC	multi physics constraints
NLP	nonlinear programming
OCP	optimal control problem
ODE	ordinary differential equation
OP	optimization problem
PDE	partial differential equation
PI	proportional-integral controller
PSD	power spectral density
RMS	root mean square
SISO	single input single output
SQP	sequential quadratic programming
SVD	singular value decomposition
VPD	virtual product development

*[...] I think, myself, it's very helpful, too
that one can take back home, and use,
what someone's penned in black and white.*

Johann Wolfgang von Goethe, Faust Part I

Chapter 1
Introduction

Multibody systems (MBS) are an essential part in the discipline of technical and computational mechanics. Simulation of technical systems and processes is getting more and more important due to decreasing development times and simultaneously increasing quality standards [157]. Models of mechanical or mechatronical systems become more accurate in order to reproduce the real physical behavior. At the same time, efficiency and computational effort play an important role [140]. Nowadays a MBS software has to fulfill two major tasks. The first challenge is the appropriate modeling of the real system and the derivation of a mechanical or mathematical model by a systematic formulation of the equations of motion. The second purpose is the solution of these differential equations [134]. Due to the high complexity of the systems and the resulting equations, numerical methods are applied. Analytical solutions can just be found in special cases.

Historically, MBS were used in order to simulate rigid bodies, which are connected by different joints and massless springs and dampers. The method was developed for relatively large translational and rotational displacements [157]. Deformable structures were treated within the finite element method (FEM) . Nowadays rigid bodies as well as flexible bodies can be integrated in a system. Applications of multibody simulation can be found in the automotive- and railroad industry, in the aeronautics and space technology, in robotics, biomechanics and general mechanical or mechatronical mechanisms.

1.1 Overview

The typical task in a multibody simulation is a forward dynamics problem, where the equations of motion have to be integrated numerically. Specific output variables $\mathbf{y}(t) \in \mathbb{R}^k$ and states $\mathbf{x}(t) \in \mathbb{R}^{2n}$ are calculated based on given input variables $\mathbf{u}(t) \in \mathbb{R}^{m_c}$ (e.g.: forces, motions, ...) and initial conditions $\mathbf{x}(0) \in \mathbb{R}^{2n}$.

The focus of this dissertation is in the field of inverse dynamics. The output variables $\mathbf{y}(t)$ (e.g.: positional coordinates, accelerations, strains, spring deflections, ...) are either analytically defined or given by measurements and the unknown inducing inputs variables $\mathbf{u}(t)$ have to be computed.

1.1.1 Multibody Systems

In the literature a range of definitions for the expression 'multibody system' can be found. Selected textbooks that deal with multibody systems and corresponding numerical methods are [9, 50, 114, 132, 133, 139, 157, 163], without claim of completeness. Textbooks that are focused on numerical methods for MBS, specially on methods for solving ordinary differential equations (ODEs) and differential-algebraic equations (DAEs) are for example [4, 35, 50, 78, 79, 92].
In [70] the most important definitions are summarized from several authors.
Generally, in the literature it is distinguished between *continuous systems*, *finite element systems*, *multibody systems* and *hybrid multibody systems*. In this dissertation multibody systems are considered in chapter 7, a finite element system is considered in chapter 8.1 and hybrid multibody systems are considered in chapters 8.2 and 8.3. Therefore the difference of these mechanical models and their specific definitions are worked out in the following paragraph:

Continuous systems: "[...] consist of elastic bodies, for which mass and elasticity are continuously distributed throughout the body. The action of forces is also continuous along the body's volume resp. surface." [157] The equations of motion can only be formulated for infinitesimal small volumetric elements and are partial differential equations, which depend on the spacial location and the time [133].

Finite element systems: "[...] bodies are assumed to have nonzero mass and to be elastic with forces and moments acting at discrete points." [157] The basic idea is to consider inertia forces, elasticity and forces in a discrete element of simple geometry. Based on local equations of motion, which are formulated for a single finite element, the global equations are assembled [133]. Detailed descriptions can be found for instance in [6].

Multibody systems: "[...] bodies are assumed to have nonzero mass and to be rigid with forces and moments acting at discrete points." [157] Springs, dampers and servo-motors are assumed to be massless. Bodies are interconnected by rigid bearings or supports. Friction and contact forces can also be included in the model [133].

Hybrid multibody systems: "[...] both elastic and rigid bodies are used to model a mechanical system." [157] Applications of hybrid multibody systems originated from vehicle-, robot- and satellite-dynamics [133].

Nowadays hybrid multibody systems are state-of-the-art and several methods are known in order to implement elastic and even plastic structures in a rigid multibody system.

1.1.2 Brief Historical Review

The historical overview of analytical mechanics, which has been essential for the field of MBS, is summarized from [3] and [133]. The equations of motion of unconstrained mechanical systems were already known from the beginning of mechanics. *Sir Isaac Newton* (1643-1727) published 1687 his famous three universal laws in his *Philosophiae Naturalis Principia Mathematica*: (1) the law of inertia ("lex prima"), (2) the law of motion or the law of linear momentum ("lex secunda") and (3) the action-reaction law ("lex tertia"). The law of motion is mathematically described by differential equations, which were also introduced by Newton. The law of linear momentum provides the equations of motion for a point mass. The equations of motion for a rigid body, i.e. the laws of linear and angular momentum were presented by *Leonhard Euler* (1707-1783) in his *Mechanica corporum solidorum* in the year 1776. To set up the Newton-Euler equations of motion a free body diagram has to be prepared first. Constraint forces like joint forces are also considered, although they are often not of interest.

Jean le Rond D'Alembert (1717-1783) published a seminal theorem for the dynamic behavior of interacting bodies in his *Traitè de dynamique* in 1743. In his work he distinguished between applied and reaction forces. In 1788 this principle was reformulated by *Joseph Louis Lagrange* (1736-1813) in his *Mècanique analytique*. This version is the one which we call "D'Alembert's principle, the principle of virtual work" today. Furthermore, Lagrange made important findings in the field of the calculus of variations. This lead to Lagrange's equations of the first kind, which represents a set of DAEs. Later Lagrange introduced generalized coordinates, which underlay his equations of motion of second kind, published 1811. This formulation resulted in a minimal set of ODEs. During this period the theory of small-amplitude oscillations and the theory of linear systems of differential equations were developed. Furthermore, fundamental terms in linear algebra were introduced (e.g.: eigenvalues and eigenvectors in an n-dimensional case).

Generalizations of the principle of D'Alembert were published by *Johann Carl Friedrich Gauß* (1777-1855) in 1829 and by *Philip Edward Bertrand Jourdain* (1879-1919) in 1908. Their principles are known as Jourdain's principle (the principle of virtual power) and Gauß' principle of least constraint. Laplace, Lagrange and Gauß made also important contributions in perturbation theory.

Lagrange's equations of second kind were extended to nonholonomic systems from *Josiah Willard Gibbs* (1839-1903) in 1879 and from *Paul Émile Appell* (1855-1930) in 1900.

Besides the state of the art, at that time, differential-principles a new integral-principle, the principle of least action, was introduced by *Sir William Rowan Hamilton* (1805-1865) in 1834.

The theory of stability of motion started with classical works from *Aleksandr Mikhailovich Lyapunov* (1857-1918). Russian mathematicians had a huge amount in further developments, e.g.: *Lev Semenovich Pontryagin* (1908-1988) in structural stability.

The beginning of computer simulation in multibody dynamics started after 1965. The simulation of satellites was of great importance for space flight. From that point on computerized formalisms have been developed [133]. 1977 multibody dynamics was set up as new branch at a IUTAM symposium held in Munich and chaired by *Kurt Magnus* [134]. Since the 1980's MBS-software has been available for modeling, simulation and animation [70]. 1990 computational aspects of multibody dynamics were highlighted at the second world congress on computational mechanics in Stuttgart, chaired by *John H. Argyris* [134].

1.1.3 State of the Art

Research:
Due to increasing challenges in simulation and the the merging fields of dynamics, continuum mechanics, control engineering, optimization, etc. MBS is a current field of research [134]. State-of-the-art topics that are discussed in MBS-conferences are listed below:

- Theoretical and computational methods
- Flexible multibody systems
- Contact and impact problems
- Control and mechatronics
- Multidisciplinary approaches
- Coupled multi-physics problems
- Algorithms, integration codes and software
- Efficient methods and real-time simulations
- Virtual reality
- Experiments and numerical verifications
- Optimization and sensitivity analysis
- Dynamics of machines and rotating structures
- Dynamics of vehicles (aerospace, automotive, railway engineering) and tire dynamics
- Robotic systems
- Biomechanics
- Nano technology in MBS
- Education in multibody dynamics

1.1. OVERVIEW

Due to the increasing complexity, which can be handled in a multibody simulation, several topics can be combined.

Conferences:
Specific international conferences are organized which are focuses on actual developments in multibody dynamics. The well-known conferences in Europe, America and Asia are representatively listed:

Europe:

- ECCOMAS (European Community on Computational Methods in Applied Sciences) Thematic Conference on Multibody Dynamics
- Joint International Conference on Multibody System Dynamics (IMSD)

America:

- International Conference on Multibody Systems, Nonlinear Dynamics, and Control (MSNDC), ASME (American Society of Mechanical Engineers)

Asia:

- ACMD (Asian Conference on Multibody Dynamics)

Other conferences which are strongly related to technical mechanics are listed in alphabetical order:

- Annual Meeting of the International Association of Applied Mathematics and Mechanics (GAMM)
- CISM-IFToMM Symposium on Robot Design, Dynamics, and Control
- EUROMECH Colloquium Advanced Applications and Perspectives of Multibody System Dynamics
- European Conference on Computational Mechanics Solids, Structures and Coupled Problems in Engineering
- European Conference on Mechanism Science (EUCOMES)
- European Congress on Computational Methods in Applied Sciences and Engineering
- IFToMM Asian Conference on Mechanism and Machine Science
- International Conference on Rotor Dynamics
- International Symposium on Mechanism and Machine Science (ISMMS)
- International Workshop on Underactuated Grasping
- IUTAM - Symposium on Multiscale Problems in Multibody System Contacts

- IUTAM Symposium on Computational Methods in Contact Mechanics
- World Congress in Mechanism and Machine Science (CFP)
- World Congress on Computational Mechanics (WCCM)

Journals:
Since 1997 the Journal *"Multibody System Dynamics"* has been published as a journal, which is fully devoted to multibody dynamics. Of course, many contributions in the field of MBS are also published in other journals due to the wide scope. The following alphabetical list shows selected journals that are related to mechanical research.

- Acta Mechanica
- Computational Methods in Applied Sciences
- Computer methods in applied mechanics and engineering
- Computers and Structures
- International Journal for Numerical Methods in Engineering
- International Journal of Non-Linear Mechanics
- International Journal of Solids and Structures
- Journal of Computational and Nonlinear Dynamics
- Journal of Mechanical Science and Technology
- Journal of Sound and Vibration
- Journal of Theoretical and Applied Mechanics
- Meccanica
- Nonlinear Dynamics
- Structural and Multidisciplinary Optimization

Optimization is as an additional field, which is also relevant in this dissertation. Hence, two journals, which are related to optimization theory, are listed here as well:

- Optimal Control Applications and Methods
- Journal of Optimization Theory and Applications

Software:
Nowadays specific software is developed for special applications. However, individual commercial MBS software tools have been establishing during the last few years. At the present stage the market leaders are:

- Adams^{TM} (www.mscsoftware.com)
- Simpack^{TM} (www.simpack.com)
- RecurDyn^{TM} (http://functionbay.de)

Historically, *Adams* (Automated Dynamic Analysis of Mechanical Systems) was developed especially for vehicle dynamics simulation [27]. The roots of *Simpack* are located in railroad applications. *RecurDyn* (Recursive Dynamics) was developed later in the 90's of the previous century and was strongly focused on integrated FEM capabilities.

Other MBS software packages are (list in alphabetical order, no claim to completeness):

- Altair Motion Solve^{TM} (www.altairhyperworks.de/Product,18,MotionView.aspx)
- Ansys Rigid Dynamics^{TM} (http://www.ansys.com/products/rigid-dynamics.asp)
- CarSim^{TM}, TruckSim^{TM}, BikeSim^{TM} (www.carsim.com)
- CASCaDE
- cosin/mbs (www.cosin.eu)
- Dymola/Modelica^{TM} (www.3ds.com/products/catia/portfolio/dymola)
- FEDEM^{TM} (www.fedem.com)
- Hotint (http://tmech.mechatronik.uni-linz.ac.at/staff/gerstmayr/hotint.html)
- LMS Virtual.Lab Motion^{TM} (previously DADS^{TM}) (http://www.lmsintl.com/simulation/virtuallab/motion)
- madymo^{TM} (www.advancedsimtech.com/software/madymo)
- MBSim (http://mbsim.berlios.de)
- Mesa Verde (Mechanism, Satellite, Vehicle, Robot Dynamics Equations)
- Neweul (Neweul-M^2) (www.itm.uni-stuttgart.de/research/neweul)
- Robotran (www.robotran.be)
- Samcef^{TM} (www.samtech.com)

- veDyna^{TM} (http://dynaware.tesis.de)

Each software is different regarding to the formalisms for generating and solving the equations of motion. While *Adams* uses absolute coordinates, *Simpack* uses relative coordinates [132]. In *Adams* or *LMS Virtual.Lab Motion* the equations of motion are always formulated with redundant coordinates, which results in a set of DAEs [106]. In *Simpack* ODEs are obtained for chain and tree topologies due to a minimal coordinates formulation and DAEs are obtained for closed-loop systems [132]. *RecurDyn* makes use of recursive formalisms, as its name already implies.

Differences also occur in the mechanical principle, which is used to formulated the equations of motion. Lagrange's equations of the first kind are used in *Adams* and d'Alembert's principle is implemented in *Simpack* [157].

Generally, most MBS packages use pure numerical methods for the generation of the equations. Exceptions are, for example, *Neweul*, *Robotran* or *Mesa Verde*, which compute the equations of motion symbolically. *Simpack* use a mixture of symbolic and numerical formalisms for the generation of the equations of motion [70, 157].

In addition, FEM codes are necessary for the implementation of flexible bodies. If a finite element solver is not directly integrated in the MBS software, an external program has to be used to compute e.g. the eigenvectors for a modal implementation. *Abaqus*, *Ansys* or *Nastran* can be used as finite element solvers in connection with *I-Deas*, *Hypermesh*, *Ansa*, *Patran* etc. as preprocessors, just to mention a few packages.

1.1.4 Types of Problems

The types of problems in the field of multibody dynamics can roughly be classified into three categories: kinematic problems, dynamics problems and optimization problems. In kinematic problems the motion of a MBS is studied without involving the forces that act on the system. These are purely geometric problems and will not be discussed here. Details can be found in [157].

Dynamic Problems:

In dynamic problems the relation between forces and motion of the system is studied. Actuating forces are considered, as well as inertia forces. The following problems are the main tasks in a dynamic computation of a MBS:

Forward Dynamics: Forces and torques, which are applied to the MBS, are known. The motion of the system should be simulated for given initial states. As a consequence, an initial value problem (IVP) has to be solved. Furthermore, velocities, accelerations and reaction forces (-torques) are of interest. To solve this problem, typically a nonlinear system of differential equations (equations of motion) has to be solved. Reaction forces and -torques are calculated afterwards. Traditionally, the forward dynamics problem has been the heart of a dynamic simulation [157].

Inverse Dynamics: The goal in an inverse dynamics problem is to compute the forces that are necessary to produce a specific motion. In typical technical systems it can occur that the motion is (partly) known from physical measurements or it is specified in order that the mechanical systems fulfills a specific task as e.g. in cranes or industrial robots. If the motion should be specified, the trajectories of discrete points have to be described by spatial coordinates. If the motion of the real system can be measured, typically specific variables like positions, velocities or accelerations are known at certain measuring points. The input variables have to be determined in a way that the outputs of the model coincide with the specified motion or the measurement data, respectively. Actuating forces or torques in a servomotor can be used as input variables as well as dynamic positions of hydraulic or pneumatic cylinders.

The inverse dynamics problem is easier to solve than the forward dynamics problem, if the equation of motion are known in a symbolic form and if the system is fully actuated. A system is called fully actuated, if the number of inputs $\mathbf{u}(t)$ is identical to the number of degrees of freedom (DOFs) . In such a case only an algebraic problem has to be solved. The inverse dynamics problem becomes much more complicated, if the MBS is underactuated. In an underactuated system the number of inputs variables is less than the number of DOFs [21]. Another challenging task for an inverse dynamics problem appears, if the equations of motion are not given in a symbolic form. This problem typically arises in commercial MBS software, where complex systems with flexible bodies are modeled. Such problems cannot be solved with standard techniques and are therefore considered in this dissertation.

Static equilibrium: For this type of problem the task is to find the position of the MBS where all forces are balanced. In many applications a static equilibrium analysis is carried out to initialize a dynamic analysis [157].

Modal dynamics (linearized dynamics problem): The model is linearized at a stationary point (e.g.: static equilibrium). Eigenfrequencies, eigenvectors and state space matrices are computed. As a result the linearized model can be exported to an external software to design for example a controller.

Optimization Problems:

During the construction phase of a mechatronical system an engineer is interested in parameters that influence the system behavior or in the determination of input variables so that the system achieves a predefined task. Several mathematical methods exist for these static or dynamic optimization problems, which are classified in [157]:

Parameter identification: Unknown system parameters have to be found so that the model behaves as close as possible to specific observations (measurements).

Parameter optimization: The model behavior can be optimized by tuning a set of parameters. The goal is to adjust parameters in a way that some objectives are fulfilled. This problem is also called "optimal design".

Optimal control: The aim of an optimal control problem is to find a control law such that a specific optimality criterion is fulfilled. This method can be applied for completely designed and optimized systems that come into operation. Optimality criteria might be minimum time or minimum energy consumption of a system, which is moved from one point to another. These problems are hard to solve and address the ultimate goal in the product development procedure.

1.2 Aim of the Study

The aim of the study is to compute outer excitations (i.e. input variables) of underactuated multibody systems based on measured or predefined target signals (output variables). The considered multibody system can either be given in a form where the equations of motion can be derived analytically or it can be modeled in commercial software where the equations are not available. Different mathematical methods are considered and new approaches are developed. The specific methods are classified by their applicability to the different types of model.

The motivation for this kind of problem results from industrial applications. In several technical systems output variables can be measured with standardized measurement techniques. In contrast, measurements of input variables are either technically not possible or financially not affordable. Such problems typically occur in the automotive and especially in the agricultural industry. Output variables like accelerations or strains can be measured at different points on a vehicle. By comparison, input variables like wheel forces can only be measured with high technical effort. In the automotive industry measuring wheels exist, which can measure the forces and torques at the wheel hub, where the load is introduced into the structure. However, the costs for such measuring wheels can reach a hundred thousand Euro and more, depending on the type of the wheel. For trucks and heavy machinery such measuring equipment is very limited and in the agricultural industry it does not exist up to now.

1.2. AIM OF THE STUDY

For simulations of complete vehicles and for durability tests it is necessary to find system-invariant excitations. These dynamic loads can be generated on a test rig in order to reproduce a test drive on a real track. Load data are called invariant if they are independent of the system under consideration [39]. Invariant variables of a test drive are the road profiles, if it is assumed that the track is a solid terrain. Digital road profiles are available for standard maneuvers in the automotive industry, but unfortunately such data are very limited in the agricultural industry up to now.

Due to these difficulties it is of great interest to compute the wheel forces or generally the outer excitations without such measuring wheels.

Another application which is investigated in this thesis is an AOD steel converter. Here the task is to compute the inducing excitation forces and torques which cause unwanted vibrations.

Similar problems also occur in the control tasks of cranes where a load should be moved along a specific trajectory.

In all these applications a similar problem appears in the computation of input variables. Therefore, different methods for the solution of the inverse problems, which are based on physical measurements of specific target signals, have to be found.

Requirements for virtual product development

Virtual product development consists of a range of individual simulation steps. The general procedure is shown in Fig. 1.1 which should be called "building of virtual product development (VPD)". In Fig. 1.1 it can be seen that the knowledge of

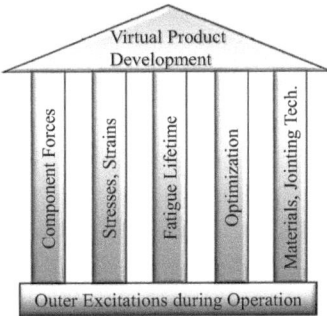

Figure 1.1: Building of virtual product development

the outer excitations are the basis of the VPD-building. If the outer excitations of a MBS are known, component forces can be computed in a multibody simulation. Based on component forces of individual parts stresses, strains, etc. can be computed in a finite element analysis. In a subsequent simulation the fatigue lifetime can be predicted. These data can be used to optimize the shape and the materials of specific parts. Additional information about jointing technology and materials is necessary to optimize the design of the construction. All these steps are important

1.2. AIM OF THE STUDY

requirements for lightweight construction.
Fig. 1.1 should illustrate that all computations are based on the outer excitations. If the inputs contain errors, all further steps are not correct anymore. Hence, the focus of this dissertation is put exactly in the first step, i.e. the computation of the outer excitations.

Mathematical description of the problem

The mathematical problem can be illustrated by the block diagram in Fig. 1.2. The

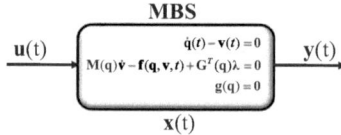

Figure 1.2: Block diagram with input, output and state variables of a MBS

mathematical model of a given multibody system can either be formulated by ODEs or DAEs, depending on the formulation with generalized or redundant coordinates, cf. chapter 2. The equations of motion can be formulated with state variables, i.e. positional coordinates and the corresponding velocities $\mathbf{x}(t) = [\mathbf{q}, \dot{\mathbf{q}}]^T \in \mathbb{R}^{2n}$. The aim is to compute the input variables $\mathbf{u}(t) \in \mathbb{R}^{m_c}$ in a way that the output variables $\mathbf{y}(t) \in \mathbb{R}^k$ are identical to predefined target signals $\tilde{\mathbf{y}}(t) \in \mathbb{R}^k$. Furthermore, underactuated systems are considered. Underactuated systems are characterized by less control inputs $\mathbf{u}$ than degrees of freedom, i.e. $m_c < n$, if n denotes the number of generalized coordinates. This type of problem is classified as a generally nonlinear inverse problem of an underactuated multibody system.

Representative problems in industrial applications

The problem, which is stated in Fig. 1.2, typically occurs in automotive or agricultural test rigs. In such a test rig the goal is to find the input signals of servo-hydraulic cylinders in order that the system is excited in the same way as during a test drive. Such a full-vehicle test rig and a suspension test rig are shown in Fig. 1.3. Nowadays full-vehicle test rigs as shown in Fig. 1.3(a) and suspension test rigs as shown in Fig. 1.3(b) are designed to excite each wheel with up to all six DOFs.
In Fig. 1.4 two typical test rigs for agricultural machines are shown. The entire service life of trailed machines like a silage trailer in Fig. 1.4(a) is tested on a 4-poster with up to four vertical excitations. Mounted machines like a plough are tested on a multi-axis shaker table (MAST) , which simulates all six DOFs at the mounting point.

1.2. AIM OF THE STUDY

(a) Full vehicle test rig, Opel Insignia (Source: http://www.insignia-blog.de/wp-content/uploads/2008/07/vlcsnap-421481.jpg)

(b) Suspension test rig, allowing six-DOF testing (simulation of F_x, F_y, F_z and M_x, M_y, M_z) [47]

Figure 1.3: Test rigs in the automotive industry

(a) 4-poster

(b) multi-axis shaker table (MAST)

Figure 1.4: Test rigs in the agricultural industry (Source: Pöttinger)

1.3 Outline of the Present Work

The following paragraph presents a short overview of each chapter and the main issues that are addressed.

Chapter 2: The basis for all further computations are the equations of motion from the considered multibody system. In this chapter the equations of motion of constrained mechanical systems are presented. Different formalisms are compared and their specific advantages and disadvantages are highlighted. An overview of numerical procedures for the solution of DAEs is given. Some basic concepts for the implementation of flexible bodies are discussed.

Chapter 3: The first considered method regarding to inverse problems is the method of virtual iteration. This method is based on a linearization of the model and the inverse computation in the frequency domain. The equations of motion do not have to be available in a symbolic form. This method is best suited for detailed multibody systems including flexible bodies, but only with moderate nonlinearities.

Chapter 4: The second method is an extension of the DAEs by so called servo or control constraints. This formalism results in high index DAEs which are more complicated to solve. A well suited index reduction procedure and an appropriate implicit solver are derived for this kind of systems. The procedure is an excellent method for systems where the equations of motion are given in a symbolic form.

Chapter 5: In this chapter optimal control methods are discussed. Optimization procedures for unconstrained and constrained static problems are stated. Formalisms of dynamic optimization problems are presented. Optimal control theory is discussed which leads to Pontryagin's maximum principle. For the numerical solution of optimal control problems an overview of direct and indirect methods is given. The indirect methods are based on the solution of the optimality conditions that results in a boundary value problem (BVP) that is challenging to solve. Other approaches are presented based on a direct optimization method. One method is qualified for MBS where the equations of motion are not given in a symbolic form. The system is treated as a black box. In the second approach state and costate equations are formulated and a gradient method is used to minimize an appropriate cost functional.

Chapter 6: The fourth method is a feed forward control design, known from automation and control engineering. The formalism for differentially flat systems is presented. This method results in an analytical control law, which is a big advantage. However, the desired trajectories must be sufficiently smooth, i.e. continuously differentiable up to a certain order.

Chapter 7: The approaches under consideration are applied to three academic examples. The first example is a nonlinear oscillator consisting of two masses that

1.3. OUTLINE OF THE PRESENT WORK

are connected by spring and damper elements. The example represents a fully actuated system, i.e. the number of inputs is equal to the number of degrees of freedom. The second example is a planar overhead crane. The problem can either be formulated with independent or dependent coordinates. Here the differences in the resulting equations of motion can be seen clearly. The example represents an underactuated system and is therefore more complicated to solve. The goal is to compute the inputs (force at the trolley and torque at the winch) that a mass follows a trajectory, which is given by a polynomial. The third example illustrates a three-dimensional rotary crane. In this larger multibody system big differences can be seen regarding to the formulation of the equations of motion. Again, the control inputs of this underactuated system are computed in a way that the load follows a desired trajectory.
Based on these academic examples the considered approaches are compared. Specific advantages and disadvantages of each method are discussed.

Chapter 8: The method of virtual iteration is applied to three examples from industrial applications. The first problem is an AOD-converter, which is modeled in the finite element software *Abaqus*. The vibrations of this converter are physically measured by acceleration sensors and strain gauges. The goal of this problem is to compute the actuating forces and torques in the vessel. The model is a completely linear model and therefore the inverse calculation can be done in one single step.
The second and the third industrial examples represent agricultural machines. The first machine is a trailed cultivator, called *Synkro 6003T*. The second machine is a plough, called *Servo 6.50*. Both machines are modeled in the MBS-software *Adams*. During a test drive on a real track accelerations and strains are measured. Afterwards the measured accelerations and strains should be reproduced on a test rig. This is firstly done on a real test rig in the laboratory and secondly on a virtual test rig on the computer. For both machines the excitation signals in servo-hydraulic cylinders are computed in a way that the resulting outputs coincide with the measured target signals.

Chapter 9: The results of the previous chapters are concluded in this chapter. A short summary of the different approaches with their specific advantages and disadvantages is given. Some considered methods have potential for further investigations that are out of the scope of this dissertation. Possible methods for future work are listed and briefly discussed.

Appendix A: Some elementary calculations, which are required for the optimal control methods in chapter 5 are performed. The gradient of a cost functional is derived.

1.4 Scope of the Present Work

The specific methods, which are presented in chapters 3 - 6, strongly depend on the type of problem. If the model is only moderately nonlinear, the method of virtual iteration is the most qualified method for large systems. It can be used for very detailed models and the equations of motion do not have to be known in a symbolic form. Furthermore, the procedure is not limited to special multibody simulation software or finite element software. The inverse computation is done in *Matlab* or can also be used as stand-alone software.

If the model includes remarkable nonlinearities, a linearization at a specific operation point does not represent the global behavior. In such a case the methods with control constraints, the optimal control approach and the feed-forward control design are better suited. All these methods are implemented in *Matlab*. In chapter 7 the methods are successfully applied to academic examples. However, the integration of these methods is another challenging task. The optimal control approach would be best suited for the implementation in a commercial MBS-software package. Unfortunately such tools do not offer an open interface for an implementation in their solver routines. The problem is that the output variables as well as all the state variables have to be saved. Based on these data the optimization procedure can calculate the input variables. A possibility would be a co-simulation between MBS-software and *Matlab*. However, a co-simulation is quite inefficient, because two different solvers work independently. Another problem is that currently the interface between *Adams* and *Matlab* does not have the capability to exchange all the state variables. In contrast, input and output variables can be defined and the interface for such variables works adequate.

Another opportunity would be to save all the state variables in an external ASCII-file. Basically this method is possible, but it is also quite inefficient.

Hence, the implementation of the optimal control approach into a commercial MBS-software is out of the scope of this dissertation. If in the future the interfaces between commercial MBS-software and numerical software like *Matlab* will be improved, it would be possible to implement an efficient algorithm.

> *Making the simple complicated is commonplace; making the complicated simple, awesomely simple, that's creativity.*
>
> Charles Mingus

Chapter 2
Dynamics of Multibody Systems

In this chapter the most important parts of kinematics of rigid bodies are discussed in order to formulate the dynamics of multibody systems. Furthermore, the principles of mechanics are summarized. The equations of motion for constrained MBS are derived from variational principles. An overview regarding the numerical solution of DAEs is given. Finally, the implementation of flexible bodies is briefly discussed. Most parts regarding to computational dynamics, which include kinematics, variational principles and Lagrangian dynamics as well as constrained dynamics, are taken from [140, 142]. Parts of flexible multibody systems are an excerpt from [139].

2.1 Kinematics of Rigid Bodies

Motions of rigid bodies are described with respect to a coordinate system, which is fixed to the body. Material points of rigid bodies remain constant, if their coordinates are formulated in this fixed-body coordinate system. An unconstrained body has six DOFs in space. These DOFs consist of three positional coordinates, i.e. the spatial coordinates of the fixed-body coordinate system and three rotational coordinates, where the orientation of the fixed-body coordinate system is related to an inertial frame. If absolute coordinates are used, all positional and rotational variables are expressed with respect to an inertial frame. This formulation is used for example in the MBS-software *Adams* [27]. If relative coordinates are used, positional and rotational coordinates of a specific body are formulated with respect to another body, i.e. a tree topology is used. Such a formulation is used e.g. in the MBS-tool *Simpack*. Relative coordinates are especially advantageous in simulations of chain drives, where the coordinates of a chain link are related to the previous and the next one, respectively.

The position of a material point of a single rigid body is expressed by

$$\mathbf{r} = \mathbf{u} + \mathbf{AR} \tag{2.1}$$

which is also shown in Fig. 2.1. The blue coordinate system $(x\,y\,z)$ in Fig. 2.1 is an inertial frame and the red coordinate system $(X\,Y\,Z)$ represents the body-fixed coordinate system. The absolute position of the point P expressed with respect to

2.1. KINEMATICS OF RIGID BODIES

the inertial frame is given by $\mathbf{r} = [x, y, z]^T$. The material point P expressed with respect to the fixed-body coordinate system is given by $\mathbf{R} = [X, Y, Z]^T$. The position of the fixed-body coordinate system is given by the vector $\mathbf{u}$ and the orientation regarding to the inertial frame is given by the orthogonal rotation matrix $\mathbf{A}$. In the majority of cases it is advantageous, if the fixed-body reference frame is located in the center of mass of the rigid body.

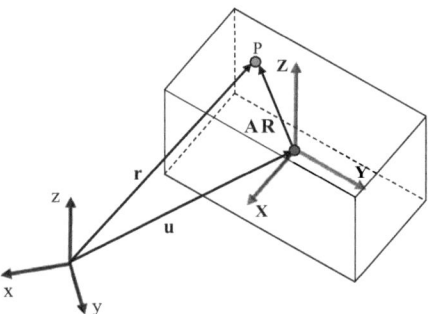

Figure 2.1: Position of a material point with respect to a fixed-body coordinate system

2.1.1 Parametrization of the Rotation Matrix

The 3×3 rotation matrix $\mathbf{A}$ in the three-dimensional Euclidean space can be parameterized in different ways. The most important property of all rotation matrices in $\mathbb{R}^3$ is the orthogonality with a determinant of $\det(\mathbf{A}) = +1$, considering Cartesian coordinates and a right hand coordinate system [74, 139]. Therefore, rotation matrices are a member of a *Lie-group* and they are called SO(3), which stands for *special orthogonal group of order 3*.

The elements of the 3×3 rotation matrix are not linearly independent. Rather the orientation can be described by three linearly independent variables. These variables are e.g. rotational angles of elementary rotations. In such a case the rotational axes are coincident with the coordinate axes. Confirming to the three basis vectors, three elementary rotation matrices are known [133].

In the following section the general terms $\cos \varphi$ and $\sin \varphi$ are abbreviated by $\cos \varphi \rightarrow c\varphi$ and $\sin \varphi \rightarrow s\varphi$, respectively. If φ denotes a rotation about x-, y- or z-axis according to the right hand rule, following rotation matrices can be found:

$$\mathbf{A}_x(\varphi) = \begin{bmatrix} 1 & 0 & 0 \\ 0 & c\varphi & -s\varphi \\ 0 & s\varphi & c\varphi \end{bmatrix}, \mathbf{A}_y(\varphi) = \begin{bmatrix} c\varphi & 0 & s\varphi \\ 0 & 1 & 0 \\ -s\varphi & 0 & c\varphi \end{bmatrix}, \mathbf{A}_z(\varphi) = \begin{bmatrix} c\varphi & -s\varphi & 0 \\ s\varphi & c\varphi & 0 \\ 0 & 0 & 1 \end{bmatrix} \quad (2.2)$$

In the following section the main parts concerning to the rotation matrix $\mathbf{A}$ are taken from [133, 139].

2.1. KINEMATICS OF RIGID BODIES

Euler angles ϕ, ψ, θ (rotation about the axes $z_0 - x_1 - z_2$)

The rotation matrix $\mathbf{A}$ is formulated by a rotation about the z_0-axis followed by the resulting x_1-axis and the new z_2-axis. The corresponding Euler angles are called ϕ, ψ, θ and the rotations are expressed by the matrix product

$$\mathbf{A} = \mathbf{A}_z(\phi)\mathbf{A}_x(\psi)\mathbf{A}_z(\theta) \tag{2.3}$$

However, it should be mentioned that Euler angles are not unique and sometimes other conventions are used [139]. If the elementary rotation matrices (2.2) are inserted into (2.3), the rotation matrix $\mathbf{A}$ results in

$$\mathbf{A} = \begin{bmatrix} c\theta c\phi - c\psi s\theta s\phi & -c\phi s\theta - c\theta c\psi s\phi & s\phi s\psi \\ c\phi c\psi s\theta + c\theta s\phi & c\theta c\phi c\psi - s\theta s\phi & -c\phi s\psi \\ s\theta s\psi & c\theta s\psi & c\psi \end{bmatrix} \tag{2.4}$$

The Euler angles can also be found from the rotation matrix (2.4). Therefore it is convenient to use the sparse coordinates.

$$\psi = \arccos(A_{33}), \quad \phi = \arcsin\left(\frac{A_{13}}{\sin\psi}\right), \quad \theta = \arccos\left(\frac{A_{32}}{\sin\psi}\right) \tag{2.5}$$

Euler angles are used to describe the orientation of a rigid body in *Adams* [27].

Tait-Bryan angles (Cardan angles, Nautical angles, yaw-pitch-roll) α, β, γ (rotation about the axes $x_0 - y_1 - z_2$)

Especially in aeronautics Tait-Bryan angles are used in connection with the maneuvers yaw, pitch and roll. Rotations are performed via x_0, y_1 and z_2.

$$\mathbf{A} = \mathbf{A}_x(\alpha)\mathbf{A}_y(\beta)\mathbf{A}_z(\gamma) \tag{2.6}$$

If the elementary rotations (2.2) are inserted in (2.6), the rotation matrix $\mathbf{A}$ yields

$$\mathbf{A} = \begin{bmatrix} c\beta c\gamma & -c\beta s\gamma & s\beta \\ c\alpha s\gamma + s\alpha s\beta c\gamma & c\alpha c\gamma - s\alpha s\beta s\gamma & -s\alpha c\beta \\ s\alpha s\gamma - c\alpha s\beta c\gamma & s\alpha c\gamma + c\alpha s\beta s\gamma & c\alpha c\beta \end{bmatrix} \tag{2.7}$$

The Tait-Bryan angles can be calculated from the rotation matrix (2.7) as well.

$$\beta = \arcsin(A_{13}), \quad \alpha = \arccos\left(\frac{A_{33}}{\cos\beta}\right), \quad \gamma = \arccos\left(\frac{A_{11}}{\cos\beta}\right) \tag{2.8}$$

Parametrization via a rotation vector

The rotation matrix $\mathbf{A}$ can also be interpreted as a linear mapping from an initial

2.1. KINEMATICS OF RIGID BODIES

vector $\mathbf{r}_0$ to a final vector $\mathbf{r}_1 = \mathbf{A}\mathbf{r}_0$. Then the rotation matrix $\mathbf{A}$ can be expressed by

$$\mathbf{A} = \mathbf{I}_{33} + \sin\theta\, \tilde{\mathbf{a}} + (1 - \cos\theta)\, \tilde{\mathbf{a}}^2 \tag{2.9}$$

Eq. (2.9) is known as *Rodriguez formula*. The term $\mathbf{I}_{33}$ denotes a 3×3 identity matrix. The rotation angle θ is defined as a rotation of the initial vector $\mathbf{r}_0$ about a rotation axis defined by the unit vector $\mathbf{a}$. $\tilde{\mathbf{a}}$ is formed by the skew-symmetric matrix

$$\tilde{\mathbf{a}} = \begin{bmatrix} 0 & -a_3 & a_2 \\ a_3 & 0 & -a_1 \\ -a_2 & a_1 & 0 \end{bmatrix} \tag{2.10}$$

Details and the derivation of the Rodriguez formula (2.9) can be found in [139].

Euler parameters e_0, e_1, e_2, e_3

The parameters e_0, e_1, e_2 and e_3 are called Euler parameters and can be derived from the Rodriguez formula (2.9) [139]:

$$e_0 = \cos\frac{\theta}{2},\ e_1 = a_x \sin\frac{\theta}{2},\ e_2 = a_y \sin\frac{\theta}{2},\ e_3 = a_z \sin\frac{\theta}{2} \tag{2.11}$$

The four Euler parameters (also called quaternions) express the rotation θ and the orientation of the rotation vector. However, the Euler parameters (2.11) are not independent of each other and therefore the constraint equation

$$e_0^2 + e_1^2 + e_2^2 + e_3^2 = 1 \tag{2.12}$$

must be fulfilled. As a consequence, the rotation matrix is again specified by three *independent* parameters.

$$\mathbf{A} = \begin{bmatrix} e_0^2 + e_1^2 - e_2^2 - e_3^2 & 2e_1e_2 - 2e_0e_3 & 2e_1e_3 + 2e_0e_2 \\ 2e_1e_2 + 2e_0e_3 & e_0^2 + e_2^2 - e_3^2 - e_1^2 & 2e_2e_3 - 2e_0e_1 \\ 2e_1e_3 - 2e_0e_2 & 2e_2e_3 + 2e_0e_1 & e_0^2 + e_3^2 - e_1^2 - e_2^2 \end{bmatrix} \tag{2.13}$$

Rodriguez parameters γ_1, γ_2, γ_3

Rodgriguez parameters can be derived by normalization of the Euler parameters (2.11).

$$\gamma_1 = \frac{e_1}{e_0},\ \gamma_2 = \frac{e_2}{e_0},\ \gamma_3 = \frac{e_3}{e_0} \tag{2.14}$$

By using definitions (2.14), the rotation matrix (2.13) can be transformed to

$$\mathbf{A} = \frac{1}{1 + \gamma_1^2 + \gamma_2^2 + \gamma_3^2} \begin{bmatrix} 1 + \gamma_1^2 - \gamma_2^2 - \gamma_3^2 & 2\gamma_1\gamma_2 - 2\gamma_3 & 2\gamma_1\gamma_3 + 2\gamma_2 \\ 2\gamma_1\gamma_2 + 2\gamma_3 & 1 + \gamma_2^2 - \gamma_3^2 - \gamma_1^2 & 2\gamma_2\gamma_3 - 2\gamma_1 \\ 2\gamma_1\gamma_3 - 2\gamma_2 & 2\gamma_2\gamma_3 + 2\gamma_1 & 1 + \gamma_3^2 - \gamma_1^2 - \gamma_2^2 \end{bmatrix} \tag{2.15}$$

2.1. KINEMATICS OF RIGID BODIES

Comparison of the different parameterizations

It can be shown that the rotation matrices $\mathbf{A}$, which are formulated by Euler angles (2.4), Tait-Bryan angles (2.7) or by a rotation vector (2.9), can run into singularities. As a consequence the relation between rotation angles and rotation matrix is not bijective.

In the case of Euler angles the critical values of the second angle $\psi = 0$ or $\psi = \pi$ must be avoided. In Eq. (2.5) it can be seen that the denominator is then equal to zero ($\sin \psi = 0$) and the angles ϕ and θ are singular. In this case the z-axes have the same or the opposite direction. If the z-axis and the Z-axis coincide, ϕ and θ can be added and the individual values are not unique any more. In the configuration $\phi + \theta = 0$ the rotation matrix yields the identity matrix $\mathbf{A} = \mathbf{I_{33}}$.

In the case of Tait-Bryan angles the value $\beta = \pi/2$ is not allowed for a unique definition. In Eq. (2.8) it can be seen that denominator would be equal to zero ($\cos \beta = 0$) and the angles α and γ would be singular.

If the rotation angle θ is equal to zero, the rotation matrix (2.9) is equal to the identity matrix $\mathbf{A} = \mathbf{I_{33}}$ and furthermore the rotation vector $\mathbf{a}$ is not uniquely defined.

Euler parameters (2.11) have the benefit that no singularities can occur. This effect results due to the implementation of four parameters instead of three parameters, which are used in the previous methods. As a consequence the constraint equation (2.12) must hold at any time of the simulation.

Rodriguez parameters (2.14) can be handled without a constraint equation. However, singularities occur if the angle $\theta = \pi$. By considering the definitions (2.14) and (2.11) it can be seen that $\gamma_{x,y,z} = a_{x,y,z} \cdot \tan \frac{\theta}{2}$, which violates the domain of the tangent function $D_{\tan} = \mathbb{R} \backslash \{\pm \frac{\pi}{2}, \pm \frac{3\pi}{2}, \pm \frac{5\pi}{2}, \ldots\}$.

Generally, it can be shown that all parameterizations that use three parameters can have singular problems. However, a compensation with complementary angles is possible in principle [133]. Because of all these reasons Euler parameters are mostly used in computational multibody dynamics.

2.1.2 Velocity of a Rigid Body

The velocity of a material point can be calculated by the total derivative of Eq. (2.1).

$$\dot{\mathbf{r}} = \mathbf{v} = \dot{\mathbf{u}} + \dot{\mathbf{A}}\mathbf{R} \qquad (2.16)$$

The vector $\mathbf{R}$ is constant in the case of a rigid body and therefore the derivative $\dot{\mathbf{R}}$ vanishes. It can be shown that the derivative of the rotation matrix $\dot{\mathbf{A}}$ is equal to [139]

$$\dot{\mathbf{A}} = \mathbf{A}\tilde{\mathbf{\Omega}} \qquad (2.17)$$

2.1. KINEMATICS OF RIGID BODIES

where the skew symmetric matrix $\tilde{\mathbf{\Omega}} = \mathbf{A}^T\dot{\mathbf{A}}$ defines the angular velocity vector $\mathbf{\Omega}$ in the body-fixed coordinate system.

$$\tilde{\mathbf{\Omega}} = \begin{bmatrix} 0 & -\Omega_z & \Omega_y \\ \Omega_z & 0 & -\Omega_x \\ -\Omega_y & \Omega_x & 0 \end{bmatrix}, \quad \mathbf{\Omega} = \begin{bmatrix} \Omega_x \\ \Omega_y \\ \Omega_z \end{bmatrix} \quad (2.18)$$

If Eq. (2.17) and (2.18) are inserted into (2.16), the velocity $\mathbf{v}$ can be expressed as

$$\mathbf{v} = \dot{\mathbf{u}} + \mathbf{A}\tilde{\mathbf{\Omega}}\mathbf{R} = \dot{\mathbf{u}} + \mathbf{A}\left(\mathbf{\Omega} \times \mathbf{R}\right) \quad (2.19)$$

which is a combination of translational and rotational velocity.

2.1.3 Formulation of the Angular Velocity

The skew symmetric matrix $\tilde{\mathbf{\Omega}}$ is formed by $\tilde{\mathbf{\Omega}} = \mathbf{A}^T\dot{\mathbf{A}}$ [139]. If the total derivative of $\dot{\mathbf{A}}$ is calculated, the equation extends to

$$\tilde{\mathbf{\Omega}} = \mathbf{A}^T\dot{\mathbf{A}} = \sum_{i=1}^{n_r} \mathbf{A}^T \frac{\partial \mathbf{A}}{\partial q_i^r} \dot{q}_i^r \quad (2.20)$$

All terms where the rotation matrix and its derivative appear are summarized in a skew-symmetric matrix $\tilde{\mathbf{H}}^i$.

$$\tilde{\mathbf{H}}^i = \mathbf{A}^T \frac{\partial \mathbf{A}}{\partial q_i^r} \quad (2.21)$$

Only the elements $\mathbf{H}^i = \begin{bmatrix} \tilde{H}^i_{32}, & \tilde{H}^i_{13}, & \tilde{H}^i_{21} \end{bmatrix}^T$ have to be considered due to the fact that $\tilde{\mathbf{\Omega}}$ is a skew symmetric matrix. Eq. (2.20) can now be written as matrix-vector product.

$$\mathbf{\Omega} = \mathbf{H}\dot{\mathbf{q}}^r \quad (2.22)$$

The matrix $\mathbf{H}$ depends on $\mathbf{q}^r$ and maps the derivatives of the rotational DOFs $\dot{\mathbf{q}}^r$ to the angular velocity vector $\mathbf{\Omega}$.

Now the goal is to find the matrices $\mathbf{H}$ for the specific parameterizations of the rotation matrix. Therefore, the multiplication of the rotation matrix and its partial derivatives by Eq. (2.21) is applied to the rotation matrices for Euler angles (2.4), Tait-Bryan angles (2.7), Euler parameters (2.13) and Rodriguez parameters (2.15).

Euler angles

$$\mathbf{H} = \begin{bmatrix} s\theta s\psi & c\theta & 0 \\ c\theta s\psi & -s\theta & 0 \\ c\psi & 0 & 1 \end{bmatrix} \quad (2.23)$$

Tait-Bryan angles

$$\mathbf{H} = \begin{bmatrix} c\beta c\gamma & s\gamma & 0 \\ -c\beta s\gamma & c\gamma & 0 \\ s\beta & 0 & 1 \end{bmatrix} \quad (2.24)$$

Euler parameters

$$\mathbf{H} = 2 \begin{bmatrix} -e_1 & e_0 & e_3 & -e_2 \\ -e_2 & -e_3 & e_0 & e_1 \\ -e_3 & e_2 & -e_1 & e_0 \end{bmatrix} \quad (2.25)$$

Eq. (2.25) can be written in the compact form

$$\mathbf{H} = 2\left[-\mathbf{e}, -\tilde{\mathbf{e}} + e_0 \mathbf{I}_{33}\right] \quad (2.26)$$

where the vector $\mathbf{e}$ and the skew symmetric matrix $\tilde{\mathbf{e}}$ are given by

$$\mathbf{e} = \begin{bmatrix} e_1 \\ e_2 \\ e_3 \end{bmatrix}, \quad \tilde{\mathbf{e}} = \begin{bmatrix} 0 & -e_3 & e_2 \\ e_3 & 0 & -e_1 \\ -e_2 & e_1 & 0 \end{bmatrix} \quad (2.27)$$

Rodriguez parameters

$$\mathbf{H} = \frac{2}{1+\gamma_1^2+\gamma_2^2+\gamma_3^2} \begin{bmatrix} 1 & \gamma_3 & -\gamma_2 \\ -\gamma_3 & 1 & \gamma_1 \\ \gamma_2 & -\gamma_1 & 1 \end{bmatrix} \quad (2.28)$$

These basics of kinematics are used to formulate the kinetic equations of motion for rigid multibody systems. The extension to flexible MBS will be given in section 2.9.

2.2 Newton-Euler equations

Newton's second law *(lex secunda)* is the fundamental law of mechanics. Newton's equations in combination with Euler's equations form the laws for linear and angular momentum. For a MBS with p rigid bodies the Newton-Euler equations reads [133]:

$$m_i \, \mathbf{a}_i(t) = \mathbf{f}_i(t), \quad (2.29a)$$
$$\mathbf{I}_i \, \dot{\mathbf{\Omega}}_i(t) + \tilde{\mathbf{\Omega}}_i(t) \, \mathbf{I}_i \, \mathbf{\Omega}_i(t) = \mathbf{M}_i(t), \quad i = 1 \ldots p \quad (2.29b)$$

To establish Eq. (2.29) a free body diagram has to be created. Thereby, forces and torques that act on each body can either be classified into outer and inner forces and torques or into applied and reaction forces and torques.

$$\mathbf{f}_i = \mathbf{f}_i^o + \mathbf{f}_i^i = \mathbf{f}_i^a + \mathbf{f}_i^r$$

$$\mathbf{M}_i = \mathbf{M}_i^o + \mathbf{M}_i^i = \mathbf{M}_i^a + \mathbf{M}_i^r$$

It is important to note that Newton's equations (2.29a) have to be formulated with respect to the inertial frame. The variable m_i denotes the mass of body i, $\mathbf{a}_i(t)$ the translational acceleration ($\mathbf{a}_i = \ddot{\mathbf{q}}^t$) and $\mathbf{f}_i$ summarizes all forces that act at body i. The small letters indicate that forces and accelerations are related to the inertial frame.

Euler's equations (2.29b) can either be formulated with respect to a fixed-body reference frame or the inertial frame. Typically, they are expressed in a fixed body reference frame, which is in contrast to Newton's equations. The advantage of that coordinate system is that the tensor of inertia $\mathbf{I}_i$ is constant at any time. The capital letters of the angular velocity vector $\mathbf{\Omega}_i(t)$, the skew-symmetric matrix $\tilde{\mathbf{\Omega}}_i(t)$ and the torques $\mathbf{M}_i$ illustrate the formulation regarding to the fixed body reference frame.

Newton-Euler equations are used e.g. in the MBS-codes *Neweul* or *LMS Virtual.Lab Motion* to formulate the equations of motion [157].

The big disadvantage of Eq. (2.29) is that six equations are required for each body. As a consequence lots of unknowns are introduced in the system of equations. Eventually, all the forces are not of interest. In order to describe the motion of a MBS, reaction forces are not needed. Hence, variational principles based on *d'Alemberts principle* are of great interest. In these formulations the reaction forces are a priori eliminated. The resulting system of equations consists of as many equations as generalized coordinates are introduced.

2.3 Types of Constraints

Typically the dynamics of MBS are not formulated with generalized (minimal, independent) coordinates but rather with redundant (dependent) coordinates. This means that the number of coordinates exceeds the number of DOFs in the system. Redundant coordinates are more convenient with regard to an automatic procedure for the formalism of the equations of motion [157]. As a consequence algebraic constraint equations have to be introduced to fulfill the kinematic boundary conditions. Further on, constraint equations are needed, if a closed loop appears in the topology of the MBS. The constraints can be classified into following groups:

Holonomic Constraints:

The constraints only depend on the positions $\mathbf{q}$ of the MBS. Holonomic constraints are always independent of the velocities $\mathbf{v} = d\mathbf{q}/dt$.

If the time t does not appear in the constraint equations, then they are called *scleronomous holonomic constraints*:

$$\mathbf{g}(\mathbf{q}) = \mathbf{0}, \quad \mathbf{g} : \mathbb{R}^n \to \mathbb{R}^{n_\lambda} \tag{2.30}$$

This type of constraints is most commonly used in MBS. If the constraints also depend explicitly on the time t, they are called *rheonomous holonomic constraints*:

$$\mathbf{g}(\mathbf{q}, t) = \mathbf{0}, \quad \mathbf{g} : \mathbb{R}^{n+1} \to \mathbb{R}^{n_\lambda} \tag{2.31}$$

Nonholonomic Constraints:
Nonholonomic constraints additionally depend explicitly on the velocities **v** of the MBS. They cannot be transformed into holonomic constraints by integration. This means that there are more DOFs in the positions than in the velocities [157]. The following form illustrates *nonholonomic constraints*, which include both, the scleronomous as well as the rheonomous type:

$$\mathbf{g}(\mathbf{q}, \dot{\mathbf{q}}, t) = \mathbf{0}, \quad \mathbf{g} : \mathbb{R}^{2n+1} \rightarrow \mathbb{R}^{n_\lambda} \tag{2.32}$$

2.4 Variational Principles

In this section variational principles starting with d'Alembert's principle of virtual work, Jourdain's principle of virtual power, Gauß' principle of least constraint and Hamilton's principle are discussed. Further on, Lagrange's equations of first and second kind are derived from Hamilton's principle.

2.4.1 Principles of d'Alembert, Jourdain and Gauß

Firstly, the variational principles of d'Alembert, Jourdain and Gauß are given for point systems, which are furthermore extended for multibody systems.

The positional coordinates for a specific point or particle are given by the vector **r**, which is a geometrical function of n generalized coordinates $q_1, \ldots, q_n$. If a dynamical system is considered, it also depends explicitly on the time t.

$$\mathbf{r} = \mathbf{r}(q_1, \ldots, q_n, t) \tag{2.33}$$

At a fixed time t the virtual variations $\delta q_1, \ldots, \delta q_n$ can be considered. The variations are called virtual variations, because the changes do not have to coincide with the real dynamics of the system. However, the variations have to be geometrically possible, i.e. no constraints are violated. Furthermore, only small variations are considered. The positional variation of a point is then given by:

$$\delta \mathbf{r} = \sum_{j=1}^{n} \frac{\partial \mathbf{r}}{\partial q_j} \delta q_j \tag{2.34}$$

Principle of d'Alembert:
The principle of d'Alembert is based on virtual works, which result from the forces acting on a particle times the virtual displacements. The variations of the positional coordinates are performed at a fixed time t while variations of velocities and accelerations are equal to zero.

$$\delta \mathbf{r} \neq \mathbf{0}, \quad \delta \mathbf{v} = \mathbf{0}, \quad \delta \mathbf{a} = \mathbf{0}, \quad \delta t = 0 \tag{2.35}$$

Hence, the principle of d'Alembert in the Lagrangian version [133] reads as:

$$\delta W = \sum_{i=1}^{p} (m_i \mathbf{a}_i - \mathbf{f}_i^a) \cdot \delta \mathbf{r}_i = 0 \tag{2.36}$$

2.4. VARIATIONAL PRINCIPLES

The sum of all virtual works for p points resulting from inertia forces $m_i \mathbf{a}_i$ and *applied forces* $\mathbf{f}_i^a$ (impressed forces) is equal to zero. The virtual works of the reaction forces $\mathbf{f}_i^r$ (constraint forces) is zero. Therefore, constraint forces are a priori not considered.

The principle of d'Alembert is only suitable for all holonomic systems. It can be shown [133] that Lagrange's equations of the second kind can be derived from Eq. (2.36). The general equation of d'Alembert principle (2.36) is called principle of virtual work, if static systems are considered. In such a case the inertia forces $m_i \mathbf{a}_i$ vanish. d'Alembert's principle is used to formulate the equations of motion in the commercial software *Simpack* [157].

Principle of Jourdain:
Jourdain's principle is based on virtual power, which is calculated by forces times virtual velocities. At a fixed time t the velocities are variated, while the variations of positional coordinates and accelerations are kept identical to zero. The variations of the velocities does not have to be small but they must be compatible with the system.

$$\delta \mathbf{r} = \mathbf{0}, \quad \delta \mathbf{v} \neq \mathbf{0}, \quad \delta \mathbf{a} = \mathbf{0}, \quad \delta t = 0 \qquad (2.37)$$

Jourdain's principle can be defined as:

$$\delta P = \sum_{i=1}^{p} (m_i \mathbf{a}_i - \mathbf{f}_i^a) \cdot \delta \mathbf{v}_i = 0 \qquad (2.38)$$

The virtual power of the constraint forces disappear in Jourdain's principle, i.e. constraint forces are not taken into account. Jourdain's principle is suitable for nonholonomic systems and is implemented in the MBS-code *Mesa Verde* [157]. Jourdain's principle leads to Kane's equations of motion [130].

Principle of Gauß:
In the principle of Gauß, a variation of the accelerations is performed. Positions and velocities are not varied as well as the time.

$$\delta \mathbf{r} = \mathbf{0}, \quad \delta \mathbf{v} = \mathbf{0}, \quad \delta \mathbf{a} \neq \mathbf{0}, \quad \delta t = 0 \qquad (2.39)$$

Thus, the Principle of Gauß reads as follows:

$$\delta C = \sum_{i=1}^{p} (m_i \mathbf{a}_i - \mathbf{f}_i^a) \cdot \delta \mathbf{a}_i = 0 \qquad (2.40)$$

Gauß' principle can be interpreted as principle that minimizes the constraint due to the averaged acceleration divergence. Hence, it is also called principle of least constraint. Nonholonomic systems can be considered as well. However, Gauß' principle has not been reached greater technical relevance till now [133].

D'Alembert's principle for rigid MBS:

2.4. VARIATIONAL PRINCIPLES

The mechanical principles above can all be extended from point systems to multibody systems. For illustrative purposes, it is only shown for d'Alembert's principle. Based on Newton-Euler's equations (2.29) d'Alembert's principle for MBS reads as follows:

$$\delta W = \sum_{i=1}^{p} \left[(m_i \mathbf{a}_i - \mathbf{f}_i^a) \cdot \delta \mathbf{r}_i + \left(\mathbf{I}_i \cdot \dot{\mathbf{\Omega}}_i + \tilde{\mathbf{\Omega}}_i \cdot \mathbf{I}_i \cdot \mathbf{\Omega}_i - \mathbf{M}_i^a \right) \cdot \delta \boldsymbol{\varphi}_i \right] = 0 \qquad (2.41)$$

Beside the virtual displacements $\delta \mathbf{r}$ also virtual rotations $\delta \boldsymbol{\varphi}$ have to be considered. Applied forces $\mathbf{f}^a$ are regarded as well as applied torques $\mathbf{M}^a$ in Eq. (2.41).

2.4.2 Hamilton's Principle

Hamilton's principle of stationary action reads as follows:
Amongst all possible motions, which can be performed by a conservative system from a given initial position within a given time into a given end position, the one is going to occur in the nature for which the action integral

$$S = \int_{t_1}^{t_2} (L + W_{nc}) dt \qquad (2.42)$$

results in a stationary value (extremum).
A derivation of Hamilton's principle can be found in [133, 157]. The Lagrangian L for constrained mechanical systems is defined as:

$$L(\mathbf{q}, \dot{\mathbf{q}}, \boldsymbol{\lambda}) = T(\mathbf{q}, \dot{\mathbf{q}}) - V(\mathbf{q}) - \mathbf{g}(\mathbf{q})^T \boldsymbol{\lambda} \qquad (2.43)$$

Here, the functional $T = T(\mathbf{q}, \dot{\mathbf{q}})$, $T : \mathbb{R}^{2n} \to \mathbb{R}$ denotes the kinetic energy and the functional $V = V(\mathbf{q})$, $V : \mathbb{R}^n \to \mathbb{R}$ the potential energy. Each of the n_λ constraints (2.30) is associated to a corresponding Lagrange multiplier λ_i. Furthermore, W_{nc} denotes the work done by non-conservative forces $\mathbf{Q}$.
In order that the action integral (2.42) vanishes, the variation δS also has to be zero. As a consequence a variation with respect to each coordinate is carried out:

$$\begin{aligned}\delta S &= \int_{t_1}^{t_2} (\delta L + \delta W_{nc}) dt \\ &= \int_{t_1}^{t_2} \left[\frac{\partial T}{\partial \dot{\mathbf{q}}} \delta \dot{\mathbf{q}} + \frac{\partial T}{\partial \mathbf{q}} \delta \mathbf{q} - \frac{\partial V}{\partial \mathbf{q}} \delta \mathbf{q} - \left(\frac{\partial \mathbf{g}}{\partial \mathbf{q}} \right)^T \cdot \boldsymbol{\lambda} \delta \mathbf{q} - \mathbf{g}^T \delta \boldsymbol{\lambda} + \delta W_{nc} \right] dt = 0\end{aligned}$$
(2.44)

Eq. (2.44) is also known as constrained Lagrange-d'Alembert principle [84, 96, 104]. By using the calculus of variations, the equations of motion can be derived from the above principle (2.44). For constrained mechanical systems this results in Lagrange's equations of the first kind, for unconstrained systems, in Lagrange's equations of the second kind.

2.5 Lagrange's Equations of the First Kind, Descriptor Form

In this section constrained multibody systems are considered. Lagrange's equations are derived from Hamilton's principle.
A partial integration of the first term in Eq. (2.44) is performed [148]:

$$\int_{t_1}^{t_2} \frac{\partial T}{\partial \dot{\mathbf{q}}} \delta \dot{\mathbf{q}} \, dt = \left(\frac{\partial T}{\partial \dot{\mathbf{q}}} \delta \mathbf{q} \right) \bigg|_{t_1}^{t_2} - \int_{t_1}^{t_2} \frac{d}{dt} \left(\frac{\partial T}{\partial \dot{\mathbf{q}}} \right) \delta \mathbf{q} \, dt \qquad (2.45)$$

The first term of the right side in Eq. (2.45) can be canceled due to the fact that the variation at the boarders vanishes: $\delta \mathbf{q}(t_0) = \delta \mathbf{q}(t_1) = 0$. This intermediate result is inserted in Eq. (2.44). Furthermore the Lagrangian $L(\mathbf{q}, \dot{\mathbf{q}}) = T(\mathbf{q}, \dot{\mathbf{q}}) - V(\mathbf{q})$ is recalled.

$$\int_{t_1}^{t_2} \left[-\frac{d}{dt}\left(\frac{\partial L}{\partial \dot{\mathbf{q}}}\right) + \frac{\partial L}{\partial \mathbf{q}} - \left(\frac{\partial \mathbf{g}}{\partial \mathbf{q}}\right)^T \cdot \boldsymbol{\lambda} \right] \delta \mathbf{q} \, dt - \int_{t_1}^{t_2} \mathbf{g}^T \delta \boldsymbol{\lambda} \, dt + \int_{t_1}^{t_2} \delta W_{nc} \, dt \stackrel{!}{=} 0 \qquad (2.46)$$

Regarding to the fundamental lemma of variational calculus, the expression inside the brackets in Eq. (2.46) must be zero for all times t [148]. This leads to *Lagrange's equations of the first kind*, also known as *Euler-Lagrange* equations.

$$\frac{d}{dt}\left(\frac{\partial L}{\partial \dot{\mathbf{q}}}\right) - \frac{\partial L}{\partial \mathbf{q}} + \left(\frac{\partial \mathbf{g}(\mathbf{q})}{\partial \mathbf{q}}\right)^T \cdot \boldsymbol{\lambda} = \mathbf{Q}(\mathbf{q}, \dot{\mathbf{q}}, t) \qquad (2.47a)$$

$$\mathbf{g}(\mathbf{q}) = \mathbf{0} \qquad (2.47b)$$

In Eq. (2.47a) the product $\left(\frac{\partial \mathbf{g}(\mathbf{q})}{\partial \mathbf{q}}\right)^T \cdot \boldsymbol{\lambda}$ denotes the constraint forces, which ensure that the constraints are fulfilled. $\mathbf{Q}(\mathbf{q}, \dot{\mathbf{q}}, t)$ is the vector of non-conservative forces that results from W_{nc}. The constraint equations $\mathbf{g}(\mathbf{q}) = \mathbf{0}$ have to be fulfilled at all times, which can cause numerical difficulties, cf. section 2.8.
For scleronomous systems the kinetic energy can be written in the quadratic form:

$$T(\mathbf{q}, \dot{\mathbf{q}}) = \frac{1}{2} \dot{\mathbf{q}}^T \mathbf{M}(\mathbf{q}) \dot{\mathbf{q}} \qquad (2.48)$$

In Eq. (2.48) $\mathbf{M}(\mathbf{q})$ denotes the symmetric positive definite mass matrix, where mass and inertia terms are included. The derivatives of the Lagrangian in Eq. (2.47) can be written as:

$$\frac{d}{dt}\left(\frac{\partial L(\mathbf{q}, \dot{\mathbf{q}})}{\partial \dot{\mathbf{q}}}\right) = \frac{d}{dt}\left(\frac{\partial T(\mathbf{q}, \dot{\mathbf{q}})}{\partial \dot{\mathbf{q}}}\right) = \frac{\partial^2 T(\mathbf{q}, \dot{\mathbf{q}})}{\partial \dot{\mathbf{q}}^2} \ddot{\mathbf{q}} + \frac{\partial^2 T(\mathbf{q}, \dot{\mathbf{q}})}{\partial \mathbf{q} \partial \dot{\mathbf{q}}} \dot{\mathbf{q}}$$

$$\frac{\partial L(\mathbf{q}, \dot{\mathbf{q}})}{\partial \mathbf{q}} = \frac{\partial T(\mathbf{q}, \dot{\mathbf{q}})}{\partial \mathbf{q}} - \frac{\partial V(\mathbf{q})}{\partial \mathbf{q}}$$

2.5. LAGRANGE'S EQUATIONS OF THE FIRST KIND, DESCRIPTOR FORM

and the specific components can further be defined as follows:

$$\frac{\partial^2 T(\mathbf{q},\dot{\mathbf{q}})}{\partial \dot{\mathbf{q}}^2} = \mathbf{M}(\mathbf{q})$$

$$\frac{\partial^2 T(\mathbf{q},\dot{\mathbf{q}})}{\partial \mathbf{q} \partial \dot{\mathbf{q}}} = \frac{\partial \mathbf{M}(\mathbf{q})\dot{\mathbf{q}}}{\partial \mathbf{q}} = \frac{d}{dt}\mathbf{M}(\mathbf{q}) = \dot{\mathbf{M}}(\mathbf{q},\dot{\mathbf{q}})$$

$$\frac{\partial T(\mathbf{q},\dot{\mathbf{q}})}{\partial \mathbf{q}} = \frac{1}{2}\dot{\mathbf{q}}^T \dot{\mathbf{M}}(\mathbf{q},\dot{\mathbf{q}})$$

Hence, the vector of generalized forces $\mathbf{f}(\mathbf{q},\dot{\mathbf{q}},t)$ can be defined. It includes applied external forces, non-conservative forces, generalized Coriolis- and centrifugal forces as well as conservative (potential) forces.

$$\mathbf{f}(\mathbf{q},\dot{\mathbf{q}},t) := \mathbf{Q}(\mathbf{q},\dot{\mathbf{q}},t) - \dot{\mathbf{M}}(\mathbf{q},\dot{\mathbf{q}})\dot{\mathbf{q}} + \frac{1}{2}\left(\dot{\mathbf{q}}^T \dot{\mathbf{M}}(\mathbf{q},\dot{\mathbf{q}})\right)^T - \left(\frac{\partial V(\mathbf{q})}{\partial \mathbf{q}}\right)^T \quad (2.49)$$

By summing up all these terms and the constraint forces $\mathbf{G}^T \boldsymbol{\lambda}$ the equations of motion for constrained MBS can be formulated. $\mathbf{G}^T$ denotes the constraint Jacobian and is calculated by the partial derivatives

$$\mathbf{G}(\mathbf{q})^T = D\left(\mathbf{g}(\mathbf{q})\right)^T = \left(\frac{\partial \mathbf{g}(\mathbf{q})}{\partial \mathbf{q}}\right)^T \quad (2.50)$$

$$\mathbf{M}(\mathbf{q})\ddot{\mathbf{q}} - \mathbf{f}(\mathbf{q},\dot{\mathbf{q}},t) + \mathbf{G}(\mathbf{q})^T \boldsymbol{\lambda} = \mathbf{0} \quad (2.51a)$$
$$\mathbf{g}(\mathbf{q}) = \mathbf{0} \quad (2.51b)$$

Eq. (2.51a) consist of n differential equations for for the coordinates $q_1 \ldots q_n$ and (2.51b) of m algebraic equations for the Lagrange multipliers $\lambda_1 \ldots \lambda_m$. Hence, the system (2.51) is a set of differential-algebraic equations (DAEs). The form (2.51) of the equation of motion is called *descriptor form*. The descriptor form can also be found by applying other principles of mechanics. Lagrange's equations of the first kind are used to formulate the equations of motion in *Adams* [157]. Mathematical properties and numerical solutions of DAEs are discussed in chapter 2.8.

The principles of d'Alembert and Jourdain can also be formulated for constrained MBS. If the general form

$$\bar{\mathbf{f}} = \sum_{i=1}^{p} \begin{bmatrix} m_i \mathbf{a}_i - \mathbf{f}_i^a \\ \mathbf{I}_i \dot{\boldsymbol{\Omega}}_i + \tilde{\boldsymbol{\Omega}}_i \mathbf{I}_i \boldsymbol{\Omega}_i - \mathbf{M}_i^a \end{bmatrix}$$

is introduced and the variations $\delta \mathbf{r}$ and $\delta \boldsymbol{\varphi}$ are summarized in $\delta \mathbf{q}^T = [\delta \mathbf{r}, \delta \boldsymbol{\varphi}]^T$, Eq. (2.41) read as [157]:

$$\text{d'Alembert}: \quad \delta W = \delta \mathbf{q}^T \bar{\mathbf{f}} = 0, \quad \delta \mathbf{q}^T \mathbf{G}^T \boldsymbol{\lambda} = 0 \quad (2.52a)$$
$$\text{Jourdain}: \quad \delta P = \delta \mathbf{v}^T \bar{\mathbf{f}} = 0, \quad \delta \mathbf{v}^T \mathbf{G}^T \boldsymbol{\lambda} = 0 \quad (2.52b)$$

2.6 Lagrange's Equations of the Second Kind

For MBS, where minimal coordinates are used, Eq. (2.44) simplifies to:

$$\begin{aligned}\delta S &= \int_{t_1}^{t_2} (\delta L + \delta W_{nc})\,dt \\ &= \int_{t_1}^{t_2} \left(\frac{\partial T}{\partial \dot{\mathbf{q}}}\delta\dot{\mathbf{q}} + \frac{\partial T}{\partial \mathbf{q}}\delta\mathbf{q} - \frac{\partial V}{\partial \mathbf{q}}\delta\mathbf{q} + \delta W_{nc} \right) dt = 0 \end{aligned} \quad (2.53)$$

Furthermore, Eq. (2.46) simplifies to:

$$\int_{t_1}^{t_2} \left[-\frac{d}{dt}\left(\frac{\partial L}{\partial \dot{\mathbf{q}}}\right) + \frac{\partial L}{\partial \mathbf{q}} \right] \delta\mathbf{q}\,dt + \int_{t_1}^{t_2} \delta W_{nc}\,dt \stackrel{!}{=} 0 \quad (2.54)$$

As a consequence, Lagrange's equation of the second kind read as:

$$\frac{d}{dt}\left(\frac{\partial L}{\partial \dot{\mathbf{q}}}\right) - \frac{\partial L}{\partial \mathbf{q}} = \mathbf{Q}\left(\mathbf{q},\dot{\mathbf{q}},t\right) \quad (2.55)$$

By using the definition of the Lagrangian for unconstrained systems $L = T - V$, Eq. (2.55) can be written as:

$$\frac{d}{dt}\left(\frac{\partial T\left(\mathbf{q},\dot{\mathbf{q}}\right)}{\partial \dot{\mathbf{q}}}\right) - \frac{\partial T\left(\mathbf{q},\dot{\mathbf{q}}\right)}{\partial \mathbf{q}} + \frac{\partial V\left(\mathbf{q}\right)}{\partial \mathbf{q}} = \mathbf{Q}\left(\mathbf{q},\dot{\mathbf{q}},t\right) \quad (2.56)$$

If the definitions for the mass matrix, its derivative and the definition for the generalized forces are used from section 2.5, Eq. (2.55), (2.56) can be written as:

$$\mathbf{M}(\mathbf{q})\ddot{\mathbf{q}} - \mathbf{f}(\mathbf{q},\dot{\mathbf{q}},t) = \mathbf{0} \quad (2.57)$$

Eq. (2.57) can be seen as generalization of Newton's second law. It should be recalled that in Eq. (2.51) the vector $\mathbf{q}$ denotes dependent (redundant) coordinates, while in Eq. (2.57) it denotes independent (minimal, generalized) coordinates.

In commercial MBS software packages different mechanical principles are applied to derive the equations of motion. Newton-Euler equations are used in *Neweul* or in *LMS Virtual.Lab Motion (DADS)*. Jourdain's principle is used e.g. in *Mesa Verde* and d'Alembert's principle in *Simpack*. *Adams* makes use of Lagrange's equations of the first kind [157].

2.7 Kinetic and Potential Energy of Rigid Bodies

In order to formulate Lagrange's equations of the first kind (2.47) or second kind (2.55), it is necessary to find the kinetic energy $T(\mathbf{q},\dot{\mathbf{q}})$ as well as the potential

2.7. KINETIC AND POTENTIAL ENERGY OF RIGID BODIES

energy $V(\mathbf{q})$ of the rigid bodies.
The kinetic energy of a rigid body is given by

$$T = \frac{1}{2} \int_m \mathbf{v}^T \mathbf{v} \, dm \qquad (2.58)$$

where the velocity vector $\mathbf{v}$ is defined in Eq. (2.19). Eq. (2.58) can be split into

$$T = T^t + T^r = \frac{1}{2} m \mathbf{v}_M^2 + \frac{1}{2} \mathbf{\Omega} \mathbf{I}_M \mathbf{\Omega} \qquad (2.59)$$

where $\mathbf{v}_M$ denotes the velocity of the center of mass and $\mathbf{I}_M$ denotes the tensor of inertia. The first term T^t denotes the translational part and the second term T^r the rotational term of the kinetic energy. It should be noted that the translational part is formulated with respect to the inertial frame and the rotational part with respect to the fixed body reference frame.

$$\mathbf{I}_M = \int_m \tilde{\mathbf{R}}^T \tilde{\mathbf{R}} \, dm = \int_m \begin{bmatrix} Y^2 + Z^2 & -XY & -ZX \\ -XY & Z^2 + X^2 & -YZ \\ -ZX & -YZ & X^2 + Y^2 \end{bmatrix} dm \qquad (2.60)$$

The skew symmetric matrix $\tilde{\mathbf{R}}$ is defined in the same way as Eq. (2.18) or (2.27). If Eq. (2.22), i.e. $\mathbf{\Omega} = \mathbf{H}\dot{\mathbf{q}}^r$ is inserted into Eq. (2.59), it results in

$$T = \frac{1}{2} m \left(\dot{\mathbf{q}}^t \right)^T \dot{\mathbf{q}}^t + \left(\dot{\mathbf{q}}^r \right)^T \mathbf{H}^T \mathbf{I}_M \mathbf{H} \dot{\mathbf{q}}^r \qquad (2.61)$$

As a consequence, the symmetric positive definite regular mass matrix $\mathbf{M}$ can be written in the form [139]

$$\mathbf{M} = \begin{bmatrix} m\mathbf{I} & 0 \\ 0 & \mathbf{H}^T \mathbf{I}_M \mathbf{H} \end{bmatrix} \qquad (2.62)$$

if the generalized coordinates are arranged in the order $\mathbf{q} = [\mathbf{q}^t, \mathbf{q}^r]$.
The potential energy can be calculated by

$$V = \int_m [0, 0, g] \mathbf{r} \, dm = [0, 0, g] \mathbf{u} \, m \qquad (2.63)$$

if the gravity field is in the negative e_z-direction. The vector $\mathbf{r}$ is a material point expressed in the coordinates of an inertial frame and the vector $\mathbf{u}$ is the position of the body-fixed reference frame, which is located in the center of mass of the rigid body, cf. Fig. 2.1.
Furthermore, the potential energy can also be expressed for any (nonlinear) spring with the spring force $f_c(s)$.

$$V_c = \int_{s_0}^{s} f_c(\xi) \, d\xi \qquad (2.64)$$

31

In the case of a linear spring with a spring force of $f_c = cs$ Eq. (2.64) results in

$$V_c = \frac{1}{2}c\left(s - s_0\right)^2 \tag{2.65}$$

2.8 Numerical Solution of DAEs

The dynamics of constrained MBS is usually described by a set of DAEs (2.51), known as descriptor system. These equations result from an augmented formulation where redundant coordinates are used. The challenge is to solve the differential equations (2.51a) while the constraint equations (2.51b) have to be fulfilled at all times. In the following sections different numerical approaches are discussed to solve the set of DAEs.

2.8.1 Index of the Descriptor Form

The index of a DAE can be seen as a measure as how different the DAE is from an ODE [157]. In the literature several definitions exist about the index. Unfortunately, they are not equivalent for all classes of DAEs. In this section DAEs for MBS are considered, i.e. the descriptor form of the type (2.51). Index definitions and numerical solution techniques are taken from [157].
The implicit differential system

$$\mathbf{F}(\mathbf{x}, \dot{\mathbf{x}}, \mathbf{u}, t) = \mathbf{0} \tag{2.66}$$

is a generalization of the descriptor system (2.51). The Jacobian matrix $\partial \mathbf{F}/\partial \dot{\mathbf{x}}$ may be singular. Generally, a *semi-explicit system of DAEs* or an *ODE with constraints* can also be written in the following form [4, 92]:

$$\dot{\mathbf{x}} = \mathbf{f}(\mathbf{x}, \mathbf{u}, t) \tag{2.67a}$$
$$\mathbf{0} = \mathbf{g}(\mathbf{x}, \mathbf{u}, t) \tag{2.67b}$$

The state of the system is represented by a vector $\mathbf{x}(t) \in \mathbb{R}^{2n}$ (differential variables). $\mathbf{u}(t) \in \mathbb{R}^m$ denotes the vector of control-inputs of the system (algebraic variables). If $\mathbf{u}$ is unknown, this equation is underdetermined [39]. The index of a DAE is defined as follows: [157]
Eq. (2.66) has differential index $i = k$ if k is the minimum number of analytical differentiations

$$\begin{aligned} \mathbf{F}(\mathbf{x}, \dot{\mathbf{x}}, \mathbf{u}, t) &= \mathbf{0}, \\ \dot{\mathbf{F}}(\mathbf{x}, \dot{\mathbf{x}}, \mathbf{u}, t) &= \mathbf{0}, \\ &\cdots, \\ \mathbf{F}^{(k)}(\mathbf{x}, \dot{\mathbf{x}}, \mathbf{u}, t) &= \mathbf{0} \end{aligned} \tag{2.68}$$

such that Eq. (2.68) can be transformed by algebraic manipulations into an explicit ODE $\dot{\mathbf{x}} = \mathbf{F}(\mathbf{x})$ (which is called the underlying ODE). Such a definition is very useful in understanding the mathematical structure of the DAE and hence in selecting an appropriate numerical method [4].

2.8.2 Index Reduction

The descriptor system (2.51) can be transformed into a system that includes only second order differential equations. Therefore, the constraint equations (2.51b) have to be differentiated with respect to time. The first and the second derivative are given by:

$$\mathbf{G}\dot{\mathbf{q}} = 0 \qquad (2.69a)$$
$$\mathbf{G}\ddot{\mathbf{q}} + \dot{\mathbf{G}}\dot{\mathbf{q}} = 0 \qquad (2.69b)$$

Eq. (2.69a) summarizes the constraints at velocity level and Eq. (2.69b) are the constraints at acceleration level [140].
The *index 2 system* can now be written in the form:

$$\mathbf{M}\ddot{\mathbf{q}} - \mathbf{f} + \mathbf{G}^T\boldsymbol{\lambda} = 0 \qquad (2.70a)$$
$$\mathbf{G}\dot{\mathbf{q}} = 0 \qquad (2.70b)$$

Furthermore, Eq. (2.69b) can be written in the form:

$$\mathbf{G}\ddot{\mathbf{q}} = \boldsymbol{\gamma}(\mathbf{q},\dot{\mathbf{q}}) \Rightarrow \boldsymbol{\gamma}(\mathbf{q},\dot{\mathbf{q}}) := -\dot{\mathbf{G}}\dot{\mathbf{q}} \qquad (2.71)$$

The constraint Jacobian $\mathbf{G}(\mathbf{q})$ and the constraints at acceleration level (2.71) can be described analytically for every type of joint. Now the constraints at positional level (2.51b) can be replaced by the constraints at acceleration level (2.71) due to the fact that both equations are mathematically equivalent. As a consequence, the descriptor system (2.51) is transformed to the *index 1 system*:

$$\mathbf{M}\ddot{\mathbf{q}} - \mathbf{f} + \mathbf{G}^T\boldsymbol{\lambda} = 0 \qquad (2.72a)$$
$$\mathbf{G}\ddot{\mathbf{q}} = \boldsymbol{\gamma} \qquad (2.72b)$$

The *index 1 system* (2.72) can also be written in matrix form:

$$\begin{bmatrix} \mathbf{M} & \mathbf{G}^T \\ \mathbf{G} & 0 \end{bmatrix} \begin{bmatrix} \ddot{\mathbf{q}} \\ \boldsymbol{\lambda} \end{bmatrix} = \begin{bmatrix} \mathbf{f} \\ \boldsymbol{\gamma} \end{bmatrix} \qquad (2.73)$$

The system (2.73) is called *index 1 equation*, because it is fully determined by a set of first order differential equations. The original descriptor system (2.51) is an *index 3 system*. It can be seen that the *index 1 system* results from two differentiations of the *index 3 system*.
The matrix on the left side of (2.73) can be summarized:

$$\mathcal{A}(\mathbf{q}) := \begin{bmatrix} \mathbf{M}(\mathbf{q}) & \mathbf{G}(\mathbf{q})^T \\ \mathbf{G}(\mathbf{q}) & 0 \end{bmatrix} \qquad (2.74)$$

2.8. NUMERICAL SOLUTION OF DAES

It is assumed that the constraint matrix $\mathbf{G}(\mathbf{q})$ has full rank and that the mass matrix $\mathbf{M}(\mathbf{q})$ is positive definite in the null space $\ker(\mathbf{G})$ of $\mathbf{G}(\mathbf{q})$, i.e.

$$\mathbf{y}^T \mathbf{M}(\mathbf{q}) \mathbf{y} > 0 \quad \forall \; y \in \ker(\mathbf{G})$$

Then the system matrix $\boldsymbol{\mathcal{A}}(\mathbf{q})$ in (2.73) is non-singular [157]. If $\boldsymbol{\mathcal{A}}(\mathbf{q})$ is non-singular, (2.73) can be solved for $\ddot{\mathbf{q}}$ and $\boldsymbol{\lambda}$:

$$\begin{bmatrix} \ddot{\mathbf{q}} \\ \boldsymbol{\lambda} \end{bmatrix} = \boldsymbol{\mathcal{A}}^{-1}(\mathbf{q}) \cdot \begin{bmatrix} \mathbf{f}(\mathbf{q}, \dot{\mathbf{q}}, t) \\ \boldsymbol{\gamma}(\mathbf{q}, \dot{\mathbf{q}}) \end{bmatrix} \qquad (2.75)$$

The second order differential equation for $\ddot{\mathbf{q}}$ can be solved by transforming the system into a system of first order differential equations. With $\mathbf{v} = \dot{\mathbf{q}}$ the state vector results in $\mathbf{x} = [\mathbf{q}, \mathbf{v}]^T$. The initial states $\mathbf{x}(t_0) = [\mathbf{q}_0, \mathbf{v}_0]^T$ must satisfy the constraint equations for positions $\mathbf{g}(\mathbf{q}_0) = \mathbf{0}$ and velocities $\mathbf{G}(\mathbf{q}_0) \cdot \mathbf{v}_0 = \mathbf{0}$. The initial value problem (IVP) for the state vector $\mathbf{x}(t)$ can be solved by standard methods. Based on the solution of $\mathbf{x}(t)$ the vector of Lagrange multipliers $\boldsymbol{\lambda}(t)$ can be calculated by the second equation in (2.75).

Several numerical methods for the solution of IVPs are known. However, the best suited method always depends on the type of the problem. A general algorithm for all types of problems cannot be selected. Single-step methods are e.g. the explicit and implicit Euler method or different Runge-Kutta methods. Extrapolation methods are also single-step methods that can be explicit or implicit and are also available for stiff systems. ADAMS-methods are multi-step methods, which are based on predictor-corrector iterations. ADAMS-methods are not suited for stiff systems. Backward-differentiation formula (BDF)-methods are implicit multistep-methods that are well suited for stiff systems [17]. A comparison of different numerical algorithms is given in Table 2.1.

2.8.3 Drift Problem of the Index 1 System

Although the system of index 1 (2.73) is mathematically equivalent to the original descriptor system of index 3 (2.51), numerical problems can occur. With standard integration methods only the constraints at acceleration level $\mathbf{G}\ddot{\mathbf{q}} = \boldsymbol{\gamma}$ can be observed. The discretized system can drift off from the original system and lower-order constraint equations (positional and velocity constraints) are violated due to accumulation of integration truncation errors [20]. The errors usually tend to increase in time and as a consequence the simulation results are not reliable any more. The drift effect is graphically illustrated in Fig. 2.2. To avoid this drift, either a stabilization method can be applied or a projection method can be used.

2.8.4 Stabilization Methods

Gear-Gupta-Leimkuhler (GGL) stabilization:
The index 2 system (2.70) is stabilized by using an additional term at velocity level.

2.8. NUMERICAL SOLUTION OF DAES

	advantages	disadvantages
Runge-Kutta methods	• robust • small overhead • well suited for small accuracies	• high effort • many function evaluations • fixed order
Extrapolation methods	• variable order and step size • highly accurate	• large overhead • too expensive for small accuracies • sparsely robust (problems with unsteady systems)
ADAMS methods	• less effort (2-3 function evaluations per step) • variable order and step size • well suited for complex functions	• high overhead (step size control)
BDF methods	• variable order and step size • stable (also for stiff systems)	• relatively high effort due to implicit character (step size control, solution of implicit equations) • high overhead

Table 2.1: Comparison of different numerical methods for the solution of IVPs [17]

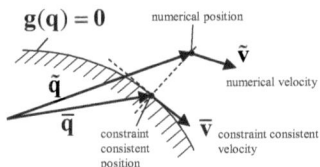

Figure 2.2: Constraint violation by the numerical solutions $\tilde{\mathbf{q}}(t)$ and $\tilde{\mathbf{v}}(t)$ [20]

A new Lagrange multiplier $\boldsymbol{\nu}$, which vanishes in the exact solution, is introduced and hence the *stabilized index 2 formulation* [35, 50, 78, 92] yields

$$\dot{\mathbf{q}} - \mathbf{v} + \mathbf{G}^T \boldsymbol{\nu} = \mathbf{0} \tag{2.76a}$$

$$\mathbf{M}\dot{\mathbf{v}} - \mathbf{f} + \mathbf{G}^T \boldsymbol{\lambda} = \mathbf{0} \tag{2.76b}$$

$$\mathbf{g} = \mathbf{0} \tag{2.76c}$$

$$\mathbf{G}\mathbf{v} = \mathbf{0} \tag{2.76d}$$

The original constraint equation $\mathbf{g} = \mathbf{0}$ is added to the system as well. As a consequence the number of equations is increased, i.e. an overdetermined system of DAEs occurs. However, the advantage of this method is that position-level constraints and velocity-level constraints are automatically enforced. Hence, the drift problem is eliminated for these constraints [35]. The solver *DASSL* or the revised version *DDASKR* provide implicit multi-step methods for the solution of such DAEs.

2.8. NUMERICAL SOLUTION OF DAES

The original *DASSL*-code is based on a backward differentiation formula (BDF) and is designed to solve index 1 systems [35]. A modified *DASSL*-integrator is e.g. used in *Simpack* [130]. Furthermore, commercial MBS-solvers as implemented e.g. in *Adams* have the possibility to solve either the index 3 equations or the stabilized index 2 equations (e.g.: *GSTIFF I3/SI2*).

Baumgarte stabilization:
In this formulation the constraint equations of different index-levels are combined with weighting factors $\alpha > 0$ and $\beta > 0$ [140].

$$\ddot{\mathbf{g}} + \alpha \dot{\mathbf{g}} + \beta \mathbf{g} = \mathbf{0} \qquad (2.77)$$

The first and second time derivative of the holonomic constraints $\mathbf{g}(\mathbf{q})$ are known from Eqs. (2.69)

$$\dot{\mathbf{g}} = \mathbf{G}\dot{\mathbf{q}} = \mathbf{0} \qquad (2.78a)$$
$$\ddot{\mathbf{g}} = \mathbf{G}\ddot{\mathbf{q}} - \boldsymbol{\gamma} = \mathbf{0} \qquad (2.78b)$$

Theoretically, the weighting factors can optimally be adapted to a certain step size by $\alpha = 1/h$ and $\beta = 1/h^2$ [130]. Eq. (2.77) can be compared with a one-mass-oscillator.

$$\ddot{g} + 2\zeta\omega_0 \dot{g} + \omega_0^2 g = 0 \qquad (2.79)$$

The dynamics of a one-mass oscillator is characterized by the undamped eigenfrequency ω_0 and the viscous damping ζ. Hence, the weighting factors can be chosen to $\alpha = 2\zeta\omega_0$ and $\beta = \omega_0^2$. By combining Eqs. (2.78) and (2.79), the system (2.73) results in

$$\begin{bmatrix} \mathbf{M} & \mathbf{G}^T \\ \mathbf{G} & \mathbf{0} \end{bmatrix} \begin{bmatrix} \ddot{\mathbf{q}} \\ \boldsymbol{\lambda} \end{bmatrix} = \begin{bmatrix} \mathbf{f} \\ \boldsymbol{\gamma} - 2\zeta\omega_0 \mathbf{G}\dot{\mathbf{q}} - \omega_0^2 \mathbf{g} \end{bmatrix} \qquad (2.80)$$

A disadvantage of Baumgarte's stabilization method is that artificial stiffness is introduced into the system if the parameters are not chosen properly [157].

Projection method:
In a projection method the state vector $\mathbf{x}(t)$ has to be projected back onto the constraint manifold from time to time. The goal is that the necessary update for $\mathbf{q}$ and $\mathbf{v}$ is minimal. If $\mathbf{q}_{mod}$ denotes the updated vector of coordinates, then $\Delta \mathbf{q} = \mathbf{q}_{mod} - \mathbf{q}$ is the change of $\mathbf{q}$. The projection algorithm should furthermore be designed in order that the norm $||\Delta \mathbf{q}||$ is a minimum.
This yields an optimization problem with a constraint condition (nonlinear constrained least squares problem) [157].

$$\mathbf{g}(\mathbf{q} + \Delta \mathbf{q}) = \mathbf{0} \qquad (2.81a)$$
$$||\Delta \mathbf{q}||_\mathbf{A} = \frac{1}{2} \Delta \mathbf{q}^T \mathbf{A} \Delta \mathbf{q} \to \min \qquad (2.81b)$$

2.8. NUMERICAL SOLUTION OF DAES

$$\mathbf{G} \cdot (\mathbf{v} + \Delta\mathbf{v}) = \mathbf{0} \quad (2.82a)$$

$$||\Delta\mathbf{v}||_{\mathbf{A}} = \frac{1}{2}\Delta\mathbf{v}^T \mathbf{A} \Delta\mathbf{v} \to \min \quad (2.82b)$$

The velocity update algorithm is executed after the the position update algorithm. The projected values fulfill both, position and velocity constraints. Then these values are used to advance the numerical solution. Details about projection methods can be found e.g. in [4, 50, 157].

2.8.5 Index 3 Solver

By using single-step methods of small order, the nonlinear constraint equations (2.51b) can also be solved within an integration step [50, 130]. If an Euler-step with the step size h is applied to the equations of motion (2.51a), it yields:

$$\mathbf{v}_{t+h} = \mathbf{v}_t + h \cdot \mathbf{M}^{-1}\left(\mathbf{f} - \mathbf{G}^T \boldsymbol{\lambda}\right) \quad (2.83)$$

The new velocity state $\mathbf{v}_{t+h}$ depends on the constraint forces $\mathbf{G}^T \boldsymbol{\lambda}$, which are initially unknown. Applied and gyroscopic forces $\mathbf{f}$ can be calculated based on the actual positional coordinates $\mathbf{q}_t$ and the actual velocities $\mathbf{v}_t$. Another Euler-step applied to the kinematic equations $\dot{\mathbf{q}} = \mathbf{v}$ results in:

$$\mathbf{q}_{t+h} = \mathbf{q}_t + h \cdot \mathbf{v}_{t+h} = \mathbf{q}_t + h \cdot \left(\mathbf{v}_t + h \cdot \mathbf{M}^{-1}\left(\mathbf{f} - \mathbf{G}^T \boldsymbol{\lambda}\right)\right) \quad (2.84)$$

In Eq. (2.84) the new velocity state $\mathbf{v}_{t+h}$ is an implicit part for the calculation of the new positional coordinates $\mathbf{q}_{t+h}$. Certainly, the new state $\mathbf{q}_{t+h} = \mathbf{q}_{t+h}(\boldsymbol{\lambda})$ has to fulfill the constraint equations:

$$\mathbf{g}\left(\mathbf{q}_{t+h}(\boldsymbol{\lambda})\right) = \mathbf{g}\left(\boldsymbol{\lambda}\right) = \mathbf{0} \quad (2.85)$$

The constraint equations (2.85) are typically nonlinear. Therefore, they can only be resolved iteratively with regard to $\boldsymbol{\lambda}$. If an approximated value $\boldsymbol{\lambda}_i$ is known from the previous integration step, the improved value $\boldsymbol{\lambda}_{i+1}$ can be calculated by Newton's method [130].

$$\boldsymbol{\lambda}_{i+1} = \boldsymbol{\lambda}_i - \mathbf{g}(\boldsymbol{\lambda}_i) \cdot \left(\frac{\partial \mathbf{g}(\boldsymbol{\lambda}_i)}{\partial \boldsymbol{\lambda}_i}\right)^{-1} \quad (2.86)$$

The partial derivative $\frac{\partial \mathbf{g}}{\partial \boldsymbol{\lambda}}$ is calculated by:

$$\frac{\partial \mathbf{g}}{\partial \boldsymbol{\lambda}} = \frac{\partial \mathbf{g}}{\partial \mathbf{q}} \cdot \frac{\partial \mathbf{q}}{\partial \boldsymbol{\lambda}} \quad (2.87)$$

and from Eq. (2.84) and the definition of the constraint Jacobian $\mathbf{G} = \frac{\partial \mathbf{g}}{\partial \mathbf{q}}$ it follows:

$$\frac{\partial \mathbf{g}}{\partial \boldsymbol{\lambda}} = -h^2 \, \mathbf{G} \, \mathbf{M}^{-1} \, \mathbf{G}^T \quad (2.88)$$

Therefore, the derivative $\frac{\partial \mathbf{g}}{\partial \boldsymbol{\lambda}}$ can be calculated with variables that are already known. Consequently, the algorithm consists of three main parts. Firstly, the Lagrange multipliers $\boldsymbol{\lambda}_{i+1}$ have to be computed by using Eq. (2.86). Secondly, the velocities $\mathbf{v}_{t+h}$ can be calculated based on these Lagrange multipliers and the variables from the previous step with Eq. (2.83). Thirdly, the positional coordinates $\mathbf{q}_{t+h}$ are computed based on the velocities $\mathbf{v}_{t+h}$ and the position from the previous step $\mathbf{q}_t$ by (2.84).

It should be noted that the Euler-method is a method of first order $\mathcal{O}(h)$, i.e. it only provides accuracies in the dimension of the step size h. As a consequence, this method should only be used for simulations with small accuracy demands [130].

Commercial MBS software tools use e.g. the Hilber-Hughes-Taylor (HHT) method or the implicit Newmark formula to integrate the index 3 system [130]. The implicit Runge-Kutta code $RADAU5$ is designed to integrate DAEs of index 1, 2 or 3 [78]. These methods are basically similar to the method above, but a higher accuracy of order $\mathcal{O}(h)^2$ is used to solve the index 3 DAEs [157].

2.9 Flexible Multibody Systems

In the industrial examples in chapters 8.2 and 8.3 flexible bodies are included in the MBS as well. This yields a hybrid multibody system. Therefore, the implementation of flexible bodies should be briefly discussed in this section.

2.9.1 Overview

Deformations can either be elastic or inelastic. Several inelastic material models like elasto-plastic, viscous or combined material behavior exist [6]. For this reason flexible MBS are considered, which is more general than elastic MBS. From the field of finite elements the distinction between (a) small deformations, (b) large deformations with small strains and (c) large deformations with large strains is well known.

For the implementation of flexible bodies in MBS several formulations are used, e.g.:

- floating frame of reference formulation
- incremental formulation
- large rotation vector formulation
- absolute nodal coordinate formulation

For the implementation of plastic material behavior especially the absolute nodal coordinate formulation has inspired researchers for the last years. Details about the formulation can be found in several publications [70, 139], just to mention some key works.

However, in the industrial application in chapters 8.1, 8.2 and 8.3 only small deformations are considered due to the type of problem. Hence, the focus in this section

2.9. FLEXIBLE MULTIBODY SYSTEMS

is in the implementation of linear elastic models and therefore in the formulation of a floating reference frame. This reference frame moves along with the body and if the body is rigid, it coincides with the fixed body frame. The position of the reference frame is described by the rigid body coordinates $\mathbf{q}^t$ (translational part) and $\mathbf{q}^r$ (rotational part). The elastic coordinates $\mathbf{q}^f$ describe the deformation with respect to the floating reference frame. The disadvantage of the formulation with a floating reference frame is that the mass matrix is much more complicated and not constant.

In a numerical analysis the continuum problem has to be discretized in space. From the finite element theory the well known *Ritz-Ansatz* is used to find spatial shape functions [139].

$$\mathbf{w}(x,y,z,t) \approx \begin{bmatrix} a_1(t)\Phi_1(x,y,z) + \cdots + a_k(t)\Phi_k(x,y,z) \\ b_1(t)\Psi_1(x,y,z) + \cdots + b_l(t)\Psi_l(x,y,z) \\ c_1(t)\Theta_1(x,y,z) + \cdots + c_m(t)\Theta_m(x,y,z) \end{bmatrix} \quad (2.89)$$

In Eq. (2.89) $\mathbf{w}(x,y,z,t)$ denotes the displacement field of a deformable body. The displacement field is defined by pre-defined shape functions $\Phi_1(x,y,z), \cdots, \Phi_k(x,y,z)$, $\Psi_1(x,y,z), \cdots, \Psi_l(x,y,z)$ and $\Theta_1(x,y,z), \cdots, \Theta_m(x,y,z)$ and is furthermore scaled by coefficients $a_1(t), \cdots, a_k(t)$, $b_1(t), \cdots, b_l(t)$ and $c_1(t), \cdots, c_m(t)$. The vector

$$\mathbf{q}^f(t) := \begin{bmatrix} a_1(t) \ldots a_k(t), & b_1(t) \ldots b_l(t), & c_1(t) \ldots c_m(t) \end{bmatrix}^T \quad (2.90)$$

forms the vector of degrees of freedom (DOFs) of the flexible body.

The floating reference frame should be defined in a way that the displacements of the material points can be assumed to be small with regard to this frame. The vector $\mathbf{R} = [X, Y, Z]^T$ denotes the position of a material point *with respect to the floating frame* in a predefined reference configuration. Furthermore, $\mathbf{W}(x,y,z,t)$ denotes the displacement field of the flexible body. Now Eq. 2.89 can be split into a matrix of shape functions $\mathbf{N}(X,Y,Z)$ and the vector of elastic DOFs $\mathbf{q}^f(t)$.

$$\mathbf{W}(X,Y,Z,t) = \mathbf{N}(X,Y,Z) \cdot \mathbf{q}^f(t) \quad (2.91)$$

The position of a material point *with respect to an inertial frame* is then given by:

$$\mathbf{r} = \mathbf{u} + \mathbf{A}\left(\mathbf{R} + \mathbf{W}\right) \quad (2.92)$$

The term $\mathbf{u} + \mathbf{AR}$ describes a rigid body motion while the term $\mathbf{AW}$ expresses the deformation of the body. Fig. 2.3 illustrates the description of a flexible body with respect to a floating reference frame. The blue coordinate systems symbolizes an inertial frame and the red coordinate system the floating reference frame of the body. The position of the floating reference frame is described by the vector $\mathbf{u}$. The position in the deformed state is described by the vector $\mathbf{r}$, which is formulated by Eq. (2.92).

The kinetics of a single flexible body can be described e.g. by Jourdain's principle

2.9. FLEXIBLE MULTIBODY SYSTEMS

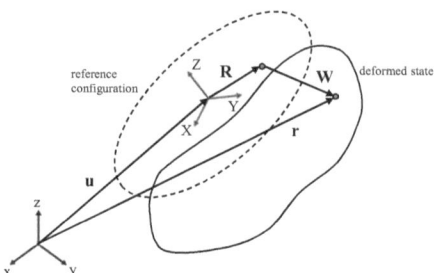

Figure 2.3: Floating reference frame

[133] (principle of virtual power):

$$\delta P = \int_V \delta \mathbf{v} \cdot \mathbf{a}\, dm + \int_V \delta \boldsymbol{\varepsilon} \cdot \boldsymbol{\sigma}\, dm - \int_V \delta \mathbf{v} \cdot \mathbf{k}\, dm - \int_V \delta \mathbf{v} \cdot \mathbf{p}\, dm - \int_V \delta \mathbf{v}^k \cdot \mathbf{f}^k + \delta \mathbf{v}^k \cdot \mathbf{m}^k\, dm \tag{2.93}$$

The first integral in Eq. (2.93) denotes inertia forces and the second integral inner forces. The third, fourth and fifth integral express volumetric forces, surface forces and single forces and torques, respectively.

In the calculation of the kinetic energy, which is necessary for the equations of motion, care must be taken. The kinetic energy is generally calculated by $T = \frac{1}{2} \int_m \mathbf{v}^T \mathbf{v}\, dm$, Eq. (2.58). However, the velocity vector $\mathbf{v}$ cannot be calculated by Eq. (2.19) any more, because of the flexibility in the body. Rather, the expression for the velocity has to be extended.

$$\begin{aligned}
\mathbf{v} &= \dot{\mathbf{u}} + \dot{\mathbf{A}}\left(\mathbf{R} + \mathbf{W}\right) + \mathbf{A}\dot{\mathbf{W}} \\
&= \mathbf{A}\left[\mathbf{A}^T \dot{\mathbf{u}} + \boldsymbol{\Omega} \times \left(\mathbf{R} + \mathbf{W}\right) + \dot{\mathbf{W}}\right] \\
&= \mathbf{A}\left[\mathbf{A}^T \dot{\mathbf{u}} - \left(\tilde{\mathbf{R}} + \tilde{\mathbf{W}}\right)\boldsymbol{\Omega} + \dot{\mathbf{W}}\right]
\end{aligned} \tag{2.94}$$

Now the Eq. (2.22) and the Ritz-Ansatz (2.91), i.e.

$$\dot{\mathbf{u}} = \dot{\mathbf{q}}^t, \quad \boldsymbol{\Omega} = \mathbf{H}\dot{\mathbf{q}}^r, \quad \mathbf{W} = \mathbf{N}(X, Y, Z)\mathbf{q}^f$$

can be inserted into Eq. (2.94). This results in

$$\mathbf{v} = \mathbf{A}\left[\mathbf{A}^T \dot{\mathbf{q}}^t - \left(\tilde{\mathbf{R}} + \tilde{\mathbf{W}}\right)\mathbf{H}\dot{\mathbf{q}}^r + \mathbf{N}\dot{\mathbf{q}}^f\right] \tag{2.95}$$

By inserting the velocity (2.95) into the kinetic energy, the mass matrix can be calculated. If the coordinates are arranged in $\mathbf{q} = \begin{bmatrix}\mathbf{q}^t, \mathbf{q}^r, \mathbf{q}^f\end{bmatrix}^T$, the mass matrix is formed by:

$$\mathbf{M} = \begin{bmatrix} \mathbf{M}_{tt} & \mathbf{M}_{tr} & \mathbf{M}_{tf} \\ & \mathbf{M}_{rr} & \mathbf{M}_{rf} \\ symm. & & \mathbf{M}_{ff} \end{bmatrix} \tag{2.96}$$

2.9. FLEXIBLE MULTIBODY SYSTEMS

The specific terms in Eq. (2.96) can be found in [139]. In (2.96) only the submatrices $\mathbf{M}_{tt}$ and $\mathbf{M}_{ff}$ are constant. All other submatrices depend on the rotational DOFs $\mathbf{q}^r$ or the elastic DOFs $\mathbf{q}^f$ [139].

2.9.2 Determination of Shape Functions

Different methods can be used to describe the shape functions $\mathbf{N}(X, Y, Z)$ for the Ritz-Ansatz (2.91). A short overview should be given in the following list.

- Eigenfunctions and static functions of continuum models like beams or plates
- Finite element discretization
- Eigenfunctions and static functions of finite element discretizations
- Spline functions, assumed modes, frequency response modes

The direct implementation of a finite element model is not discussed here. Methods that keep the number of DOFs small are of interest. Hence, a mapping to a lower dimensional space of the elastic coordinates $\mathbf{q}^f$ is of interest. In the following section the static reduction, the modal reduction and the Craig-Bampton reduction are discussed, which are needed for the component mode synthesis (CMS).

2.9.3 Guyan Reduction (Static Reduction)

The deformation of a flexible body can be described in many cases, if the displacements of just a few points are known. In the Guyan reduction specific *master nodes* and *slave nodes* are defined. Furthermore, the slave-DOFs are expressed by the master-DOFs. The displacements of the slave nodes are calculated from the static deformation, which results from the displacements of the master nodes, even in the dynamic case. The nodal DOFs are split into master and slave components $\mathbf{u} = [\mathbf{u}_M, \mathbf{u}_S]^T$, Fig. 2.4. The goal is to calculate $\mathbf{u}_S$, if $\mathbf{u}_M$ is known. Therefore the

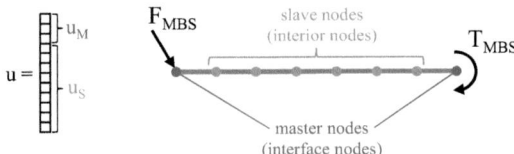

Figure 2.4: Master nodes and slave nodes in a FEM of a beam

static FEM-problem $\mathbf{Ku} = \mathbf{f}$ is split into:

$$\begin{bmatrix} \mathbf{K}_{MM} & \mathbf{K}_{MS} \\ \mathbf{K}_{SM} & \mathbf{K}_{SS} \end{bmatrix} \cdot \begin{bmatrix} \mathbf{u}_M \\ \mathbf{u}_S \end{bmatrix} = \begin{bmatrix} \mathbf{f}_M \\ \mathbf{f}_S \end{bmatrix} = \begin{bmatrix} \mathbf{f}_M \\ \mathbf{0} \end{bmatrix} \quad (2.97)$$

The term $\mathbf{f}_S$ in Eq. (2.97) is set to zero $\mathbf{f}_S = \mathbf{0}$, because the slave nodes should only follow the motion of the master nodes, i.e. no external forces act on the slave nodes.

2.9. FLEXIBLE MULTIBODY SYSTEMS

The forces $\mathbf{f}_M$ have to be applied in order to deflect the master nodes. The motion of the slave nodes can be calculated from Eq. (2.97):

$$\mathbf{u}_S = -\mathbf{K}_{SS}^{-1}\mathbf{K}_{SM}\mathbf{u}_M \qquad (2.98)$$

The vector of all nodal DOFs is given by:

$$\mathbf{u} = \begin{bmatrix} \mathbf{u}_M \\ \mathbf{u}_S \end{bmatrix} = \begin{bmatrix} \mathbf{I}_{MM} \\ -\mathbf{K}_{SS}^{-1}\mathbf{K}_{SM} \end{bmatrix} \mathbf{u}_M \qquad (2.99)$$

In Eq. (2.99) $\mathbf{I}_{MM}$ denotes the $(n_M \times n_M)$ identity matrix. Furthermore, (2.99) is given in the form

$$\mathbf{u} = \mathbf{\Phi}\mathbf{q}^f \qquad (2.100)$$

where the matrix $\mathbf{\Phi}$ and the vector $\mathbf{q}^f$ are formulated by:

$$\mathbf{\Phi} = \begin{bmatrix} \mathbf{I}_{MM} \\ -\mathbf{K}_{SS}^{-1}\mathbf{K}_{SM} \end{bmatrix}, \quad \mathbf{q}^f = \mathbf{u}_M \qquad (2.101)$$

To attain the kth column of $\mathbf{\Phi}$, the kth component of $\mathbf{u}_M$ has to be set to one and all other components to zero. If a static FEM-analysis is performed, the resulting displacements are used to fill the column.

2.9.4 Modal Reduction

It is assumed that the flexible body is modeled with finite elements. For small displacements the equations of motion are given by the linear differential equation

$$\mathbf{M}\ddot{\mathbf{u}} + \mathbf{C}\dot{\mathbf{u}} + \mathbf{K}\mathbf{u} = \mathbf{f} \qquad (2.102)$$

The vector of nodal DOFs is given by $\mathbf{u} \in \mathbb{R}^n$, the mass matrix by $\mathbf{M} \in \mathbb{R}^{n \times n}$, the damping matrix by $\mathbf{C} \in \mathbb{R}^{n \times n}$ and the stiffness matrix by $\mathbf{K} \in \mathbb{R}^{n \times n}$. The force vector $\mathbf{f} \in \mathbb{R}^n$ contains externally applied forces and reaction forces due to its connection to adjacent components at boundary DOFs. The basis of a modal reduction is an undamped autonomous system

$$\mathbf{M}\ddot{\mathbf{u}} + \mathbf{K}\mathbf{u} = \mathbf{0} \qquad (2.103)$$

The harmonic approach $\mathbf{u} = \boldsymbol{\phi}\cos(\omega t)$ or $\mathbf{u} = \boldsymbol{\phi}\sin(\omega t)$ leads to the eigenvalue problem [140]:

$$\left(\mathbf{K} - \omega^2\mathbf{M}\right)\boldsymbol{\phi} = \mathbf{0} \qquad (2.104)$$

It is well known that a nontrivial solution $\boldsymbol{\phi} \neq \mathbf{0}$ only exists if the condition $\det\left(\mathbf{K} - \omega^2\mathbf{M}\right) = 0$ is fulfilled. This characteristic equation results in n independent undamped eigenfrequencies $\omega = \pm\omega_k$, $k = 1, ..., n$ (ω^2 are the eigenvalues).
If $\mathbf{K}$ and $\mathbf{M}$ are positive definite, all eigenfrequencies are greater than zero $\omega_k > 0$, $k = 1, ..., n$ and if they are positive semidefinite, they are greater or equal to zero $\omega_k \geq 0$, $k = 1, ..., n$. The number of zero eigenfrequencies is equal to the number of

rigid body modes in the system [6].
Each eigenvalue ω_k^2 is associated to an eigenvector ϕ_k, $k = 1, ..., n$ and the eigenvectors are linearly independent. Due to the fact that the number of eigenvectors is equal to the number of nodal DOFs, the vector $\mathbf{u}(t)$ can be written as:

$$\mathbf{u}(t) = \sum_{k=1}^{n} \phi_k q_k(t) \quad (2.105)$$

Eq. (2.105) is called *modal transformation* and can be applied to linear or linearized systems. It is well known that the eigenvectors, which correspond to the lowest eigenfrequencies ω_k, have the highest influence in the dynamical movement of the system. Higher eigenfrequencies affect the system much less and therefore they can be neglected [53]. However, the cut-off frequency strongly depends on the system and the excitation. If the eigenfrequencies are sorted according to their magnitudes $0 \leq \omega_1 \leq \omega_2 \leq \cdots \leq \omega_n$, the sum in (2.105) can be cut at the specific cut-off frequency:

$$\mathbf{u}(t) \approx \sum_{k=1}^{p} \phi_k q_k(t) \quad (2.106)$$

Note, that $p \ll n$, i.e. the number of DOFs can be decreased dramatically. Eq. (2.106) can also be written in matrix form:

$$\mathbf{u} = \mathbf{\Phi} \mathbf{q}^f \quad (2.107)$$

The matrix $\mathbf{\Phi}$ contains the eigenvectors ϕ_k as column vectors and the vector $\mathbf{q}^f$ the modal DOFs $q_k(t)$.

$$\mathbf{\Phi} = [\phi_1, \ldots, \phi_p], \quad \mathbf{q}^f = [q_1, \ldots, q_p]^T \quad (2.108)$$

2.9.5 Craig-Bampton Reduction

The Craig-Bampton reduction is a combination of static and modal reduction [43]. The idea of the Craig-Bampton reduction is that the internal dynamics of the slave nodes is described by eigenvectors for fixed master nodes. First of all, it is distinguished between normal modes $\mathbf{\Phi}_N$ and constraint modes $\mathbf{\Phi}_M$ [42].

Normal modes, dynamic modes $\mathbf{\Phi}_N$:

The matrix $\mathbf{\Phi}_N$ is formed by the first n_N eigenvectors, if all master DOFs are locked. These eigenvectors are called normal modes, which define the modal expansion of the interior DOFs.

Constraint modes, static modes $\mathbf{\Phi}_M$:

2.9. FLEXIBLE MULTIBODY SYSTEMS

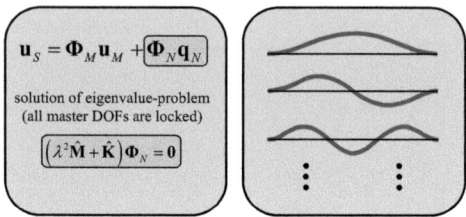

Figure 2.5: Normal modes, sketched for a Hermite beam element

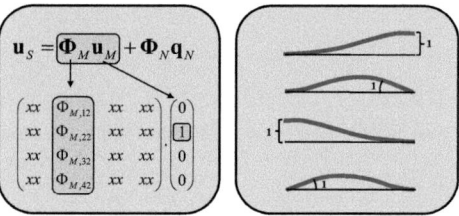

Figure 2.6: Constraint modes, sketched for a Hermite beam element

The matrix $\boldsymbol{\Phi}_M = -\mathbf{K}_{SS}^{-1}\mathbf{K}_{SM}$ from Eq. (2.99) maps the motion of the master-DOFs to the slave-DOFs. The columns of the matrix $\boldsymbol{\Phi}_M$ are the constraint modes. These modes are static shapes obtained by giving each master DOF a unit displacement while holding all other master DOFs fixed. The basis of constraint modes completely spans all possible motions of the master DOFs [109]. The displacements of the slave-DOFs are then described by the displacements of the master-DOFs $\mathbf{u}_M$ and the modal coordinates $\mathbf{q}_N$ of the normal modes:

$$\mathbf{u}_S = \boldsymbol{\Phi}_M \mathbf{u}_M + \boldsymbol{\Phi}_N \mathbf{q}_N \qquad (2.109)$$

Now the motion of master and slave nodes can be written as follows:

$$\mathbf{u} = \begin{bmatrix} \mathbf{u}_M \\ \mathbf{u}_S \end{bmatrix} = \begin{bmatrix} \mathbf{I}_{MM} & \mathbf{0} \\ \boldsymbol{\Phi}_M & \boldsymbol{\Phi}_N \end{bmatrix} \cdot \begin{bmatrix} \mathbf{u}_M \\ \mathbf{q}_N \end{bmatrix} \qquad (2.110)$$

It can be seen that the Craig-Bampton reduction (2.110) is given in the form

$$\mathbf{u} = \boldsymbol{\Phi} \mathbf{q}^f \qquad (2.111)$$

where the matrix $\boldsymbol{\Phi}$ and the vector $\mathbf{q}^f$ are formulated by:

$$\boldsymbol{\Phi} = \begin{bmatrix} \mathbf{I}_{MM} & \mathbf{0} \\ \boldsymbol{\Phi}_M & \boldsymbol{\Phi}_N \end{bmatrix}, \quad \mathbf{q}^f = \begin{bmatrix} \mathbf{u}_M \\ \mathbf{q}_N \end{bmatrix} \qquad (2.112)$$

Eq. (2.110) is a relation between physical DOFs and Craig-Bampton modes with their modal coordinates [109]. The generalized stiffness and mass-matrices corresponding to the Craig-Bampton modal basis are given by following modal transformation:

$$\bar{\mathbf{K}} = \mathbf{\Phi}^T \mathbf{K} \mathbf{\Phi} = \begin{bmatrix} \mathbf{I}_{MM} & 0 \\ \mathbf{\Phi}_M & \mathbf{\Phi}_N \end{bmatrix}^T \begin{bmatrix} \mathbf{K}_{MM} & \mathbf{K}_{MN} \\ \mathbf{K}_{NM} & \mathbf{K}_{NN} \end{bmatrix} \begin{bmatrix} \mathbf{I}_{MM} & 0 \\ \mathbf{\Phi}_M & \mathbf{\Phi}_N \end{bmatrix}$$
$$= \begin{bmatrix} \bar{\mathbf{K}}_{MM} & 0 \\ 0 & \bar{\mathbf{K}}_{NN} \end{bmatrix} \qquad (2.113)$$

$$\bar{\mathbf{M}} = \mathbf{\Phi}^T \mathbf{M} \mathbf{\Phi} = \begin{bmatrix} \mathbf{I}_{MM} & 0 \\ \mathbf{\Phi}_M & \mathbf{\Phi}_N \end{bmatrix}^T \begin{bmatrix} \mathbf{M}_{MM} & \mathbf{M}_{MN} \\ \mathbf{M}_{NM} & \mathbf{M}_{NN} \end{bmatrix} \begin{bmatrix} \mathbf{I}_{MM} & 0 \\ \mathbf{\Phi}_M & \mathbf{\Phi}_N \end{bmatrix}$$
$$= \begin{bmatrix} \bar{\mathbf{M}}_{MM} & \bar{\mathbf{M}}_{MN} \\ \bar{\mathbf{M}}_{NM} & \bar{\mathbf{M}}_{NN} \end{bmatrix} \qquad (2.114)$$

Eq. (2.113) and (2.114) have some noteworthy properties [109].

- $\bar{\mathbf{M}}_{NN}$ and $\bar{\mathbf{K}}_{NN}$ are diagonal matrices because they are associated with eigenvectors.

- $\bar{\mathbf{K}}$ is block diagonal, i.e. there is no stiffness coupling between constraint modes and normal modes.

- $\bar{\mathbf{M}}$ is not block diagonal, i.e. there is inertia coupling between constraint modes and normal modes.

The Craig-Bampton method has become one of the most popular methods for the implementation of flexible bodies into MBS as long as small deformations are existent.

2.9.6 Component Mode Synthesis (CMS)

The disadvantage of the Craig-Bampton method is that it is usually applied in a slightly different form in order to describe the problem of a floating reference frame. The component mode synthesis is used in many MBS-tools for the implementation of flexible bodies and can be divided in four steps [43].

1. Generation of a finite element model
   The flexible body has to be modeled as a finite element model. An appropriate mesh is created as well as mass- and stiffness properties. In this step no boundary conditions are introduced.

2. Definition of interface nodes
   The interface nodes are the nodes where the flexible body interacts with other rigid or flexible bodies in the MBS. Forces and/or torques can be introduced directly or via joints at the interface nodes. Hence, these nodes determine the master DOFs for the Craig-Bampton method. Accordingly, no loads can be applied to interior DOFs [71]. The number of master DOFs (interface DOFs) influence the number of DOFs in the MBS.

3. Craig Bampton reduction
   In a finite element solver (e.g.: *Nastran, Abaqus, ...*) the constraint modes and the normal modes are calculated. For the constraint modes (static modes) a unit displacement is applied for one master-DOF while all other master-DOFs are locked. As a consequence at least six master DOFs are needed in order that the system is not statically underdetermined. For the computation of the normal modes all master DOFs are set to zero. Then the eigenfrequencies and eigenvectors can be calculated. It is the decision of the user how many eigenfrequencies are taken into consideration. Therefore, the frequency range of the excitation of the MBS has to be known.

4. Mode shape orthonormalization
   Due to the fact that no boundary conditions are introduced till now, the system is statically underdetermined. It contains six rigid body modes. As a consequence, the system is not decoupled as in a classic modal analysis. The mode shape orthonormalization is used to remove the rigid body DOFs [138].

For the mode shape orthonormalization one more transformation to the Craig-Bampton vector $\mathbf{q}^f = [\mathbf{u}_M, \quad \mathbf{q}_N]^T$ is applied. The six rigid body modes of the reduced model correspond to zero eigenfrequencies. By carrying out a modal transformation of the reduced system, the rigid body modes can be eliminated. Hence, the undamped autonomous system

$$\bar{\mathbf{M}}\ddot{\mathbf{q}}^f + \bar{\mathbf{K}}\mathbf{q}^f = \mathbf{0} \tag{2.115}$$

with the reduced mass-matrix $\bar{\mathbf{M}} = \mathbf{\Phi}^T \mathbf{M} \mathbf{\Phi}$ and the reduced stiffness-matrix $\bar{\mathbf{K}} = \mathbf{\Phi}^T \mathbf{K} \mathbf{\Phi}$. The new eigenvalue-problem reads as

$$\left(\bar{\mathbf{K}} - \omega^2 \bar{\mathbf{M}}\right) \bar{\phi} = \mathbf{0} \tag{2.116}$$

The n_f eigenvectors $\bar{\phi}_1, \ldots, \bar{\phi}_{n_f}$ can be used as base vectors to represent $\mathbf{q}^f$ by new coordinates $\bar{q}_i^f$.

$$\mathbf{q}^f = \sum_{i=1}^{n_f} \bar{\phi}_i \bar{q}_i^f. \tag{2.117}$$

Eq. (2.117) is not a further dimension reduction, but rather a projection of $\mathbf{q}^f$ onto a new basis. If $\bar{\phi}_1, \ldots, \bar{\phi}_6$ are the six zero eigenvectors, the rigid body modes can then be removed by:

$$\mathbf{q}^f = \sum_{i=7}^{n_f} \bar{\phi}_i \bar{q}_i^f. \tag{2.118}$$

Eq. (2.118) can be written in matrix form:

$$\mathbf{q}^f = \bar{\mathbf{\Phi}} \bar{\mathbf{q}}^f \tag{2.119}$$

with the matrix $\bar{\mathbf{\Phi}}$ and the vector $\bar{\mathbf{q}}^f$:

$$\bar{\mathbf{\Phi}} = \left[\bar{\phi}_7, \ldots, \bar{\phi}_{n_f}\right], \quad \bar{\mathbf{q}}^f = \left[\bar{q}_7, \ldots, \bar{q}_{n_f}\right]^T \tag{2.120}$$

The new coordinates $\bar{\mathbf{q}}^f$ have only an abstract meaning. They can easily be transformed back to $\mathbf{q}^f$ by Eq. (2.119) and to the nodal displacements $\mathbf{u}$ by Eq. (2.110).

Pick battles big enough to matter, small enough to win.

Jonathan Kozol

Chapter 3
Virtual Iteration

This section is focused on the method of virtual iteration, also called iterative learning control (ILC) [39, 40]. This method is already state-of-the-art in the industry. Applications of real test rigs and partially also of virtual test rigs are published in [18, 19, 45, 47, 62, 63, 80, 97, 107, 113, 116, 117, 123, 124, 125, 128, 153, 158, 161, 162, 164]. Its roots can be found especially in test rigs in the automotive industry, Fig. 1.3, 1.4.

Firstly, the method was developed for real servo-hydraulic test rigs. Nowadays it is also used in a virtual environment, where the real physical test object is replaced by a multibody system. The method is based on a linearization either of the real system or the virtual model and an inverse computation in the frequency domain. Hence, the first part of the method of virtual iteration is the *system identification*, i.e. the computation of the transfer matrix $\mathbf{G}(i\omega)$. The second part is called *target simulation*. During this phase the drive signals $\mathbf{u}(t)$ are computed in an inverse way based on the transfer matrix and specific measured target signals $\mathbf{y}(t)$. This procedure is an iterative algorithm due to nonlinearities in the considered system. The two parts (i) system identification and (ii) target simulation are explicitly described in the following chapters.

3.1 System Identification

The aim of system identification is to find individual transfer functions between specific inputs and outputs. Transfer functions are calculated in the frequency domain. This fact implies that the system is linear. If the system is nonlinear, it has to be linearized at a specific operating point.

Input signals on a real test rig (cf. Fig. 1.3 or 1.4) are typically strokes of the servo-hydraulic cylinders. They are realized by a subsidiary control, because the direct interaction between control and test rig is realized by the oil pressure and hence the forces of the cylinders. In a virtual test rig these control inputs are realized by a motion in a joint or a single marker.

Output signals are typically accelerations, strains or spring deflections, both for the real system and also for the virtual model.

Transfer functions between specific inputs and outputs can be computed with dif-

3.1. SYSTEM IDENTIFICATION

ferent methods, which are described in the following sections.

3.1.1 System Identification via Noise Excitations

Non-parametric system identification methods which have been developed for a real test rig are based on an excitation by a white/pink noise signal $\mathbf{u}(t)$. The responses $\mathbf{y}(t)$ of the noise excitations are measured at suitable measuring points. Both input and output signals are transformed into the frequency domain $\mathbf{U}(i\omega)$, $\mathbf{Y}(i\omega)$ via a Fourier transform or a fast Fourier transform (FFT) . With these signals in the frequency domain it is possible to calculate the individual transfer functions and hence the transfer matrix $\mathbf{G}(i\omega)$.

The noise signals are typically generated in the frequency domain. The amplitudes can be defined with specific shape functions while the phase data are uniformly distributed random sequences over the relevant frequency band. For the generation of a noise signal it is important that the excitation level is as close as possible to the operational excitation level [47]. In a virtual environment the noise signal can have a constant amplitude over the whole frequency range, i.e. a white noise signal can be used. However, on a real test rig the amplitudes at higher frequencies must decrease because of physical restrictions, i.e. the available energy in the hydraulic cylinders. Therefore, the test rig is protected from load levels that are too high. Typically, a pink noise signal with a low-pass characteristic is used. Such signal is characterized by a constant amplitude up to a specific cut-off frequency. At higher frequencies the amplitude is decreased indirect proportional to the frequency $A \propto 1/f$. This means that the noise power is divided by two, if the frequency is doubled. In other words the power density is decreased by $3dB$/octave [44].

The noise signal is applied either to the real system or the multibody system in the time domain and the responses are measured. Then the input and output signals have to be transformed from the time domain into the frequency domain. Therefore, a Fourier transform is used [36].

$$F(i\omega) = \mathcal{F}\left\{f(t)\right\} = \int\limits_{-\infty}^{\infty} f(t)e^{-i\omega t}\,dt \qquad (3.1)$$

Nowadays a FFT-algorithm [53] is implemented in typical numerical software. Hence, a FFT procedure is used in order to reduce the computational time. Later in the virtual iteration algorithm the inverse Fourier transform is also needed which transforms the data from the frequency domain back into the time domain.

$$f(t) = \mathcal{F}^{-1}\left\{F(i\omega)\right\} = \frac{1}{2\pi}\int\limits_{-\infty}^{+\infty} F(i\omega)e^{i\omega t}\,d\omega \qquad (3.2)$$

Nowadays the inverse fast Fourier transform (IFFT) is used in typical numerical software. Based on the input and output data in the frequency domain $U(i\omega)$,

3.1. SYSTEM IDENTIFICATION

$Y(i\omega)$, the auto spectral density (ASD) or power spectral density (PSD) can be computed:

$$S_{xx}(\omega) = \lim_{T\to\infty} \frac{1}{2T} |\mathcal{F}\{x(t)\}|^2 = \lim_{T\to\infty} \frac{1}{2T} \{X^*(i\omega)X(i\omega)\} \qquad (3.3)$$

The power spectral density is greater or equal to zero $S_{xx}(\omega) \geq 0$ and is always a real function, i.e. phase information is not included [105]. The related time signal cannot be reproduced from the PSD. The power spectral density is the Fourier transform of the auto correlation function

$$R_{xx}(\tau) = \lim_{T\to\infty} \frac{1}{2T} \int_{-T}^{T} x(t)\,x(t+\tau)\,d\tau \qquad (3.4)$$

which converges towards zero for a noise signal, if $\tau \to \infty$.
Furthermore, the cross spectral density (CSD) can be defined for two signals $x(t)$ and $y(t)$.

$$S_{xy}(i\omega) = \lim_{T\to\infty} \frac{1}{2T} |\mathcal{F}^*\{x(t)\}\,\mathcal{F}\{y(t)\}|^2 = \lim_{T\to\infty} \frac{1}{2T} \{X^*(i\omega)Y(i\omega)\} \qquad (3.5)$$

The CSD is a complex function that includes phase information between $x(t)$ and $y(t)$. Another characteristic is that $S_{xy}(i\omega) = S^*_{yx}(i\omega)$. The cross spectral density is the Fourier transform of the cross correlation function

$$R_{xy}(\tau) = \lim_{T\to\infty} \frac{1}{2T} \int_{-T}^{T} x(t)\,y(t+\tau)\,d\tau \qquad (3.6)$$

which is a measure of the statistical relation between $x(t)$ and $y(t)$.
By using the power spectral density and the cross spectral density, a coherence function can be calculated.

$$\gamma^2_{xy}(\omega) = \frac{S^*_{xy}(i\omega)S_{xy}(i\omega)}{S_{xx}(\omega)S_{yy}(\omega)} \qquad (3.7)$$

The coherence function γ^2 is used to estimate the quality of measured transfer functions. It is defined between 0 and 1. If both signals $x(t)$ and $y(t)$ are statistically independent, the coherence function is equal to zero and if they are linearly dependent, it is equal to one.
In a real measurement the condition $T \to \infty$ cannot be fulfilled and hence a specific measurement time T_{mea} is used. Furthermore, the integral is replaced by a finite sum. Therefore the PSD $S_{xx}(\omega)$ has to be replaced by the approximation $\hat{S}_{xx}(\omega)$ and the CSD $S_{xy}(i\omega)$ by $\hat{S}_{xy}(i\omega)$ [53].

By using the definitions of the PSD (3.3) and the CSD (3.5), the transfer function between input $u(t)$, $U(i\omega)$ and output $y(t)$, $Y(i\omega)$ can be computed by [53, 105]

$$\hat{G}(i\omega) = \frac{\hat{S}_{yu}(i\omega)}{\hat{S}_{uu}(\omega)} \qquad (3.8)$$

3.1. SYSTEM IDENTIFICATION

$\hat{S}_{yu}(i\omega)$ denotes the estimated input-output cross spectral density and $\hat{S}_{uu}(i\omega)$ the estimated input power spectral density at frequency ω.

In MIMO-systems a transfer matrix can be computed as well [53, 105]. Therefore, all input channels $\mathbf{u}(t) \in \mathbb{R}^{m_c}$ have to be simultaneously excited by uncorrelated noise signals. This means that the motion of one actuator is fully independent from the motion of another actuator. The responses of the noise excitations are measured in specific output channels $\mathbf{y}(t) \in \mathbb{R}^k$.

$$\hat{\mathbf{G}}_{yu}(i\omega) = \hat{\mathbf{S}}_{yu}(i\omega) \cdot \hat{\mathbf{S}}_{uu}(\omega)^{-1} \qquad (3.9)$$

In Eq. (3.9) the matrix $\hat{\mathbf{S}}_{yu}(i\omega) \in \mathbb{C}^{k \times m_c}$ denotes the matrix of estimated input-output cross spectrum densities and $\hat{\mathbf{S}}_{uu}(\omega) \in \mathbb{R}^{m_c \times m_c}$ denotes the matrix of estimated input power spectral densities.

Eqs. (3.8) and (3.9) are estimators for transfer functions that minimize output disturbances in an optimal way [72]. The transfer matrix $\hat{\mathbf{G}}(i\omega)$ is also called frequency response function (FRF) and the estimator (3.8) is known as H_1-technique [47].

In addition to the transfer matrix multiple coherence functions (3.7) are calculated. They indicate which energy amount of the input channels results in which energy amount in a specific output channel. The coherence functions of the outputs are measurements of the quality of the transfer matrix. Low values of the coherence functions can be caused from nonlinearities in the system or from uncorrelated external noise. Higher values indicate that the amplitudes in the input channels are high enough to obtain a good signal to noise ratio while too high excitations, which result in nonlinear behavior, are avoided [47].

In order to obtain accurate noiseless transfer functions the noise excitation has to be sufficiently long enough. As a consequence it can be guaranteed that all frequencies in the frequency range of interest are excited. Therefore, the noise excitation, which can only be done in the time domain, is very time consuming. This drawback appears in the real test rig as well as in the virtual test rig. Fig. 3.1 shows

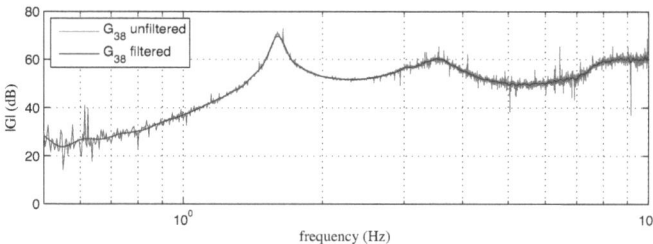

Figure 3.1: Generic magnitude plot, computed by noise excitation

an illustrative magnitude plot, which is computed by noise excitation. It is taken from the industrial example of a trailed cultivator, cf. section 8.2. It can be seen that the direct computation of the transfer matrix results in functions with noise.

3.1. SYSTEM IDENTIFICATION

These functions have to be filtered before they can be used in the virtual iteration algorithm. In Fig. 3.1 the unfiltered signal is shown as well as the filtered signal, where a *Savitzky-Golay filter* is used. After signal filtering, the transfer functions can be compared with the functions that were generated by the state matrices of the system, cf. section 3.1.3. These transfer functions are shown in Fig. 8.32. It can be seen that the magnitude plot $|G_{38}|$ of Fig. 3.1 is identical to that in Fig. 8.32.

On a real test rig the noise excitation in the time domain is the only possible method for the system identification, because the frequency domain is only a mathematical construct. However, more advantageous methods are possible in a virtual test rig, where the model's equations of motion are known. Additionally, the noise excitation is not very beneficial for a numerical solver. Because of the random numbers in the noise signal, a very short step size is needed. For these reasons more advantageous mathematical methods for the system identification are discussed in the following sections.

3.1.2 Linearization of the Nonlinear Equations of Motion

The nonlinear model (2.51) can be linearized, only if small displacements and rotations occur from an equilibrium point or a stationary operating point. Therefore, the first required step is the computation of the equilibrium point.

Calculation of an equilibrium point
The equations of motion of the multibody system can be written in the general form

$$\dot{\mathbf{x}} = \mathbf{f}(\mathbf{x}, \mathbf{u}) \tag{3.10}$$

with the state vector $\mathbf{x} = [\mathbf{q}, \dot{\mathbf{q}}]^T$, $\mathbf{x} \in \mathbb{R}^{2n}$ and the control inputs $\mathbf{u} \in \mathbb{R}^{m_c}$. The vector $\mathbf{x}_s \in \mathbb{R}^{2n}$ is called equilibrium point, if the condition

$$\mathbf{f}(\mathbf{x}_s, \mathbf{u}_s) = \mathbf{0} \tag{3.11}$$

is fulfilled. To compute the equilibrium point (3.11), the inputs are kept constant $\mathbf{u} = \mathbf{u}_s$. The pair $(\mathbf{u}_s, \mathbf{x}_s)$ is called operating point of the nonlinear system (3.10). Nowadays the computation of an equilibrium point or operating point is implemented in each commercial MBS-software. An arbitrary operating point can be computed either by an initial-conditions analysis or a static or dynamic analysis. These methods typically allow the treatment of systems with flexible bodies, friction, control elements (user-defined differential equations), non-holonomic constraints etc. [111].

Linearization at an equilibrium point
If an equilibrium point or an arbitrary operating point is calculated, the nonlinear system can be linearized at this point [60]. The basis are the equations of motion (2.57), which can be written in the general form of a time-invariant, nonlinear system

3.1. SYSTEM IDENTIFICATION

$$\dot{\mathbf{x}} = \mathbf{f}(\mathbf{x}, \mathbf{u}) \tag{3.12a}$$
$$\mathbf{y} = \mathbf{h}(\mathbf{x}, \mathbf{u}) \tag{3.12b}$$

In (3.12) $\mathbf{x} \in \mathbb{R}^{2n}$ denotes the vector of state variables, which are displacements and velocities of mass-bearing elements like parts, point masses or flexible bodies [110]. The vector $\mathbf{u} \in \mathbb{R}^{m_c}$ includes the input variables and the vector $\mathbf{y} \in \mathbb{R}^k$ the output variables of the system under consideration. If the system (3.12) is time-invariant and no inputs $\mathbf{u}$ appear, it is called *autonomous*.

The equilibrium point or the operating point fulfills the equations

$$\mathbf{f}(\mathbf{x}_s, \mathbf{u}_s) = \mathbf{0}, \quad \mathbf{h}(\mathbf{x}_s, \mathbf{u}_s) = \mathbf{y}_s \tag{3.13}$$

If the displacements and rotations from the equilibrium point are sufficiently small, the variables $\mathbf{x}$, $\mathbf{u}$ and $\mathbf{y}$ can be written as

$$\mathbf{x}(t) = \mathbf{x}_s + \Delta \mathbf{x}(t) \tag{3.14a}$$
$$\mathbf{u}(t) = \mathbf{u}_s + \Delta \mathbf{u}(t) \tag{3.14b}$$
$$\mathbf{y}(t) = \mathbf{y}_s + \Delta \mathbf{y}(t) \tag{3.14c}$$

Now the equations (3.14) can be inserted into system (3.12).

$$\dot{\mathbf{x}}_s + \Delta \dot{\mathbf{x}} = \mathbf{f}(\mathbf{x}_s + \Delta \mathbf{x}, \mathbf{u}_s + \Delta \mathbf{u}), \quad \mathbf{x}(t_0) = \mathbf{x}_s + \Delta \mathbf{x}(t_0) \tag{3.15a}$$
$$\mathbf{y}_s + \Delta \mathbf{y} = \mathbf{h}(\mathbf{x}_s + \Delta \mathbf{x}, \mathbf{u}_s + \Delta \mathbf{u}) \tag{3.15b}$$

Furthermore, $\dot{\mathbf{x}}_s = \mathbf{0}$ due to the stationary equilibrium point. The system (3.15) can be developed in a Taylor series of second order, where the remainder $\mathcal{O}^2(\mathbf{x}, \mathbf{u})$ is neglected [103].

$$\Delta \dot{\mathbf{x}} = \underbrace{\mathbf{f}(\mathbf{x}_s, \mathbf{u}_s)}_{\mathbf{0}} + \underbrace{\frac{\partial}{\partial \mathbf{x}} \mathbf{f}(\mathbf{x}_s, \mathbf{u}_s) \Delta \mathbf{x}}_{\mathbf{A}} + \underbrace{\frac{\partial}{\partial \mathbf{u}} \mathbf{f}(\mathbf{x}_s, \mathbf{u}_s) \Delta \mathbf{u}}_{\mathbf{B}} \tag{3.16a}$$

$$\mathbf{y}_s + \Delta \mathbf{y} = \underbrace{\mathbf{h}(\mathbf{x}_s, \mathbf{u}_s)}_{\mathbf{y}_s} + \underbrace{\frac{\partial}{\partial \mathbf{x}} \mathbf{h}(\mathbf{x}_s, \mathbf{u}_s) \Delta \mathbf{x}}_{\mathbf{C}} + \underbrace{\frac{\partial}{\partial \mathbf{u}} \mathbf{h}(\mathbf{x}_s, \mathbf{u}_s) \Delta \mathbf{u}}_{\mathbf{D}} \tag{3.16b}$$

By using the definitions from Eqs. (3.16)

$$\mathbf{A} = \frac{\partial}{\partial \mathbf{x}} \mathbf{f}(\mathbf{x}_s, \mathbf{u}_s), \quad \mathbf{B} = \frac{\partial}{\partial \mathbf{u}} \mathbf{f}(\mathbf{x}_s, \mathbf{u}_s)$$
$$\mathbf{C} = \frac{\partial}{\partial \mathbf{x}} \mathbf{h}(\mathbf{x}_s, \mathbf{u}_s), \quad \mathbf{D} = \frac{\partial}{\partial \mathbf{u}} \mathbf{h}(\mathbf{x}_s, \mathbf{u}_s) \tag{3.17}$$

the state matrices $\mathbf{A} \in \mathbb{R}^{2n \times 2n}$, $\mathbf{B} \in \mathbb{R}^{2n \times m_c}$, $\mathbf{C} \in \mathbb{R}^{k \times 2n}$ and $\mathbf{D} \in \mathbb{R}^{k \times m_c}$ are defined. The matrix $\mathbf{A}$ is called system matrix, $\mathbf{B}$ input matrix, $\mathbf{C}$ output matrix and $\mathbf{D}$ pass-through matrix [103]. Hence, the linearized system reads as

$$\Delta \dot{\mathbf{x}} = \mathbf{A} \Delta \mathbf{x} + \mathbf{B} \Delta \mathbf{u}, \quad \Delta \mathbf{x}(t_0) = \Delta \mathbf{x}_0 = \mathbf{x}_0 - \mathbf{x}_s \quad (3.18\text{a})$$
$$\Delta \mathbf{y} = \mathbf{C} \Delta \mathbf{x} + \mathbf{D} \Delta \mathbf{u} \quad (3.18\text{b})$$

The system (3.18) is called state-space-form. If only one input variable $m_c = 1$ and one outputs variable $k = 1$ appear, the system is called SISO-system (single input single output). Otherwise, it is called MIMO-system (multiple input multiple output).

If the system is linearized at an equilibrium point or an operating point, it results in a LTI-system (linear time-invariant system). If it is linearized about a trajectory $(\tilde{\mathbf{x}}, \tilde{\mathbf{u}})$, it results in a linear time-variant system [103]

$$\Delta \dot{\mathbf{x}} = \mathbf{A}(t) \Delta \mathbf{x} + \mathbf{B}(t) \Delta \mathbf{u}, \quad \Delta \mathbf{x}(t_0) = \Delta \mathbf{x}_0 = \mathbf{x}_0 - \tilde{\mathbf{x}}_0 \quad (3.19\text{a})$$
$$\Delta \mathbf{y} = \mathbf{C}(t) \Delta \mathbf{x} + \mathbf{D}(t) \Delta \mathbf{u} \quad (3.19\text{b})$$

Specific commercial MBS-software tools provide an automatic linearization process, which enables the linearization of the index 3 DAEs (2.51). The resulting state-space formulation (3.18) with a minimal set of states provides the state matrices $\mathbf{A}$, $\mathbf{B}$, $\mathbf{C}$ and $\mathbf{D}$ as ASCII-files, which can be further processed in numerical software like *Matlab*. The algorithm that is implemented in *Adams/Linear* inflates the governing equations and computes a set of sensitivities which provide the linearization of interest [111, 141].

If the system is given as linear time invariant system, the characteristics of an equilibrium point can be validated. The different possibilities for the equilibrium point are summarized in Table 3.1.

$\text{rank}(\mathbf{A}) = \text{rank}([\mathbf{A}, \mathbf{B}\mathbf{u}_s])$, $\det(\mathbf{A}) \neq 0$	$\mathbf{x}_s = -\mathbf{A}^{-1}\mathbf{B}\mathbf{u}_s$ is a single equilibrium point
$\text{rank}(\mathbf{A}) = \text{rank}([\mathbf{A}, \mathbf{B}\mathbf{u}_s])$, $\det(\mathbf{A}) = 0$	infinite equilibrium points exist
$\text{rank}(\mathbf{A}) \neq \text{rank}([\mathbf{A}, \mathbf{B}\mathbf{u}_s])$	no equilibrium point exists

Table 3.1: Different cases for an equilibrium point [60]

3.1.3 Calculation of the Transfer Matrix

Transfer matrix of a linear multibody system

The linear (or linearized) time invariant system can mathematically either be described by the state space formulation (3.18) in the time domain or by transfer functions between inputs and outputs in the frequency domain. Transfer functions or the transfer matrix directly describe the behavior from the inputs u_i, $i = 1, \cdots, m_c$ to the outputs y_j, $j = 1, \cdots, k$. The individual inputs and outputs are summarized in the vectors $\mathbf{u} \in \mathbb{R}^{m_c}$ and $\mathbf{y} \in \mathbb{R}^k$. If the transfer functions or the transfer matrix are known, the outputs can be calculated depending on the inputs, without knowing the specific state variables.

The linear equations (3.18) can be transformed from the time domain into the frequency domain by using the Laplace transform. The Laplace transform is defined as [36]

$$\hat{f}(s) = \mathcal{L}\{f(t)\} = \int_0^\infty f(t) e^{-st}\, dt \qquad (3.20)$$

The variable $s = \alpha + i\omega$, $i = \sqrt{-1}$ denotes the Laplace variable. Details of the Laplace transform can be found e.g. in [103].

For the sake of completeness the inverse Laplace transform should also be mentioned [36].

$$f(t) = \mathcal{L}^{-1}\{\hat{f}(s)\} = \frac{1}{2\pi i} \int_{\alpha-i\infty}^{\alpha+i\infty} \hat{f}(s) e^{st}\, ds, \quad t \geq 0 \qquad (3.21)$$

At a point of discontinuity of $f(t)$ the inverse Laplace transform provides the average value of left- and right sided boundary value, i.e. $f(t) = 1/2(f(+t) + f(-t))$.

If the Laplace transform is applied to the LTI-system (3.18), it results in

$$s\Delta\hat{\mathbf{x}}(s) - \Delta\mathbf{x}_0 = \mathbf{A}\Delta\hat{\mathbf{x}}(s) + \mathbf{B}\Delta\hat{\mathbf{u}}(s) \qquad (3.22a)$$
$$\Delta\hat{\mathbf{y}}(s) = \mathbf{C}\Delta\hat{\mathbf{x}}(s) + \mathbf{D}\Delta\hat{\mathbf{u}}(s) \qquad (3.22b)$$

The vector $\Delta\hat{\mathbf{x}}$ can be calculated from (3.22a)

$$\Delta\hat{\mathbf{x}}(s) = (s\mathbf{I} - \mathbf{A})^{-1}\mathbf{B}\Delta\hat{\mathbf{u}}(s) + (s\mathbf{I} - \mathbf{A})^{-1}\Delta\mathbf{x}_0 \qquad (3.23)$$

The matrix $\mathbf{I}$ denotes the $(2n \times 2n)$ identity matrix. Then the term (3.23) is inserted in (3.22b).

$$\Delta\hat{\mathbf{y}}(s) = \left[\mathbf{C}(s\mathbf{I} - \mathbf{A})^{-1}\mathbf{B} + \mathbf{D}\right]\Delta\hat{\mathbf{u}}(s) + \mathbf{C}(s\mathbf{I} - \mathbf{A})^{-1}\Delta\mathbf{x}_0 \qquad (3.24)$$

The second term vanishes due to $\Delta\mathbf{x}_0 = \mathbf{0}$ and hence the $(k \times m_c)$ transfer matrix $\mathbf{G}(s)$ with the individual transfer functions $G_{ij}(s) = \frac{\Delta\hat{y}_i(s)}{\Delta\hat{u}_j(s)}$, $i = 1, \ldots, k$, $j = 1, \ldots, m_c$ can be calculated by [103]

$$\mathbf{G}(s) = \mathbf{C}(s\mathbf{I} - \mathbf{A})^{-1}\mathbf{B} + \mathbf{D} \qquad (3.25)$$

It can be shown that the condition $\Delta \mathbf{x}_0 = \mathbf{0}$ is not a loss of generality. If it is not fulfilled, a state transformation can be found in order that the new states fulfill this condition [91].

The computation of the dynamic behavior by using a transfer matrix $\mathbf{G}(s)$ in the frequency domain implies a dramatic reduction of the size of a MBS. The multibody systems in chapter 8, which include flexible bodies, have hundreds of state variables. By using sufficiently enough inputs and outputs for the inverse simulation task, the dimension of the transfer matrix is approximately (10×3), i.e. the size of the model is substantially reduced.

It should be mentioned that the direct computation of the transfer matrix by (3.25) can even be accelerated. By using the eigenvectors of the model, the system (3.22) can be transformed into the modal space. This procedure is implemented for example in *Adams/Vibration* [110].

$$\Delta \hat{\mathbf{x}}(s) = \boldsymbol{\Phi} \hat{\mathbf{q}}(s) \tag{3.26}$$

The matrix $\boldsymbol{\Phi}$ includes all the eigenvectors as column vectors, cp. (2.108). The vector $\hat{\mathbf{q}}(s)$ denotes the Laplace transform of the modal coordinates $\mathbf{q}(t)$. Hence, the transfer matrix for the model in modal space reads as

$$\mathbf{G}(s) = \mathbf{C}_m(s\mathbf{I} - \mathbf{A}_m)^{-1}\mathbf{B}_m + \mathbf{D} \tag{3.27}$$

where the matrices $\mathbf{A}_m$, $\mathbf{B}_m$ and $\mathbf{C}_m$ are defined as

$$\mathbf{A}_m = \boldsymbol{\Phi}^{-1}\mathbf{A}\boldsymbol{\Phi}, \quad \mathbf{B}_m = \boldsymbol{\Phi}^{-1}\mathbf{B}, \quad \mathbf{C}_m = \mathbf{C}\boldsymbol{\Phi} \tag{3.28}$$

The computation of the transfer matrix in the modal space (3.28) is much faster than the direct solution (3.25).

Transfer matrix of a linear finite element model

The method of virtual iteration is not limited to commercial MBS-software as *Adams, Simpack, RecurDyn* etc. In section 8.1 the algorithm is also applied to a finite element model. From the knowledge of the author it is the first time that the virtual iteration approach is applied to a FEM. Therefore, the transfer matrix of the FEM has to be computed. Typically, linearized finite element models are not represented in the state-space form (3.18). Rather a linear (linearized) FEM is given in the form

$$\mathbf{M}\ddot{\mathbf{q}} + \mathbf{C}\dot{\mathbf{q}} + \mathbf{K}\mathbf{q} = \mathbf{f} \tag{3.29}$$

with the mass matrix $\mathbf{M}$, the damping matrix $\mathbf{C}$, the stiffness matrix $\mathbf{K}$ and the applied forces $\mathbf{f}$, cf. Eq. (2.102). If periodic excitations are applied to the model, they can be developed in a Fourier series of the form [48]

$$\mathbf{f}(t) = \sum_{k=-\infty}^{\infty} \hat{\mathbf{f}}_k e^{i\omega_k t} \tag{3.30}$$

The vector $\hat{\mathbf{f}}_k$ denotes the vector of complex harmonic functions of kth order.

$$\hat{\mathbf{f}}_k = \begin{bmatrix} f_{k1} \\ \vdots \\ f_{kn} \end{bmatrix} = \begin{bmatrix} |f_{k1}|\, e^{i\psi_{k1}} \\ \vdots \\ |f_{kn}|\, e^{i\psi_{kn}} \end{bmatrix} \quad (3.31)$$

The periodic excitation with the fundamental circle frequency ω_1 causes a phase-shifted periodic oscillation with the same circle frequency ω_1 in a linear system. Therefore, the vector of nodal coordinates reads as

$$\mathbf{q}(t) = \sum_{k=-\infty}^{\infty} \hat{\mathbf{q}}_k e^{i\omega_k t} \quad (3.32)$$

By inserting Eqs. (3.30) and (3.32) in the differential equation (3.29) and equating the coefficients, following equation is achieved:

$$\left(-\omega_k^2 \mathbf{M} + i\omega_k \mathbf{C} + \mathbf{K}\right) \hat{\mathbf{q}}_k = \hat{\mathbf{f}}_k \quad (3.33)$$

Hence, the $(n \times n)$ transfer matrix $\mathbf{G}(i\omega)$ can be computed by [48]

$$\mathbf{G}(i\omega_k) = \left(-\omega_k^2 \mathbf{M} + i\omega_k \mathbf{C} + \mathbf{K}\right)^{-1}, \quad -\infty \leq k \leq \infty \quad (3.34)$$

The rows and columns, which represent the behavior between specific inputs and outputs, can be extracted from (3.34).

The virtual iteration procedure is applied to multibody systems which are modeled in *Adams*, cf. section 8.2 and 8.3. The state matrices $\mathbf{A}$, $\mathbf{B}$, $\mathbf{C}$ and $\mathbf{D}$ are computed in *Adams/Linear*. The transfer matrix $\mathbf{G}(s)$ is externally computed in *Matlab* by using Eq. (3.25). The module *Adams/Vibration* [110] computes the individual transfer functions $G_{ij}(s)$, $i = 1,\ldots,k$, $j = 1,\ldots,m_c$ by (3.27), which can be exported as ASCII-files. However, the method with the state matrices has been shown as beneficial regarding to the models in section 8.2 and 8.3.

Furthermore, the virtual iteration algorithm is applied to a finite element model which is modeled in *Abaqus*. A so-called *steady-state-dynamics step* [46] is used to compute the individual transfer functions. These transfer functions are exported as ASCII-files and externally assembled to the transfer matrix $\mathbf{G}(s)$ in *Matlab*.

3.2 Target Simulation

The basic inverse computation in the virtual iteration is done in the frequency domain. Therefore, the transfer matrix of the real system or of the virtual model has to be computed by using one of the previously described methods, cf. section 3.1. Furthermore, the measured targets $\tilde{\mathbf{y}}(t) \in \mathbb{R}^k$ have to be transformed into the frequency domain $\tilde{\mathbf{Y}}(i\omega)$ via a FFT.

3.2. TARGET SIMULATION

The transfer matrix $\mathbf{G}(i\omega)$ of a MIMO-system describes the transmission behavior between inputs $\mathbf{U}(i\omega) \in \mathbb{C}^{m_c}$ and outputs $\mathbf{Y}(i\omega) \in \mathbb{C}^k$ in the forward direction

$$\mathbf{Y}(i\omega) = \mathbf{G}(i\omega)\,\mathbf{U}(i\omega) \tag{3.35}$$

This simple vector-matrix equation can be inverted at each frequency ω, if the transfer matrix is quadratic and non-singular. For that reason it is assumed that the system is fully determined, i.e. the number of inputs m_c is equal to the number of outputs k. In the sense of the virtual iteration algorithm the calculation of the so-called *first drives* $\mathbf{U}^0(i\omega)$ can be written as

$$\mathbf{U}^0(i\omega) = \mathbf{G}^{-1}(i\omega)\,\tilde{\mathbf{Y}}(i\omega) \tag{3.36}$$

The vector $\tilde{\mathbf{Y}}(i\omega)$ includes the Fourier transforms of the targets. If the system is completely linear, Eq. (3.36) already results in the final solution of the drives in the frequency domain. However, in the general case nonlinear systems are considered. Hence, Eq. (3.36) of the linearized system can only give a first guess of the inputs. Therefore, an iterative procedure is required in order to find the final solution for the drives. The first drives $\mathbf{U}^0(i\omega)$ are transformed into the time domain via an IFFT. These drives or input signals $\mathbf{u}^0(t)$ are used as excitations in a forward computation in the time domain. This means that the MBS is numerically integrated in a "standard" dynamic simulation.

The forward simulation results in outputs $\mathbf{y}(t)$ that can be compared with the measured targets $\tilde{\mathbf{y}}(t)$. Due to the nonlinearities in the model, an error between simulation outputs and targets will occur. This error can either be calculated in the time domain $\mathbf{e}(t) = \tilde{\mathbf{y}}(t) - \mathbf{y}(t)$ or in the frequency domain $\mathbf{E}(i\omega) = \tilde{\mathbf{Y}}(i\omega) - \mathbf{Y}(i\omega)$. The goal of virtual iteration is to reduce this error as much as possible. Therefore, the error is used to improve the guess of the first drives [62]. As a consequence following iterative procedure is obtained:

$$\mathbf{U}^{n+1}(i\omega) = \mathbf{U}^n(i\omega) + \alpha^{n+1}\mathbf{G}^{-1}(i\omega)\left[\tilde{\mathbf{Y}}(i\omega) - \mathbf{Y}^n(i\omega)\right] \tag{3.37}$$

Eq. (3.37) describes the central point of the virtual iteration procedure. The superscript n denotes the iteration counter. The scalar factor $0 \leq \alpha^{n+1} \leq 1$ can be used to improve the convergence behavior. Especially on a real test rig it is used because of safety reasons. With a low value of α^{n+1} the amplitudes of the drive signals can be reduced.

The iteration (3.37) is repeated until the error between targets and simulation outputs is sufficiently small. Hence, a quantifiable factor has to be calculated from the time-dependent error $\mathbf{e}(t)$. Typically, the root mean square (RMS) error is computed for each output $i = 1, ..., k$.

$$e^n_{RMS,i} = \frac{1}{T}\int_0^T (\tilde{y}_i(t) - y_i^n(t))\,dt, \quad i = 1, ..., k \tag{3.38}$$

3.2. TARGET SIMULATION

Due to the numerical discretization of the outputs the integral in Eq. (3.38) is replaced by a finite sum

$$\hat{e}_{RMS,i}^n = \frac{1}{N} \sum_{j=1}^{N} (\tilde{y}_i(j) - y_i^n(j)), \quad i = 1, ..., k \qquad (3.39)$$

where N denotes the length of the outputs $y_i(t)$ and the targets $\tilde{y}_i(t)$, respectively. In addition to the RMS-error, the maximum deviation between targets and simulation outputs can be calculated.

$$e_{MAX,i}^n = \max |\tilde{y}_i(t) - y_i^n(t)| \qquad (3.40)$$

Then, the two error indicators (3.39) and (3.40) can be combined with scalar weighting factors α and β. Furthermore, the error indicator is related to its corresponding target signal in order to obtain a relative value which can be expressed in percent.

$$e_i^n = \alpha \frac{\hat{e}_{RMS,i}^n}{\tilde{y}_{RMS,i}} + \beta \frac{e_{MAX,i}^n}{\tilde{y}_{MAX,i}} \qquad (3.41)$$

This percentage value can be calculated for each individual output channel in every iteration. The virtual iteration (3.37) is repeated until the error indicators are below a specific error tolerance $e_i^n < \varepsilon$.

3.2.1 Virtual Iteration Algorithm

The main steps of the virtual iteration algorithm are summarized in the following list:

1. Computation of the transfer matrix $\mathbf{G}(i\omega)$
2. FFT of the target signals $\tilde{\mathbf{y}}(t)$
3. Calculation of the inverse of the transfer matrix $\mathbf{G}^{-1}(i\omega)$
4. Calculation of the first drives by Eq. (3.36)
5. IFFT of the first drives
6. Forward simulation in MBS/FEM-software by numerical time integration
7. Calculation of the error indicator
8. Iteration (3.37) until $e_i^n < \varepsilon$

This procedure is also illustrated in Fig. 3.2. The green box symbolizes the MBS- or FEM-software with the virtual model. The transfer matrix $\mathbf{G}(s)$ can be computed in different ways, cf. section 3.1.3. The blue box illustrates the inverse computation in the frequency domain, which is done in the numerical software *Matlab*. The blue arrows symbolize the forward computation and the red arrows the inverse computation.

3.2. TARGET SIMULATION

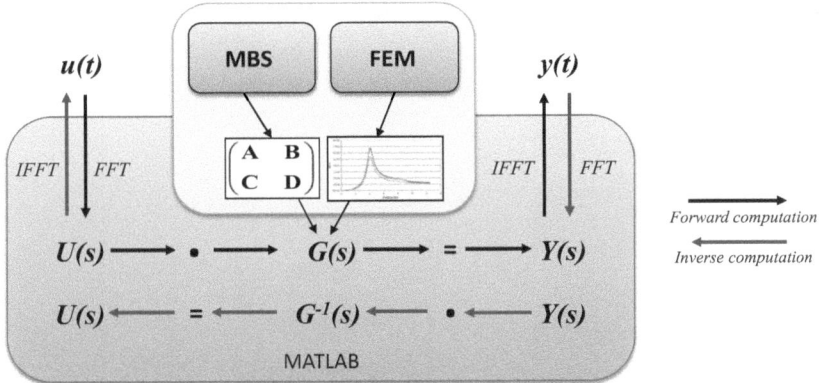

Figure 3.2: Flow chart of the virtual iteration

The individual steps in the virtual iteration algorithm are illustrated in the flow chart 3.3 in more detail. The boxes "signal filtering" and "filtering of transfer matrix" are important for specific measurements of the target signals. The targets are measured with a specific sampling frequency. Furthermore, the length of the measured targets can vary from measurement to measurement. As a consequence, the Fourier transform of the targets covers a specific frequency range. However, the computed transfer matrix is independent of the measurements. Hence, the transfer matrix has to be adapted regarding to the frequency range of the targets. The frequency resolution as well as the lower and upper border of the frequency band must match between transfer matrix and measurements. The box "filtering of transfer matrix" treats the cut-off frequencies of the transfer matrix. The box "resampling of the inverse transfer matrix" adapts the transfer functions in order that the frequency resolution matches with the measurements.

3.2.2 The Inverse of the Transfer Matrix

An important point is the computation of the inverse of the transfer matrix $\mathbf{G}^{-1}(i\omega)$ (3rd point in the procedure in section 3.2.1, box "inverse of transfer matrix" in Fig. 3.3). Until now it was assumed that the system is fully determined, i.e. $k = m_c$. In this case the transfer matrix is quadratic. If it is regular at each frequency ω, it can be inverted. However, the case of an under-determined or an over-determined system can occur, depending on the number of input and output channels.

Before the three possible cases are considered in more detail, some general definitions of a linear inverse problem should be given [61]. The continuous linear operator $\mathbf{G} : \mathcal{U} \rightarrow \mathcal{Y}$ is considered with a set of data $\mathbf{Y} \in \mathcal{Y}$. The Hilbert spaces $\mathcal{U}$ and $\mathcal{Y}$ are called solution space and data space, respectively. The task is to find $\mathbf{U} \in \mathcal{U}$ so

3.2. TARGET SIMULATION

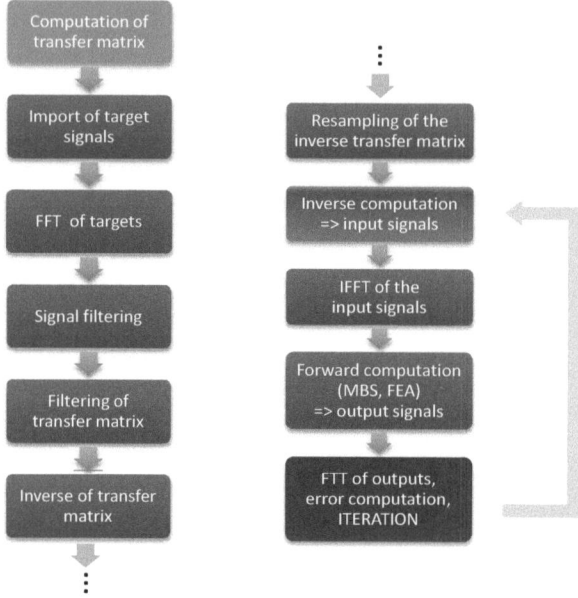

Figure 3.3: Algorithm of the virtual iteration

that
$$\mathbf{Y} = \mathbf{GU} \qquad (3.42)$$

The problem (3.42) is called well-posed in the sense of *Hadamard* if (i) the solution $\mathbf{U}$ is unique in $\mathcal{U}$, (ii) the solution $\mathbf{U} \in \mathcal{U}$ exists for any $\mathbf{Y} \in \mathcal{Y}$ and (iii) the inverse mapping $\mathbf{Y} \to \mathbf{U}$ is continuous.

The three cases of a fully-determined, an under-determined and an over-determined system are considered in the list below [61].

- **case 1:** $k = m_c$ **(fully-determined system)**
  In this case the transfer matrix $\mathbf{G}(i\omega)$ is quadratic and its rank is rank($\mathbf{G}$) = $m_c = k$. If the transfer matrix is regular at each frequency, it can be inverted and a unique solution exits for the drives $\mathbf{U}(i\omega)$.

$$\mathbf{U} = \mathbf{G}^{-1}\mathbf{Y} \qquad (3.43)$$

- **case 2:** $k < m_c$ **(under-determined system)**
  In this case the number of output channels is less than the number of input channels. The transfer matrix does not have full rank, i.e. rank($\mathbf{G}$) = k and as a consequence it cannot be inverted. The solution of this under-determined system is not unique, but exists because $\mathbf{Y}$ belongs to range

$R(\mathbf{G}) = \{\mathbf{Y} \in \mathcal{Y} \mid \mathbf{Y} = \mathbf{GU}, \mathbf{U} \in \mathcal{U}\}$. In this case a *minimum norm solution* can be computed where the norm of the drives is minimized $\|\mathbf{U}\| \to MIN$.

$$\mathbf{U} = \mathbf{G}^+\mathbf{Y} = \mathbf{G}^H \left(\mathbf{GG}^H\right)^{-1}\mathbf{Y} \tag{3.44}$$

- **case 3:** $k > m_c$ (**over-determined system**)
  In the case of an over-determined system the number of output channels exceeds the number of input channels. The rank of the transfer matrix is $\text{rank}(\mathbf{G}) = m_c$. A solution of (3.42) does not exist. However, a *least square solution* or a *pseudo-solution* can be found, which minimizes the norm $\|\mathbf{Y} - \mathbf{GU}\| \to MIN$ as well as the norm $\|\mathbf{U}\| \to MIN$.

$$\mathbf{U} = \mathbf{G}^+\mathbf{Y} = \left(\mathbf{G}^H\mathbf{G}\right)^{-1}\mathbf{G}^H\mathbf{Y} \tag{3.45}$$

The matrix $\mathbf{G}^+$ is called *Moore-Penrose pseudoinverse*, named after *Eliakim Hastings Moore* and *Roger Penrose* [50, 53]. The matrix $\mathbf{G}^H$ is the *adjoint matrix* of $\mathbf{G}$, i.e. the transpose of the conjugate complex of $\mathbf{G}$: $\mathbf{G}^H = (\mathbf{G}^*)^T$ [36]. The Moore-Penrose pseudoinverse fulfills all four Penrose axioms [92]:

1. $\mathbf{GG}^+\mathbf{G} = \mathbf{G}$
2. $\mathbf{G}^+\mathbf{GG}^+ = \mathbf{G}^+$
3. $\left(\mathbf{GG}^+\right)^H = \mathbf{GG}^+$
4. $\left(\mathbf{G}^+\mathbf{G}\right)^H = \mathbf{G}^+\mathbf{G}$

Typically, the 2nd case of an under-determined system does not occur and hence it should not be discussed further here. In typical industrial applications the 3rd case appears, i.e. an over-determined system. The number of input channels depends on the type of the test rig. On a 4-poster the number of inputs is equal to four and on a MAST it is equal to six, cf. Fig. 1.4. Typically a higher number of output channels is used, depending on the investigated system. Nowadays the measuring equipment is designed for a simultaneous recording of several channels.

Because of these reasons the Moore-Penrose pseudoinverse is implemented in the virtual iteration algorithm in *Matlab*. Therefore, the inverse $\mathbf{G}^{-1}(i\omega)$ is replaced by $\mathbf{G}^+(i\omega)$ in Eq. (3.37) and the 3rd point in the algorithm described in section 3.2.1. In *Matlab* the Moore-Penrose pseudoinverse is implemented in the command *pinv*. An efficient computation of the Moore-Penrose pseudoinverse is based on a singular value decomposition (SVD) [36].
The singular value decomposition is applied to the transfer matrix $\mathbf{G}(i\omega) \in \mathbb{C}^{k \times m_c}$ and reads as

$$\mathbf{G} = \mathbf{\Gamma \Sigma \Pi}^H \tag{3.46}$$

3.2. TARGET SIMULATION

The matrices $\mathbf{\Gamma} \in \mathbb{C}^{k \times k}$ and $\mathbf{\Pi}^H \in \mathbb{C}^{m_c \times m_c}$ are unitary matrices. The matrix $\mathbf{\Sigma} \in \mathbb{R}^{k \times k}$ is assembled by

$$\mathbf{\Sigma} = \begin{bmatrix} \mathbf{S} & \mathbf{0} \\ \mathbf{0} & \mathbf{0} \end{bmatrix}$$

where the diagonal matrix $\mathbf{S} = \text{diag}(\sigma_1, ..., \sigma_{m_c})$ contains the singular values $\sigma_1 \geq \cdots \geq \sigma_{m_c} > 0$, which are the square roots of the eigenvalues of $\mathbf{G}^H \mathbf{G}$. The number of non disappearing singular values is equal to the rank m_c of the transfer matrix $\mathbf{G}$.

By rearranging Eq. (3.46) the Moore-Penrose pseudoinverse can be computed by

$$\mathbf{G}^+ = \mathbf{\Pi} \mathbf{\Sigma}^+ \mathbf{\Gamma}^H \tag{3.47}$$

with

$$\mathbf{\Sigma}^+ = \begin{bmatrix} \frac{1}{\sigma_1} & & & \\ & \ddots & & \mathbf{0} \\ & & \frac{1}{\sigma_{m_c}} & \\ & \mathbf{0} & & \mathbf{0} \end{bmatrix}$$

The computation of $\mathbf{G}^+$ from Eq. (3.47) is more efficient than the definition in (3.45) and hence it is implemented to compute the Moore-Penrose pseudoinverse.

The method of virtual iteration works well for small as well as large multibody systems that are nearly linear, cf. the examples in chapter 8. Due to the linearization and the computation in the frequency domain sharp peaks are eventually undetected. But such sharp edges in the drive signals are important in a fatigue analysis and hence errors can be introduced [143]. Furthermore, important mathematical statements as accuracy, stability and convergence behavior cannot be given in general [40].

The following methods, namely (i) DAE approach with control constraints, (ii) optimal control and (iii) flatness-based trajectory tracking, are designed for nonlinear models, which are treated directly in the time domain.

Nothing is more practical than a good theory.
*Kurt Lewin / Gustav Robert Kirchhoff /
Todor Karman / Ludwig Boltzmann*

Chapter 4
DAE Approach with Control Constraints

This section focuses on controlled multibody systems, where the number of inputs is equal to the number of outputs. Furthermore, underactuated systems are considered, i.e. the number of control inputs m_c is less than the number of mechanical DOFs n. It is shown that the index 3 DAEs (2.51) can be extended by so-called *control or servo constraints*, which partially describe the motion of the multibody system. The implementation of control constraints in systems with independent coordinates has previously been published by [13, 21, 22, 23, 24]. Extended versions with a redundant coordinates formulation can be found, for example, in [12, 25, 26, 29, 119, 122, 149, 150, 151].
In the following considerations the DAEs (2.51) are written in the form

$$\dot{\mathbf{q}} - \mathbf{v} = \mathbf{0} \tag{4.1a}$$

$$\mathbf{M}\dot{\mathbf{v}} - \mathbf{f} + \mathbf{G}^T \boldsymbol{\lambda} = \mathbf{0} \tag{4.1b}$$

$$\mathbf{g}(\mathbf{q}) = \mathbf{0} \tag{4.1c}$$

where $\mathbf{v} \in \mathbb{R}^n$ denotes the vector of velocities, i.e. the derivations of the generalized or redundant coordinates $\mathbf{q}$ with respect to time. Furthermore, it is assumed that the geometric constraints are restricted to scleronomic holonomic constraints $\mathbf{g}(\mathbf{q}) \in \mathbb{R}^m$. The general incorporation of nonholonomic constraints $\mathbf{g}(\mathbf{q}, \dot{\mathbf{q}})$ can be found for example in [11, 151].

4.1 Formulation of the DAEs

In a straightforward way the DAEs (4.1) can be extended by additional constraints. As a consequence, the DAEs with control constraints are given by the system

$$\dot{\mathbf{q}} - \mathbf{v} = \mathbf{0} \qquad (4.2a)$$
$$\mathbf{M}\dot{\mathbf{v}} - \mathbf{f} + \mathbf{G}^T\boldsymbol{\lambda} + \mathbf{B}^T\mathbf{u} = \mathbf{0} \qquad (4.2b)$$
$$\mathbf{c}(\mathbf{q},t) = \mathbf{0} \qquad (4.2c)$$
$$\mathbf{g}(\mathbf{q}) = \mathbf{0} \qquad (4.2d)$$

The algebraic rheonomic control (servo) constraints are formulated by $\mathbf{c}(\mathbf{q},t) \in \mathbb{R}^{m_c}$ (4.2c) and read as

$$\mathbf{c}(\mathbf{q},t) = \Phi(\mathbf{q}) - \boldsymbol{\gamma}(t) \qquad (4.3)$$

The vector $\Phi(\mathbf{q}) \in \mathbb{R}^{m_c}$ summarizes the outputs of the multibody system and the vector $\boldsymbol{\gamma}(t) \in \mathbb{R}^{m_c}$ includes the requested target signals. In section 4.3 it is shown that the target signals have to be continuously differentiable up to a certain order. This effect states that the solvability of underactuated multibody systems is strongly related to differential flatness and feedback linearizability [12, 21], cf. chapter 6. Therefore, polynomial functions can be used to formulate the desired trajectory. Arbitrary targets e.g. from measurements have to be filtered in order to generate differentiable functions.

The outputs $\Phi(\mathbf{q})$ are typically given in a linear form, if dependent coordinates are used [25]. Eq. (4.3) formulate the condition that targets and system outputs must be identical. This is the ultimate aim of the trajectory tracking problem. In section 5.4.1 it is shown that in an optimal control problem this condition is formulated in a different way (5.39).

In (4.2b) a new Lagrange multiplier $\mathbf{u} \in \mathbb{R}^{m_c}$ is introduced. This Lagrange multiplier is related to the generalized actuator forces $\mathbf{f}_u = -\mathbf{B}^T\mathbf{u}$. This formulation can be compared with the geometric constraint reaction forces $\mathbf{f}_g = -\mathbf{G}^T\boldsymbol{\lambda}$ from Eq. (2.51).

The matrix $\mathbf{B} \in \mathbb{R}^{m_c \times n}$ is called input-transformation matrix. It can be derived directly from the equations of motion, either if dependent or independent coordinates are used. Generally, $\mathbf{B}$ is formulated independently from the control constraints. Typically, the entries in the input transformation matrix are constant.

If the multibody system is fully actuated, i.e. $m_c = n$, the input-transformation matrix is quadratic. Hence, it can be inverted, if it is non-singular and therefore the classical inverse dynamics methods can be used. In fully actuated systems the motion is entirely specified and index 3 integrators can be applied [150].

In contrast to that underactuated systems are more challenging to solve, because $\mathbf{B}$ is not quadratic as in the case of fully actuated systems. The motion is only partly specified and the constraint Jacobian does not span the space of control variables [13].

4.2. DEPENDENT VERSUS INDEPENDENT COORDINATES

From the geometrical point of view, the control constraint realization can be classified into an "*orthogonal realization*", a "*mixed orthogonal-tangent realization*" and a "*tangent realization*" [21]. Geometric constraints and control constraints are realized by constraint forces $\mathbf{f_g}$ and control forces $\mathbf{f_u}$, respectively. Reaction forces of ideal passive (geometric) constraints are orthogonal to the respective constraint manifold. Control constraints are typically characterized by a non-orthogonal realization, which becomes tangent in the worst case [21]. This is illustrated in Fig. 4.1. As a consequence, the DAEs of underactuated systems have an index that is higher than 3. Typically, index 5 problems (4.2) are obtained, if redundant coordinates are used to formulate the dynamic equations of motion [22]. However, DAEs with a high index produce nuisances in their numerical treatment [25] and furthermore most commercial solvers are designed for index 1 DAEs, cf. the *DASSL*-solver from section 2.8.4. Hence, such systems should not be solved directly. An appropriate projection method (section 4.3) can be applied to reduce the systems index. The resulting system can stably be integrated.

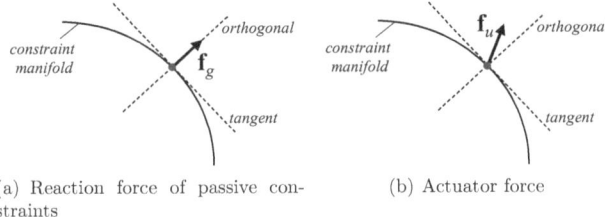

(a) Reaction force of passive constraints

(b) Actuator force

Figure 4.1: Reactions of passive geometric and control constraints [22]

4.2 Dependent versus Independent Coordinates

Control constraints can also be implemented in systems, which are formulated by independent (generalized, minimal) coordinates. As a consequence, the equations of motion, which are originally ODEs, become index 3 DAEs.

$$\dot{\mathbf{q}} - \mathbf{v} = \mathbf{0} \tag{4.4a}$$

$$\mathbf{M}\dot{\mathbf{v}} - \mathbf{f} + \mathbf{B}^T \mathbf{u} = \mathbf{0} \tag{4.4b}$$

$$\mathbf{c}(\mathbf{q}, t) = \mathbf{0} \tag{4.4c}$$

One might guess that this system is easier to solve compared to the index 5 system (4.2). However, it can be shown that the formulation with redundant coordinates is quite beneficial regarding to the inverse problem [25].
The most important advantages of redundant coordinates are shown in the following list:

- Dynamic equations of motion are considerably simpler (constant (often diagonal) mass matrix $\mathbf{M}$)

- Straightforward formulation of servo-constraints (servo-constraint matrix $\mathbf{C} = D\mathbf{\Phi}(\mathbf{q})$ is a simple sparse matrix (Boolean))

- Governing DAEs of the inverse simulation problem are less complex

- Constraint forces (e.g.: the cable tension force in crane examples) can be monitored by $\boldsymbol{\lambda}$ during the integration (more physical insight)

4.3 Index Reduction Procedure

A suitable projection method that reduces the index of the DAE-system can either be applied to systems with independent coordinates [13, 21, 22] or dependent coordinates [13, 26]. By using appropriate projection matrices $\mathbf{C}$ and $\mathbf{D}$, the index 5 system (4.2) can be projected to a constrained subspace $\mathcal{C}$ and an unconstrained subspace $\mathcal{D}$ relative to the manifold of servo constraints [26]. As a consequence, the index 5 problem is reduced to an index 3 problem and therefore the projection method can be seen as an index reduction procedure.

The control constraints (4.2c) are differentiated with respect to time.

$$\frac{d}{dt}\mathbf{c}(\mathbf{q},t) = \mathbf{C}\mathbf{v} - \dot{\boldsymbol{\gamma}} = \mathbf{0}, \quad \mathbf{C} = D\mathbf{\Phi}(\mathbf{q}) \tag{4.5}$$

Eq. (4.5) denotes the control constraints at velocity level. The matrix $\mathbf{C} \in \mathbb{R}^{m_c \times n}$ includes the partial derivations from $\mathbf{\Phi}(\mathbf{q})$ with respect to the redundant coordinates $\mathbf{q}$. Later, the matrix $\mathbf{C}$ will be used to project the differential equation (4.2b) to the constrained subspace $\mathcal{C}$. A further differentiation with respect to time affords the control constraints at acceleration level.

$$\frac{d^2}{dt^2}\mathbf{c}(\mathbf{q},t) = \mathbf{C}\dot{\mathbf{v}} + \boldsymbol{\xi} = \mathbf{0}, \quad \boldsymbol{\xi} = \dot{\mathbf{C}}\mathbf{v} - \ddot{\boldsymbol{\gamma}} \tag{4.6}$$

The control constraints at acceleration level can now be written as

$$\mathbf{C}\dot{\mathbf{v}} = -\boldsymbol{\xi} \tag{4.7}$$

The projection to a m_c-dimensional constrained (orthogonal) subspace $\mathcal{C}$ and a $(n-m_c)$-dimensional unconstrained (tangent) subspace $\mathcal{D}$ has to fulfill the conditions $\mathcal{C} \cup \mathcal{D} = \mathcal{N}$, where $\mathcal{N}$ denotes the original configuration manifold and $\mathcal{C} \cap \mathcal{D} = \emptyset$, i.e. the empty set. As a consequence, the projection matrix $\mathbf{D} \in \mathbb{R}^{n \times (n-m_c)}$ has to be formed in a way that the complementarity condition reads as [12, 22]

$$\text{range}(\mathbf{D}) = \ker(\mathbf{C}) \Leftrightarrow \mathbf{C}\mathbf{D} = \mathbf{0} \tag{4.8}$$

The orthogonal complement matrix $\mathbf{D}$ can sometimes be guessed or calculated by a coordinate partitioning method as indicated in [22, 25]. Subsequently, the projection

4.3. INDEX REDUCTION PROCEDURE

matrices **C** and **D** can be used to split the system (4.2b) to $(n - m_c)$ differential equations and m_c algebraic equations. For that reason Eq. (4.2b) is multiplied from the left side by $\mathbf{D}^T$ and $\mathbf{CM}^{-1}$, respectively. By considering also (4.7), the projection method results in

$$\begin{bmatrix} \mathbf{D}^T \\ \mathbf{CM}^{-1} \end{bmatrix} \{\mathbf{M}\dot{\mathbf{v}} - \mathbf{f} + \mathbf{G}^T\boldsymbol{\lambda} + \mathbf{B}^T\mathbf{u}\} = \\ \left\{ \begin{array}{c} \mathbf{D}^T\mathbf{M}\dot{\mathbf{v}} + \mathbf{D}^T\{-\mathbf{f} + \mathbf{G}^T\boldsymbol{\lambda} + \mathbf{B}^T\mathbf{u}\} \\ \mathbf{CM}^{-1}\{-\mathbf{f} + \mathbf{G}^T\boldsymbol{\lambda} + \mathbf{B}^T\mathbf{u}\} - \boldsymbol{\xi} \end{array} \right\} \tag{4.9}$$

The system $(4.9)_1$ are the differential equations that are projected to the unconstrained subspace $\mathcal{D}$ and $(4.9)_2$ are the algebraic equations on the constrained subspace $\mathcal{C}$. Accordingly, the DAEs with a reduced index of 3 read as

$$\dot{\mathbf{q}} - \mathbf{v} = 0 \tag{4.10a}$$
$$\mathbf{D}^T\mathbf{M}\dot{\mathbf{v}} + \mathbf{D}^T\{-\mathbf{f} + \mathbf{G}^T\boldsymbol{\lambda} + \mathbf{B}^T\mathbf{u}\} = 0 \tag{4.10b}$$
$$\mathbf{CM}^{-1}\{-\mathbf{f} + \mathbf{G}^T\boldsymbol{\lambda} + \mathbf{B}^T\mathbf{u}\} - \boldsymbol{\xi} = 0 \tag{4.10c}$$
$$\mathbf{c}(\mathbf{q},t) = 0 \tag{4.10d}$$
$$\mathbf{g}(\mathbf{q}) = 0 \tag{4.10e}$$

This new set of DAEs, which consists of the differential equations (4.10a) and (4.10b) and the algebraic equations (4.10c)-(4.10e), is then directly discretized. Hence, the discrete system is suitable for a numerical treatment.

By considering the term $\mathbf{CM}^{-1}\mathbf{B}^T$ in Eq. (4.10c), the control constraint realization can be investigated in more detail.

$$-\boldsymbol{\xi} - \mathbf{CM}^{-1}\mathbf{f} + \mathbf{CM}^{-1}\mathbf{G}^T\boldsymbol{\lambda} + \mathbf{CM}^{-1}\mathbf{B}^T\mathbf{u} = 0 \tag{4.11}$$

From the geometrical point of view, the control constraint realization can be quantified as mutual dot product of the spanning vectors of the subspaces $\mathcal{C}$ and $\mathcal{B}$ [21]. The row vectors of **C** span the m_c-dimensional constrained subspace $\mathcal{C}$ and the column vectors of $\mathbf{B}^T$ span the m_c-dimensional controlled subspace $\mathcal{B}$, if the matrices **C** and **B** have full rank m_c. The realization of control constraints can be considered as the inner product of the two subspaces $\mathcal{P} = \mathcal{C} \cap \mathcal{B}$. In differential geometry this inner product yields the matrix $\mathbf{P} = \mathbf{CM}^{-1}\mathbf{B}^T$ [21]. If **C** and **B** have full rank, the matrix **P** is quadratic ($m_c \times m_c$). The mass matrix is assumed to be regular and hence **P** is regular as well. As a consequence, **P** can be inverted and therefore the controls **u** can directly be computed from (4.11).
However, the matrices **C** and **B** do not always have full rank in general. The value

$$p = \text{rank}(\mathbf{CM}^{-1}\mathbf{B}^T) \tag{4.12}$$

is a measure of the representation of control reactions in the constrained directions [21]. It shows how many constraint conditions can directly be actuated by the control inputs **u**. If $p = m_c$, the control constraints are orthogonal to the constraint manifold. This situation is illustrated in Fig. 4.1(a) for passive constraints. The controls (actuator forces) can directly actuate all constraint conditions.

In a mixed orthogonal-tangent realization the control reactions are projected in the constrained directions as well as in the unconstrained directions. In this case the actuator forces **u** can only control p control constraints. The remaining $(m_c - p)$ directions in the constrained subspace $\mathcal{C}$ are not affected by **u**.

The worst case is a pure tangent realization $p = 0$, where the control reactions are not projected into $\mathcal{C}$. In this case all constraint conditions must be realized by tangent control reactions. Examples for the different control constraint realizations can be found in [21].

$p = m_c$	orthogonal realization
$0 < p < m_c$	mixed orthogonal-tangent realization
$p = 0$	tangent realization

Table 4.1: Possible control constraint realizations [21]

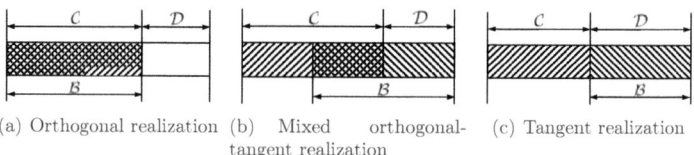

(a) Orthogonal realization (b) Mixed orthogonal-tangent realization (c) Tangent realization

Figure 4.2: Subspaces of control constraint realizations [21]

4.4 Numerical Solution

The projected index 3 DAEs (4.10) are discretized and solved by an implicit Euler algorithm [12, 26, 151].

$$\mathbf{q}_{n+1} - \mathbf{q}_n - \frac{\Delta t}{2}\left[\mathbf{v}_n + \mathbf{v}_{n+1}\right] = \mathbf{0} \quad (4.13a)$$

$$\mathbf{D}^T \mathbf{M}\left(\mathbf{v}_{n+1} - \mathbf{v}_n\right) + \mathbf{D}^T\left\{-\bar{\mathbf{f}} + \bar{\mathbf{G}}^T\bar{\boldsymbol{\lambda}} + \mathbf{B}^T\bar{\mathbf{u}}\right\}\Delta t = \mathbf{0} \quad (4.13b)$$

$$\mathbf{C}\mathbf{M}^{-1}\left\{-\bar{\mathbf{f}} + \bar{\mathbf{G}}^T\bar{\boldsymbol{\lambda}} + \mathbf{B}^T\bar{\mathbf{u}}\right\} - \bar{\boldsymbol{\xi}} = \mathbf{0} \quad (4.13c)$$

$$\mathbf{c}(\mathbf{q}_{n+1}, t_{n+1}) = \mathbf{0} \quad (4.13d)$$

$$\mathbf{g}(\mathbf{q}_{n+1}) = \mathbf{0} \quad (4.13e)$$

The abbreviations of the discrete vector of applied and gyroscopic forces $\bar{\mathbf{f}} = \mathbf{f}(\mathbf{q}_n, \mathbf{q}_{n+1})$, the discrete constraint Jacobian $\bar{\mathbf{G}}^T = \mathbf{G}^T(\mathbf{q}_n, \mathbf{q}_{n+1})$, the Lagrange multiplier $\bar{\boldsymbol{\lambda}}$, the

4.4. NUMERICAL SOLUTION

targets at acceleration level $\bar{\boldsymbol{\xi}}$ and the inputs $\bar{\mathbf{u}}$ denote an evaluation in the midpoint configuration $\mathbf{q}_{n+\frac{1}{2}} = (\mathbf{q}_n + \mathbf{q}_{n+1})/2$ [150, 152]. In [29, 148] proofs of conservation of energy and angular momentum during numerical integration are presented for the formulation (4.13). In a conservative system all external forces can be derived from a potential, i.e. $\bar{\mathbf{f}} = \bar{\mathbf{Q}} - \bar{\nabla} V$, $\bar{\mathbf{Q}} = \mathbf{0}$. If this is the case, the total energy is constant $E(\mathbf{q}, \mathbf{v}) = T + V = const.$, i.e. $E(\mathbf{q}_{n+1}, \mathbf{v}_{n+1}) = E(\mathbf{q}_n, \mathbf{v}_n)$. Energy and momentum conserving algorithms as well as stability of symplectic integrators can be found in [73, 75, 76, 85].

However, satisfying results are also achieved with an evaluation at step $n+1$, i.e. $\mathbf{f}(\mathbf{q}_{n+1})$, $\mathbf{G}(\mathbf{q}_{n+1})$, $\boldsymbol{\lambda}_{n+1}$, $\boldsymbol{\xi}_{n+1}$, $\mathbf{u}_{n+1}$. This backward Euler scheme is presented e.g. in [13, 21, 22, 119]. The one-step algorithm is denoted as "C-BEM-scheme (control basic energy-momentum scheme)" in [148]. The implicit algorithm provides the vectors $\mathbf{q}_{n+1}$, $\mathbf{v}_{n+1}$, $\boldsymbol{\lambda}_{n+1}$ and the control inputs $\mathbf{u}_{n+1}$. The main steps of the implementation without control constraints are presented in [10, 11]. In this thesis the algorithm is extended for a system with control constraints (4.13).

The following implicit procedure is shown for the kth iteration. It is repeated until convergence $\|\mathbf{R}^{(k)}\| < \varepsilon$ is reached ($\mathbf{R}$ denotes a residual vector).
The velocities at step $n+1$ results directly from (4.13a):

$$\mathbf{v}_{n+1}^{(k)} = \frac{2}{\Delta t}(\mathbf{q}_{n+1}^{(k)} - \mathbf{q}_n^{(k)}) - \mathbf{v}_n^{(k)} \tag{4.14}$$

Then the residual vectors $\mathbf{R}_\mathbf{q}, \mathbf{R}_\mathbf{v}, \mathbf{R}_\boldsymbol{\lambda}, \mathbf{R}_\mathbf{u}$ have to be calculated for the Newton-iteration, which is required in the implicit one-step algorithm.

$$\mathbf{R}_{\mathbf{q}_1}^{(k)} = \underbrace{\mathbf{D}^T \mathbf{M}\left(\frac{2}{\Delta t}(\mathbf{q}_{n+1}^{(k)} - \mathbf{q}_n^{(k)}) - 2\mathbf{v}_n^{(k)}\right) - \Delta t \mathbf{D}^T \mathbf{f}^{(k)} + \cdots}_{\mathbf{H}(\mathbf{q}_{n+1})^{(k)}} \tag{4.15a}$$
$$+ \Delta t \mathbf{D}^T \mathbf{G}^{T(k)} \boldsymbol{\lambda}_{n+1}^{(k)} + \Delta t \mathbf{D}^T \mathbf{B}^T \mathbf{u}_{n+1}^{(k)}$$

$$\mathbf{R}_{\mathbf{q}_2}^{(k)} = \mathbf{C} \mathbf{M}^{-1}\left\{-\mathbf{f}^{(k)} + \mathbf{G}^{T(k)} \boldsymbol{\lambda}_{n+1}^{(k)} + \mathbf{B}^T \mathbf{u}_{n+1}^{(k)}\right\} - \boldsymbol{\xi}_{n+1} \tag{4.15b}$$

$$\mathbf{R}_\mathbf{v}^{(k)} = \frac{\mathbf{q}_{n+1}^{(k)} - \mathbf{q}_n^{(k)}}{\Delta t} - \frac{\mathbf{v}_n^{(k)} + \mathbf{v}_{n+1}^{(k)}}{2} \tag{4.15c}$$

$$\mathbf{R}_\boldsymbol{\lambda}^{(k)} = \mathbf{g}(\mathbf{q}_{n+1})^{(k)} \tag{4.15d}$$

$$\mathbf{R}_\mathbf{u}^{(k)} = \mathbf{c}(\mathbf{q}_{n+1}, t_{n+1})^{(k)} \tag{4.15e}$$

The residual vector $\mathbf{R}_\mathbf{q}$ is split into $\mathbf{R}_{\mathbf{q}_1}$ and $\mathbf{R}_{\mathbf{q}_2}$ regarding to Eqs. (4.13b) and (4.13c). In $\mathbf{R}_{\mathbf{q}_1}$ the result from (4.14) is inserted for $\mathbf{v}_{n+1}$.
The residuum $\mathbf{R}_\mathbf{v}$ is derived from (4.13a) and the residuals $\mathbf{R}_\boldsymbol{\lambda}$ and $\mathbf{R}_\mathbf{u}$ from (4.13e) and (4.13d), respectively. Furthermore, the Jacobian $\mathbf{J}^{(k)}$ has to be calculated. Then, the Newton-iteration can be formulated as

4.4. NUMERICAL SOLUTION

$$\begin{bmatrix} \frac{\partial \mathbf{R_q}}{\partial \mathbf{q}_{n+1}} & \frac{\partial \mathbf{R_q}}{\partial \mathbf{v}_{n+1}} & \frac{\partial \mathbf{R_q}}{\partial \boldsymbol{\lambda}_{n+1}} & \frac{\partial \mathbf{R_q}}{\partial \mathbf{u}_{n+1}} \\ \frac{\partial \mathbf{R_v}}{\partial \mathbf{q}_{n+1}} & \frac{\partial \mathbf{R_v}}{\partial \mathbf{v}_{n+1}} & \frac{\partial \mathbf{R_v}}{\partial \boldsymbol{\lambda}_{n+1}} & \frac{\partial \mathbf{R_v}}{\partial \mathbf{u}_{n+1}} \\ \frac{\partial \mathbf{R_\lambda}}{\partial \mathbf{q}_{n+1}} & \frac{\partial \mathbf{R_\lambda}}{\partial \mathbf{v}_{n+1}} & \frac{\partial \mathbf{R_\lambda}}{\partial \boldsymbol{\lambda}_{n+1}} & \frac{\partial \mathbf{R_\lambda}}{\partial \mathbf{u}_{n+1}} \\ \frac{\partial \mathbf{R_u}}{\partial \mathbf{q}_{n+1}} & \frac{\partial \mathbf{R_u}}{\partial \mathbf{v}_{n+1}} & \frac{\partial \mathbf{R_u}}{\partial \boldsymbol{\lambda}_{n+1}} & \frac{\partial \mathbf{R_u}}{\partial \mathbf{u}_{n+1}} \end{bmatrix}^{(k)} \cdot \begin{bmatrix} \Delta \mathbf{q}_{n+1} \\ \Delta \mathbf{v}_{n+1} \\ \Delta \boldsymbol{\lambda}_{n+1} \\ \Delta \mathbf{u}_{n+1} \end{bmatrix} = - \begin{bmatrix} \mathbf{R_q} \\ \mathbf{R_v} \\ \mathbf{R_\lambda} \\ \mathbf{R_u} \end{bmatrix}^{(k)} \quad (4.16)$$

The Jacobian $\mathbf{J}^{(k)}$ should be investigated more in detail.

$$\mathbf{J}^{(k)} = \begin{bmatrix} \begin{bmatrix} \mathbf{N}_{[(n-m_c)\times n]} \\ \mathbf{V}_{[m_c\times n]} \end{bmatrix} & \begin{bmatrix} \mathbf{X}_{[(n-m_c)\times n]} \\ \mathbf{0}_{[m_c\times n]} \end{bmatrix} & \begin{bmatrix} \Delta t \mathbf{D}^T \mathbf{G}^T_{[(n-m_c)\times m]} \\ \mathbf{CM}^{-1}\mathbf{G}^T_{[m_c\times m)]} \end{bmatrix} & \begin{bmatrix} \Delta t \mathbf{D}^T \mathbf{B}^T_{[(n-m_c)\times m_c]} \\ \mathbf{CM}^{-1}\mathbf{G}^T_{[m_c\times m_c)]} \end{bmatrix} \\ diag\left(\frac{1}{\Delta t}\right)_{[n\times n]} & diag\left(-\frac{1}{2}\right)_{[n\times n]} & \mathbf{0}_{[n\times m]} & \mathbf{0}_{[n\times m_c]} \\ \mathbf{G}_{[m\times n]} & \mathbf{0}_{[m\times n]} & \mathbf{0}_{[m\times m]} & \mathbf{0}_{[m\times m_c]} \\ \boldsymbol{\Theta}_{[m_c\times n]} & \mathbf{0}_{[m_c\times n]} & \mathbf{0}_{[m_c\times m]} & \mathbf{0}_{[m_c\times m_c]} \end{bmatrix}^{(k)}$$
(4.17)

The sub-matrix $\mathbf{N} \in \mathbb{R}^{(n-m_c)\times n}$ includes the partial derivatives of $\mathbf{R}_{\mathbf{q}_1}$ with respect to $\mathbf{q}$. It can be computed by

$$\mathbf{N} = \frac{\partial \mathbf{R}_{\mathbf{q}_1}}{\partial \mathbf{q}_{n+1}} = D\mathbf{H}(\mathbf{q}_{n+1}) + \Delta t \mathbf{D}^T \boldsymbol{\lambda}_{n+1} \frac{\partial \mathbf{G}^T}{\partial \mathbf{q}_{n+1}} + \underbrace{\Delta t \mathbf{D}^T \mathbf{u}_{n+1} \frac{\partial \mathbf{B}^T}{\partial \mathbf{q}_{n+1}}}_{=0} \quad (4.18)$$

The abbreviation $\mathbf{H}(\mathbf{q}_{n+1})$ is given in (4.15a). The sub-matrix $\mathbf{V} \in \mathbb{R}^{m_c\times n}$ is formed by the partial derivatives of $\mathbf{R}_{\mathbf{q}_2}$ with respect to $\mathbf{q}$.

$$\mathbf{V} = \frac{\partial \mathbf{R}_{\mathbf{q}_2}}{\partial \mathbf{q}_{n+1}} = \mathbf{CM}^{-1}\boldsymbol{\lambda}_{n+1}\frac{\partial \mathbf{G}^T}{\partial \mathbf{q}_{n+1}} \quad (4.19)$$

The sub-matrix $\mathbf{X} \in \mathbb{R}^{(n-m_c)\times n}$ is formed by the partial derivatives of $\mathbf{R}_{\mathbf{q}_1}$ with respect to $\mathbf{v}$ and $\boldsymbol{\Theta}$ by the derivatives of $\mathbf{R_u}$ with respect to $\mathbf{q}$.

$$\mathbf{X} = \frac{\partial \mathbf{R}_{\mathbf{q}_1}}{\partial \mathbf{v}_{n+1}} = \mathbf{0}, \quad \boldsymbol{\Theta} = \frac{\partial \mathbf{R_u}}{\partial \mathbf{q}_{n+1}} \quad (4.20)$$

By inverting the Jacobian $\mathbf{J}^{(k)}$, the updates $\Delta \mathbf{q}_{n+1}, \Delta \mathbf{v}_{n+1}, \Delta \boldsymbol{\lambda}_{n+1}, \Delta \mathbf{u}_{n+1}$ can be calculated from (4.16). Finally, the variables $\mathbf{q}, \mathbf{v}, \boldsymbol{\lambda}, \mathbf{u}$ can be updated for iteration $(k+1)$.

$$\mathbf{q}_{n+1}^{(k+1)} = \mathbf{q}_{n+1}^{(k)} + \Delta \mathbf{q}_{n+1} \quad (4.21a)$$
$$\mathbf{v}_{n+1}^{(k+1)} = \mathbf{v}_{n+1}^{(k)} + \Delta \mathbf{v}_{n+1} \quad (4.21b)$$
$$\boldsymbol{\lambda}_{n+1}^{(k+1)} = \boldsymbol{\lambda}_{n+1}^{(k)} + \Delta \boldsymbol{\lambda}_{n+1} \quad (4.21c)$$
$$\mathbf{u}_{n+1}^{(k+1)} = \mathbf{u}_{n+1}^{(k)} + \Delta \mathbf{u}_{n+1} \quad (4.21d)$$

Then the iterative procedure starts again at Eq. (4.15) until convergence is reached.

> *The Fundamental Variational Principle*
> Namely, because the shape of the whole
> universe is the most perfect and, in fact,
> designed by the wisest creator, nothing in all
> the world will occur in which no maximum or
> minimum rule is somehow shining forth...
>
> Leonhard Euler

Chapter 5
Optimization and Optimal Control

The field of optimization occupies itself with finding an optimal solution of a given problem. Optimization is used in a range of industrial and economical applications. The mathematical description of an optimization problem (OP) yields control strategies such that a certain optimality criterion is achieved [115]. As a consequence, the problem is formulated in a way that specific variables occur, whose values can be changed from outside. Such variables are called control variables. Additionally, state variables exist that describe the system's behavior at time t. Generally, it must be distinguished between static and dynamic optimization problems.
Applications of optimization problems, which are listed below, are taken from [115].

Applications in mechanical engineering:
In mechanical engineering one requires the motion of a mechanical system to be controlled from an initial state to a final state. The goal is e.g. that the control effort or the maneuver time is minimized. Vehicle dynamics is a broad field of application. An example of an optimal control is the minimization of the time that is needed in order to drive through a given path. In space missions the challenge is to minimize the energy of a satellite or a spacecraft that is needed in order to follow a trajectory. In robotics the motion of a robot and its tool has to be steered in a way that specific requirements are fulfilled. Biological processes can be studied with optimal control. Another field of application is biomechanics. Prostheses and implants for the human body can be improved by considering biomechanical models. The movements in sports can also be investigated with optimal control theory.

Applications in economics:
In economics optimization is used to optimize a certain portfolio or the development of a company. Optimal financing or optimal investment strategies can be studied with optimization techniques.
In this thesis several methods of static optimization problems and the optimal control theory are used. Applied methods are restricted to unconstrained problems. Therefore, these methods are discussed more in detail while methods for constrained problems are just briefly described. The main parts that are discussed in the following sections are taken from [14, 15, 37, 41, 77, 87].

5.1 Static Optimization Problems

In a static optimization problem the aim is to minimize a function $f(\mathbf{x})$ while considering specific constraints. The optimization variables $\mathbf{x}$ are elements of the Euclidean space $\mathbb{R}^n$.

Static optimization is also called *mathematical programming* or *finite-dimensional optimization* [14].

5.1.1 Formulation of the Problem

Static optimization problems are typically formulated by the system (5.1), [15].

$$\min_{\mathbf{x} \in \mathbb{R}^n} f(\mathbf{x}) \tag{5.1a}$$

$$s.t.: \quad g_i(\mathbf{x}) = 0, \quad i = 1, \ldots, p \tag{5.1b}$$

$$h_i(\mathbf{x}) \leq 0, \quad i = 1, \ldots, q \tag{5.1c}$$

The general formulation of the system (5.1) consists of the *cost function* (5.1a), the *equality-constraints* (5.1b) and the *inequality-constraints* (5.1c). If the optimization problem (5.1) is given without equality constraints (5.1b) and inequality constraints (5.1c), it is called *unconstrained optimization problem*. Otherwise, the full system (5.1) is called *constrained optimization problem*.

If cost function and constraints are linear, the OP is called *linear programming*. In *quadratic programming* the cost function is quadratic while the constraints are linear. If the cost function or at least one constraint is nonlinear, the OP is in the class of *nonlinear programming*.

5.1.2 Optimality Conditions of Unconstrained Optimization Problems

It is supposed that $f : \mathbb{R}^n \to \mathbb{R}$ is twice continuously differentiable. If $f(\mathbf{x}^*)$ fulfills the condition

$$\nabla f(\mathbf{x}^*) = \mathbf{0} \tag{5.2a}$$

$$\nabla^2 f(\mathbf{x}^*) \geq \mathbf{0} \quad \text{(positive semi} - \text{definite)} \tag{5.2b}$$

then $\mathbf{x}^*$ is a local minimum of f. If (5.2b) is positive definite $\nabla^2 f(\mathbf{x}^*) > \mathbf{0}$, then $\mathbf{x}^*$ is a strict local minimum. These conditions result from a Taylor series expansion and are shown in [15]. In (5.2a) $\nabla f(\mathbf{x})$ denotes the gradient of the function $f(\mathbf{x})$

$$\nabla f(\mathbf{x}) = \frac{\partial f}{\partial \mathbf{x}} = \begin{bmatrix} \frac{\partial f}{\partial x_1} \\ \vdots \\ \frac{\partial f}{\partial x_n} \end{bmatrix} \tag{5.3}$$

5.2. NUMERICAL METHODS FOR UNCONSTRAINED STATIC PROBLEMS

and $\nabla^2 f(\mathbf{x})$ in (5.2b) denotes the symmetric Hessian matrix

$$\nabla^2 f(\mathbf{x}) = \begin{bmatrix} \frac{\partial^2 f}{\partial x_1^2} & \cdots & \frac{\partial^2 f}{\partial x_1 \partial x_n} \\ \vdots & \ddots & \vdots \\ \frac{\partial^2 f}{\partial x_n \partial x_1} & \cdots & \frac{\partial^2 f}{\partial x_n^2} \end{bmatrix} \tag{5.4}$$

5.2 Numerical Methods for Unconstrained Static Problems

In many cases the analytical solution of the stationarity condition $\nabla f(\mathbf{x}^*) = \mathbf{0}$ cannot be derived. As a consequence, numerical methods are used that start from a sub-optimal initial solution $\mathbf{x}^0$ and converge iteratively to a minimum point $\mathbf{x}^*$. In every iteration the solution is improved compared to the previous step.

$$f(\mathbf{x}^{k+1}) < f(\mathbf{x}^k), \quad k = 0, 1, 2, \ldots \tag{5.5}$$

This results in a solution that converges to a minimum, if $k \to \infty$:

$$\lim_{k \to \infty} \mathbf{x}^k = \mathbf{x}^* \tag{5.6}$$

Line search algorithms are based on finding a direction $\mathbf{d}^k$ in which Eq. (5.5) is fulfilled. Then an optimal step size α^k has to be calculated to define the optimal distance along the direction $\mathbf{d}^k$ [15].

$$\mathbf{x}^{k+1} = \mathbf{x}^k + \alpha^k \mathbf{d}^k \tag{5.7}$$

The optimal step size α^k results from a second OP:

$$\min_{\alpha^k > 0} \phi(\alpha^k), \quad \phi(\alpha^k) = f(\mathbf{x}^k + \alpha^k \mathbf{d}^k) \tag{5.8}$$

This results in a maximum descent along the search direction $\mathbf{d}^k$. For the numerical computation of the optimal step size α^k different methods like the "equal interval search", the "section search", the "golden section search", a "quadratic interpolation method" or an approximate line search based on "Armijo's rule" are used. These methods are described in detail in [15] and should not be further discussed here.
For the computation of the search direction $\mathbf{d}^k$ different methods are known as well. They differ in accuracy, convergence speed and computational effort and should be described in the following sections.
Fig. 5.1 illustrates the idea of a line search algorithm for a quadratic function $f(\mathbf{x})$ with $\mathbf{x} \in \mathbb{R}^2$. The black dotted line shows the search direction $\mathbf{d}^k$ for a given initial value $\mathbf{x}^0$.

5.2.1 Steepest Descent Method

The simplest way to define a search direction $\mathbf{d}^k$ is to evaluate the negative gradient at the point $\mathbf{x}^k$.

$$\mathbf{d}^k = -\nabla f(\mathbf{x}^k) \tag{5.9}$$

5.2. NUMERICAL METHODS FOR UNCONSTRAINED STATIC PROBLEMS

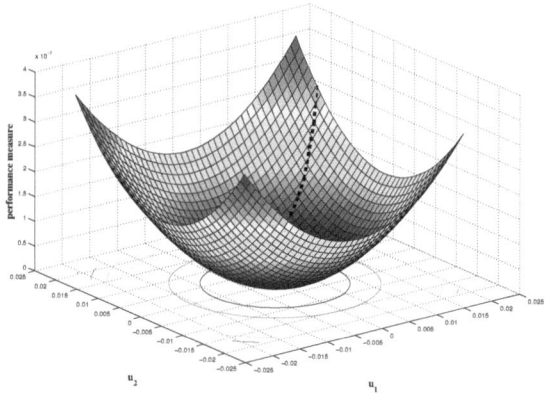

Figure 5.1: Line search algorithm for a quadratic function

The gradient is orthogonal to the contour line $f(\mathbf{x}^k) = const.$ and defines the direction of the steepest descent. If a problem is well conditioned, the gradient points into the direction of $\mathbf{x}^*$ and the steepest descent method converges rapidly. Otherwise, if the problem is badly conditioned, the negative gradient does not directly point to $\mathbf{x}^*$ and the method converges very slowly. It also depends on the chosen step size and on how close $\mathbf{x}^k$ is to $\mathbf{x}^*$. There are many schemes with adaptive step sizes α [15].

Advantages and disadvantages of the steepest descent method are summarized below [77]:

- \+   global convergence for convex problems
- \+   simple method
- \-   slow convergence if the problem is badly conditioned
- \-   limited accuracy

5.2.2 Conjugate Gradient Method

The conjugate gradient method also uses information from the previous step in order to define the search direction $\mathbf{d}^k$. The computational effort is not much higher compared to the steepest descent method while the convergence behavior is much better.

$$\mathbf{d}^0 = -\nabla f(\mathbf{x}^0) \tag{5.10a}$$
$$\mathbf{d}^k = -\nabla f(\mathbf{x}^k) + \beta^k \mathbf{d}^{k-1} \tag{5.10b}$$

For initialization in the first step the negative gradient is used as it is done in the steepest descent method, Eq. (5.10a). In the following steps a correction term is

5.2. NUMERICAL METHODS FOR UNCONSTRAINED STATIC PROBLEMS

added which depends on the previous step, Eq. (5.10b). The scalar factor β^k can be calculated by using the Fletcher-Reeves formula [15]:

$$\beta^k = \frac{[\nabla f(\mathbf{x}^k)]^T \nabla f(\mathbf{x}^k)}{[\nabla f(\mathbf{x}^{k-1})]^T \nabla f(\mathbf{x}^{k-1})} \tag{5.11}$$

or by using the Polak-Ribiere formula [15]:

$$\beta^k = \frac{[\nabla f(\mathbf{x}^{k-1}) - \nabla f(\mathbf{x}^k)]^T \nabla f(\mathbf{x}^k)}{[\nabla f(\mathbf{x}^{k-1})]^T \nabla f(\mathbf{x}^{k-1})} \tag{5.12}$$

5.2.3 Newton's Method

The basic idea behind Newton's method is to approximate the function $f(\mathbf{x}^{k+1})$ by a quadratic function, i.e. to develop a Taylor series that is terminated after the quadratic terms [15].

$$f(\mathbf{x}^{k+1}) \approx f(\mathbf{x}^k) + \nabla f(\mathbf{x}^k)^T \mathbf{d}^k + \frac{1}{2}(\mathbf{d}^k)^T \nabla^2 f(\mathbf{x}^k) \mathbf{d}^k \tag{5.13}$$

Then, Eq. (5.13) has to be differentiated with respect to $\mathbf{d}^k$ and the resulting equation has to be set to zero in order to find the minimum of $\mathbf{d}^k$.

$$\nabla f(\mathbf{x}^k) + \nabla^2 f(\mathbf{x}^k) \mathbf{d}^k = \mathbf{0} \tag{5.14}$$

Assuming that $\nabla^2 f(\mathbf{x}^k)$ is positive definite, the Newton-direction $\mathbf{d}^k$ can be calculated as follows:

$$\mathbf{d}^k = -\left[\nabla^2 f(\mathbf{x}^k)\right]^{-1} \nabla f(\mathbf{x}^k) \tag{5.15}$$

Advantages and disadvantages of the Newton-method are summarized below [77]:

- + quadratic convergence
- + high accuracy
- - local convergence
- - time-consuming calculation of the Hessian matrix $\nabla^2 f(\mathbf{x}^k)$

5.2.4 Quasi-Newton Methods

Quasi-Newton methods avoid the disadvantage of the expensive calculation and inversion of the Hessian matrix $\nabla^2 f$ in each iteration. The inverse Hessian matrix is approximated by a matrix $\mathbf{Q}^k \approx (\nabla^2 f(\mathbf{x}^k))^{-1}$. Therefore, the direction of all Quasi-Newton methods is calculated as follows:

$$\mathbf{d}^k = -\mathbf{Q}^k \nabla f(\mathbf{x}^k) \tag{5.16}$$

The Quasi-Newton methods differ in the way in which the matrix $\mathbf{Q}$ is updated. Two of them should be pointed out.

5.2. NUMERICAL METHODS FOR UNCONSTRAINED STATIC PROBLEMS

DFP (Davidon, Fletcher, Powell) Update [15]

$$\mathbf{Q}^{k+1} = \mathbf{Q}^k + \frac{\mathbf{s}^k(\mathbf{s}^k)^T}{(\mathbf{q}^k)^T \mathbf{s}^k} - \frac{(\mathbf{Q}^k \mathbf{q}^k)(\mathbf{Q}^k \mathbf{q}^k)^T}{(\mathbf{q}^k)^T \mathbf{Q}^k \mathbf{q}^k} \tag{5.17}$$

with the abbreviations

$$\mathbf{q}^k = \nabla f(\mathbf{x}^{k+1}) - \nabla f(\mathbf{x}^k)$$
$$\mathbf{s}^k = \mathbf{x}^{k+1} - \mathbf{x}^k = \alpha^k \mathbf{d}^k \tag{5.18}$$

The DFP formula is initialized with an identity matrix. It can be shown that $\mathbf{Q}^k$ is positive definite as long as $(\mathbf{q}^k)^T \mathbf{s}^k > 0$ [15].

BFGS (Broyden, Fletcher, Goldfarb, Shanon) Update [15]

$$\mathbf{Q}^{k+1} = \mathbf{Q}^k + \left(1 + \frac{(\mathbf{q}^k)^T \mathbf{Q}^k \mathbf{q}^k}{(\mathbf{q}^k)^T \mathbf{s}^k}\right) \frac{\mathbf{s}^k(\mathbf{s}^k)^T}{(\mathbf{q}^k)^T \mathbf{s}^k} - \frac{1}{(\mathbf{q}^k)^T \mathbf{s}^k} \left(\left(\mathbf{s}^k(\mathbf{q}^k)^T \mathbf{Q}^k\right)^T + \mathbf{s}^k(\mathbf{q}^k)^T \mathbf{Q}^k \right) \tag{5.19}$$

Numerical observations for updating $\mathbf{Q}$ are equal to the DFP-formula. However, the BFGS-update is numerically better qualified than the DFP-update [15].
Quasi-Newton methods show a fast convergence behavior, but do not reach the speed of pure Newton methods. However, the vector-matrix manipulations to compute $\mathbf{Q}^k$ are much faster compared to the computation of the Hessian $\nabla^2 f$ and its Inverse. Another advantage is that singularities cannot occur.
Other numerical methods for unconstrained OPs are e.g.: the *trust region method* or the *Simplex-algorithm*. These methods are described in detail in [8, 15] and will not be further discussed here.
It should be mentioned that in the *Matlab Optimization Toolbox* the function *fminunc* can be used as gradient method, Newton- and Quasi-Newton-method and the function *fminsearch* can be used as Simplex-method.

5.2.5 Levenberg-Marquardt Algorithm

The Levenberg-Marquardt algorithm is a standard procedure in nonlinear optimization. It can be classified as pseudo-second order method, i.e. only function and gradient evaluations are required. The Hessian matrix is computed by products of the gradients. Typically, this method converges faster than first order methods [108].
The method is based on a linearization and a quadratic approximation of the function $f(\mathbf{x})$ near the minimum. If an adequate approximation is used, the convergence speed is faster compared to steepest descent methods. *Kenneth Levenberg* suggested a modified Hessian matrix $\mathbf{H} + \lambda \mathbf{I}$ for the optimization purpose [94]. By using this modified Hessian, the update can be varied between the steepest descent direction (5.9)

$$\mathbf{x}^{k+1} = \mathbf{x}^k - \alpha^k \mathbf{d}^k \tag{5.20}$$

and the quadratic rule (5.15)

$$\mathbf{x}^{k+1} = \mathbf{x}^k - \left(\mathbf{H}^k\right)^{-1} \mathbf{d}^k \qquad (5.21)$$

This yields

$$\mathbf{x}^{k+1} = \mathbf{x}^k - \left[\mathbf{H}^k + \lambda^k \mathbf{I}\right]^{-1} \mathbf{d}^k \qquad (5.22)$$

If the blending factor λ^k is small ($\lambda^k \to 0$), Eq. (5.22) results in the quadratic approximation (5.21) and if λ^k is large, it results in the steepest descent method

$$\mathbf{x}^{k+1} = \mathbf{x}^k - \frac{1}{\lambda^k} \mathbf{d}^k \qquad (5.23)$$

Donald Marquardt improved the algorithm by using the diagonal of the Hessian matrix instead of the identity matrix in Eq. (5.22).

$$\mathbf{x}^{k+1} = \mathbf{x}^k - \left[\mathbf{H}^k + \lambda^k \text{diag}(\mathbf{H}^k)\right]^{-1} \mathbf{d}^k \qquad (5.24)$$

Eq. (5.24) is known as Levenberg-Marquardt method. If the minimizing function increases during the optimization, the value λ^k is increased. Hence, the procedure shows a similar behavior as the steepest descent method. If the function is decreased, λ^k is decreased and the quadratic approximation is used. As a result the convergence behavior of the algorithm is improved. Details can be found e.g. in [14, 94, 108].

5.3 Numerical Methods for Constrained Static Problems

In this section constrained problems as given in Eqs. (5.1) are briefly discussed.

5.3.1 Stationarity Condition for a Single Equality Constraint

If one equality constraint $g(\mathbf{x}) = 0$ (5.1b) is used in the problem, the Lagrangian can be formulated as

$$L(\mathbf{x}, \lambda) = f(\mathbf{x}) + \lambda g(\mathbf{x}) \qquad (5.25)$$

The Lagrange multiplier λ is used to add the equality constraint to the cost function $f(\mathbf{x})$. At the optimal point $\mathbf{x}^*$ the gradient-vectors $\nabla f(\mathbf{x}^*)$ and $\nabla g(\mathbf{x}^*)$ are co-linear, i.e. $\nabla f(\mathbf{x}^*) || \nabla g(\mathbf{x}^*)$. As a consequence, the stationarity condition can be formulated as [15]

$$\nabla_\mathbf{x} L(\mathbf{x}^*, \lambda^*) = \nabla f(\mathbf{x}^*) + \lambda^* \nabla g(\mathbf{x}^*) = 0 \qquad (5.26)$$

By considering Eq. (5.26) and the equality constraint equation $g(\mathbf{x}^*) = 0$ (5.1b) a system of equations of the order $(n+1)$ can be formulated for the $(n+1)$ unknowns $\mathbf{x}^* \in \mathbb{R}^n$ and $\lambda^* \in \mathbb{R}$.

5.3. NUMERICAL METHODS FOR CONSTRAINED STATIC PROBLEMS

5.3.2 Stationarity Condition for a Single Inequality Constraint

If a single inequality constraint $h(\mathbf{x}) \leq 0$ (5.1c) is used in the problem, the Lagrangian can be formulated in the same way than in Eq. (5.25).

$$L(\mathbf{x}, \mu) = f(\mathbf{x}) + \mu h(\mathbf{x}) \qquad (5.27)$$

In such a case it has to be distinguished whether or not the inequality constraint $h(\mathbf{x}) \leq 0$ is active in the optimal point $\mathbf{x}^*$. If it is not active ($h(\mathbf{x}^*) < 0$), the stationarity condition is identical to the unconstrained case:

$$\nabla L_{\mathbf{x}}(\mathbf{x}, \mu^*) = \nabla f(\mathbf{x}^*) = 0 \quad \text{with} \quad \mu^* = 0 \qquad (5.28)$$

If the inequality constraint is active ($h(\mathbf{x}) = 0$), then the gradients ∇f and ∇h have to be co-linear in the optimal point $\mathbf{x}^*$ [37]. As a result, the stationarity condition is formulated as

$$\nabla L_{\mathbf{x}}(\mathbf{x}^*, \mu^*) = \nabla f(\mathbf{x}^*) + \mu^* \nabla h(\mathbf{x}^*) = 0 \qquad (5.29)$$

In this case the sign of the Lagrange multiplier μ^* plays an important part. By developing the cost function $f(\mathbf{x}^*)$ and the inequality constraint function $h(\mathbf{x})$ in a Taylor series it can be shown that the condition for a minimum point at $\mathbf{x}^*$ reads as [15]

$$\exists \mu^* \geq 0: \quad \nabla L_{\mathbf{x}}(\mathbf{x}^*, \mu^*) = 0, \quad \mu^* h(\mathbf{x}^*) = 0 \qquad (5.30)$$

The equation $\mu^* h(\mathbf{x}^*) = 0$ is called *complementarity condition* [77].

5.3.3 Karush-Kuhn-Tucker Conditions

The general case (5.1) with p equality-constraints and q inequality constraints is considered. The Lagrangian function can now be written as

$$L(\mathbf{x}, \boldsymbol{\lambda}, \boldsymbol{\mu}) = f(\mathbf{x}) + \sum_{i=1}^{p} \lambda_i g_i(\mathbf{x}) + \sum_{i=1}^{p} \mu_i h_i(\mathbf{x}) \qquad (5.31)$$

with the Lagrange multipliers $\boldsymbol{\lambda} = [\lambda_1, \ldots, \lambda_p]^T$ and $\boldsymbol{\mu} = [\mu_1, \ldots, \mu_q]^T$. The necessary optimality conditions in this general case are called KKT conditions (*Karush-Kuhn-Tucker conditions*) [14, 15, 115] and are formulated in Eqs. (5.32).

$$\nabla_{\mathbf{x}} L(\mathbf{x}^*, \boldsymbol{\lambda}^*, \boldsymbol{\mu}^*) = 0 \qquad (5.32a)$$
$$g_i(\mathbf{x}^*) = 0, \quad i = 1, \ldots, p \qquad (5.32b)$$
$$h_i(\mathbf{x}^*) \leq 0, \quad i = 1, \ldots, q \qquad (5.32c)$$
$$\mu_i^* \geq 0, \quad i = 1, \ldots, q \qquad (5.32d)$$
$$\mu_i^* h_i(\mathbf{x}^*) = 0, \quad i = 1, \ldots, q \qquad (5.32e)$$

The complementarity condition $\mu_i^* h_i(\mathbf{x}^*) = 0$ describes, if the ith inequality constraint is either active $h_i(\mathbf{x}^*) = 0$ or inactive $\mu_i^* = 0$.

5.3.4 Sequential Quadratic Programming

Sequential quadratic programming (SQP) is the most widely used algorithm in the field of nonlinear optimization [14]. The algorithm solves a sequence of quadratic approximations of (5.1). The basic approach introduces a quadratic approximation to the cost function (Lagrangian) and a linear approximation to the constraints. The quadratic approximation of $f(\mathbf{x})$ yields

$$f(\mathbf{x}) \approx f(\mathbf{x}^k) + \nabla f(\mathbf{x}^k)^T \mathbf{d} + \frac{1}{2}\mathbf{d}^T \mathbf{H}^k \mathbf{d} \qquad (5.33)$$

If an initial value $\mathbf{x}^k$ is given, the update can be calculated as follows:

$$\mathbf{x}^{k+1} = \mathbf{x}^k + \alpha^k \mathbf{d}^k \qquad (5.34)$$

The vector $\mathbf{d}^k$ is the solution of a quadratic sub-problem:

$$\min_{\mathbf{d}\in\mathbb{R}^n} \quad \frac{1}{2}\mathbf{d}^T \mathbf{H}^k \mathbf{d} + \nabla f(\mathbf{x}^k)^T \mathbf{d} \qquad (5.35\text{a})$$
$$\text{s.t.:} \quad g_i(\mathbf{x}^k) + \nabla g_i(\mathbf{x}^k)^T \mathbf{d} = 0, \quad i = 1, \ldots, p \qquad (5.35\text{b})$$
$$h_i(\mathbf{x}^k) + \nabla h_i(\mathbf{x}^k)^T \mathbf{d} \leq 0, \quad i = 1, \ldots, q \qquad (5.35\text{c})$$

The term $f(\mathbf{x}^k)$ in Eq. (5.33) does not depend on $\mathbf{d}$ and hence it can be neglected in (5.35a). $\mathbf{H}^k$ is an approximation of the Hessian matrix at $\mathbf{x}^k$. As a consequence, the system (5.35) is a quadratic problem with the solution $\mathbf{d}^k$. For details it is referred to [14, 15].

5.4 Dynamic Optimization Problems

The aim of dynamic optimization is to find a function of an independent variable (e.g. the time t) that minimizes a *cost functional*. In this context the most important application of a dynamical system is to find an optimal input trajectory $u^*(t)$. This type of problem is also called *optimal control problem (OCP)*.

5.4.1 Formulation of the Optimal Control Problem

The general formulation of an OCP is given as [14, 37, 38, 59, 64, 87, 115]

$$\min_{\mathbf{u}(\cdot)} \; J(\mathbf{u}) = \Phi(\mathbf{x}(t_f), t_f) + \int_{t_0}^{t_f} \mathcal{L}(\mathbf{x}(t), \mathbf{u}(t), t)\, dt \qquad (5.36\text{a})$$
$$\text{s.t.:} \quad \dot{\mathbf{x}} = \mathbf{f}(\mathbf{x}, \mathbf{u}, t), \quad \mathbf{x}(t_0) = \mathbf{x}_0 \qquad (5.36\text{b})$$
$$\mathbf{g}(\mathbf{x}(t_f), t_f) = \mathbf{0} \qquad (5.36\text{c})$$
$$\mathbf{h}(\mathbf{x}(t), \mathbf{u}(t), t) \leq \mathbf{0} \quad \forall t \in [t_0, t_f] \qquad (5.36\text{d})$$

5.4. DYNAMIC OPTIMIZATION PROBLEMS

The OCP is characterized by the formulation of the cost functional or performance measure (5.36a), the terminal constraints (5.36c), the inequality constraints (5.36d) as well as the terminal time t_f. The dynamical system (5.36b) acts also as constraint with respect to the OCP.
In the following sections the different terms of the system (5.36) are described and classified.

Cost functional, performance measure (5.36a)
For the dynamical system (5.36b) infinite control functions $\mathbf{u}(t)$ exist that drive it from the given initial conditions $\mathbf{x}(t_0) = \mathbf{x}_0$ to the terminal constraints (5.36c). The aim is to find an optimal control $\mathbf{u}^*(t)$ that minimizes the cost functional (5.36a) [87]. The general form of (5.36a) can be subdivided into the integral part $\mathcal{L}(\mathbf{x}(t), \mathbf{u}(t), t)$ and the Mayer-term $\Phi(\mathbf{x}(t_f), t_f)$, which classifies the terminal constraints as a scrap function. Generally, the cost functional can be formulated in the following ways [14, 38]:

$$J(\mathbf{u}) = \begin{cases} \Phi(\mathbf{x}(t_f), t_f) + \int_{t_0}^{t_f} \mathcal{L}(\mathbf{x}(t), \mathbf{u}(t), t)\, dt & \ldots \text{Bolza} - \text{form} \\ \int_{t_0}^{t_f} \mathcal{L}(\mathbf{x}(t), \mathbf{u}(t), t)\, dt & \ldots \text{Lagrange} - \text{form} \\ \Phi(\mathbf{x}(t_f), t_f) & \ldots \text{Mayer} - \text{form} \end{cases} \quad (5.37)$$

Bolza-form and Lagrange-form can always be converted into the Mayer-form by using the integral part as new state

$$\dot{x}_{n+1} = \mathcal{L}(\mathbf{x}, \mathbf{u}, t), \quad x_{n+1}(t_0) = 0 \quad (5.38)$$

and add $x_{n+1}(t_f)$ to $\Phi(\mathbf{x}(t_f), t_f)$ [14].
In the special case of *trajectory tracking problems*, which are the focus of this thesis, the cost functional is formulated as (5.39b).

$$\dot{\mathbf{x}} = \mathbf{f}(\mathbf{x}, \mathbf{u}, t) \quad (5.39a)$$

$$J(\mathbf{u}) = \int_{t_0}^{t_f} \|\mathbf{y}(\mathbf{x}, t) - \tilde{\mathbf{y}}(t)\|^2\, dt \to \min \quad (5.39b)$$

where $\mathbf{y}(\mathbf{x}, \mathbf{u}, t)$ denote the outputs of the multibody system and $\tilde{\mathbf{y}}(t)$ are the target signals. Therefore, the goal is to minimize the error between the system outputs and the targets.
In comparison to that the DAE-approach (4.2) from chapter 4 should be repeated:

$$\dot{\mathbf{x}} = \mathbf{f}(\mathbf{x}, \mathbf{u}, t) \quad (5.40a)$$
$$0 = \Phi(\mathbf{q}) - \boldsymbol{\gamma}(t) \quad (5.40b)$$

5.4. DYNAMIC OPTIMIZATION PROBLEMS

The formulation of the trajectory tracking problem by the optimal control approach (5.39) instead of the DAE-approach (5.40) has two advantages [144]:

1. The system (5.39) can be solved, even if the MBS is over- or under-determined.

2. The system (5.39) can also be solved, if the initial conditions $\mathbf{x}(t_0) = \mathbf{x}_0$ violate the constraint $\mathbf{y}(\mathbf{x}, t_0) = \tilde{\mathbf{y}}(t_0)$.

Terminal Constraints (5.36c)

The Eqs. (5.36b) and (5.36c) by themselves describe a control problem [77, 87]. The goal is to drive the nonlinear system with the states $\mathbf{x} \in \mathbb{R}^n$ and the inputs $\mathbf{u} \in \mathbb{R}^m$ with a control trajectory $\mathbf{u}(t)$, $t \in [t_0, t_f]$ from the initial state $\mathbf{x}(t_0) = \mathbf{x}_0$ to the terminal constraint (5.36c). Mostly, the terminal constraints are given in a partial form

$$x_i(t_f) = x_{f,i}, \quad i \in \mathcal{I}_f \tag{5.41}$$

where the set $\mathcal{I}_f$ contains the indexes of the fixes states x_i at t_f. If the case $\mathcal{I}_f = \{\}$ occurs, it is called a system with an *open terminal state* [87].

Inequality Constraints (5.36d)

The general form of the inequality constraints (5.36d) can be reduced to constraints of the inputs $\mathbf{u}$

$$\mathbf{u}(t) \in U \subseteq \mathbb{R}^{m_c} \quad \forall t \in [t_0, t_f] \tag{5.42}$$

and/or the states $\mathbf{x}$

$$\mathbf{x}(t) \in X \subseteq \mathbb{R}^{2n} \quad \forall t \in [t_0, t_f] \tag{5.43}$$

In most applications constraints of inputs are more relevant because of physical constraints like maximum forces or maximum voltages.

Terminal Time t_f

The terminal time t_f can either be given or it can be open. If it is open, then t_f has be to be calculated as part of (5.36).

5.4.2 Unconstrained Problems

In this section unconstrained problems (5.44) are considered.

$$\min_{\mathbf{u}(\cdot)} J(\mathbf{u}) = \Phi(\mathbf{x}(t_f), t_f) + \int_{t_0}^{t_f} \mathcal{L}(\mathbf{x}(t), \mathbf{u}(t), t) \, dt \tag{5.44a}$$

$$\text{s.t.}: \quad \dot{\mathbf{x}} = \mathbf{f}(\mathbf{x}, \mathbf{u}, t), \quad \mathbf{x}(t_0) = \mathbf{x}_0 \tag{5.44b}$$

$$x_i(t_f) = x_{f,i}, \quad i \in \mathcal{I}_f \tag{5.44c}$$

To derive the optimality conditions the system equations (5.44b) are considered as (dynamical) equality constraints and are added to the cost functional by using Lagrange-multipliers.

5.4.3 Optimality Conditions

Derivation from the Calculus of Variations

The derivation of the necessary optimality conditions can either be done for a fixed or for a free terminal time t_f using the calculus of variations. In this thesis the focus is put on problems with a fixed terminal time t_f and therefore the optimality conditions are only deduced for this case. General formulations for a free final time can be found in [87].

By introducing the Lagrange multiplier $\mathbf{p}$ and adding the constraints (5.44b) the augmented functional

$$J_a = \Phi(\mathbf{x}(t_f), t_f) + \int_{t_0}^{t_f} \left\{ \mathcal{L}(\mathbf{x}, \mathbf{u}, t) + \mathbf{p}^T[\mathbf{f}(\mathbf{x}, \mathbf{u}, t) - \dot{\mathbf{x}}] \right\} dt \qquad (5.45)$$

is considered. Variations of the states $\mathbf{x}(t) = \mathbf{x}^*(t) + \varepsilon\,\delta\mathbf{x}(t)$, the multipliers $\mathbf{p}(t) = \mathbf{p}^*(t) + \varepsilon\,\delta\mathbf{p}(t)$ and the inputs $\mathbf{u}(t) = \mathbf{u}^*(t) + \varepsilon\,\delta\mathbf{u}^*(t)$ are performed about a optimal trajectory $\mathbf{x}^*(t), \mathbf{p}^*(t), \mathbf{u}^*(t),\ t \in [t_0, t_f]$ [87].

$$\begin{aligned} J_a &= \Phi\left(\mathbf{x}^*(t_f) + \varepsilon\delta\mathbf{x}(t_f)\right) + \cdots \\ &+ \int_{t_0}^{t_f} \Big\{ \mathcal{L}\left(\mathbf{x}^* + \varepsilon\delta\mathbf{x}, \mathbf{u}^* + \varepsilon\delta\mathbf{u}, t\right) + \cdots \\ &+ (\mathbf{p}^* + \varepsilon\delta\mathbf{p})^T \left[\mathbf{f}\left(\mathbf{x}^* + \varepsilon\delta\mathbf{x}, \mathbf{u}^* + \varepsilon\delta\mathbf{u}, t\right) - (\dot{\mathbf{x}}^* + \varepsilon\delta\dot{\mathbf{x}})\right] \Big\} dt \end{aligned} \qquad (5.46)$$

Now the variation $\delta J_a = \left.\dfrac{dJ_a}{d\varepsilon}\right|_{\varepsilon=0}$ is performed:

$$\begin{aligned} \delta J_a &= \left(\frac{\partial \Phi}{\partial \mathbf{x}}\right)^T\bigg|_{t_f} \delta\mathbf{x}(t_f) + \cdots \\ &+ \int_{t_0}^{t_f} \left\{ \left(\frac{\partial \mathcal{L}}{\partial \mathbf{x}} + \left(\frac{\partial \mathbf{f}}{\partial \mathbf{x}}\right)^T \mathbf{p}^*\right)^T \delta\mathbf{x} + \left(\frac{\partial \mathcal{L}}{\partial \mathbf{u}} + \left(\frac{\partial \mathbf{f}}{\partial \mathbf{u}}\right)^T \mathbf{p}^*\right)^T \delta\mathbf{u} - \mathbf{p}^{*T}\delta\dot{\mathbf{x}} \right\} dt + \cdots \\ &+ \int_{t_0}^{t_f} \left\{ (\mathbf{f}(\mathbf{x}^*, \mathbf{u}^*, t) - \dot{\mathbf{x}}^*)\,\delta\mathbf{p} \right\} dt \end{aligned}$$

$$(5.47)$$

5.4. DYNAMIC OPTIMIZATION PROBLEMS

Partial integration of the term $\mathbf{p}^{*T}\delta\dot{\mathbf{x}}$ results in

$$\delta J_a = \left(\frac{\partial \Phi}{\partial \mathbf{x}}\right)^T\bigg|_{t_f} \delta\mathbf{x}(t_f) - (\mathbf{p}^{*T}\delta\mathbf{x})\big|_{t_0}^{t_f} + \cdots$$

$$+ \int_{t_0}^{t_f} \left\{\left(\frac{\partial \mathcal{L}}{\partial \mathbf{x}} + \left(\frac{\partial \mathbf{f}}{\partial \mathbf{x}}\right)^T \mathbf{p}^* + \dot{\mathbf{p}}^*\right)^T \delta\mathbf{x} + \left(\frac{\partial \mathcal{L}}{\partial \mathbf{u}} + \left(\frac{\partial \mathbf{f}}{\partial \mathbf{u}}\right)^T \mathbf{p}^*\right)^T \delta\mathbf{u}\right\} dt + \cdots$$

$$+ \int_{t_0}^{t_f} \{(\mathbf{f}(\mathbf{x}^*, \mathbf{u}^*, t) - \dot{\mathbf{x}}^*)\delta\mathbf{p}\}\, dt$$

(5.48)

The second term can further be simplified to

$$(\mathbf{p}^{*T}\delta\mathbf{x})\big|_{t_0}^{t_f} = \mathbf{p}^*(t_f)^T \delta\mathbf{x}(t_f) - \mathbf{p}^*(t_0)^T \delta\mathbf{x}(t_0) = \mathbf{p}^*(t_f)^T \delta\mathbf{x}(t_f)$$

because of the variation $\delta\mathbf{x}(t_0) = \mathbf{0}$ in order that the initial conditions $\mathbf{x}(t_0) = \mathbf{x}_0$ are fulfilled. The variation $\delta J_a = 0$ must hold for all admissible variations $\delta\mathbf{x}(t)$, $\delta\mathbf{p}(t)$ and $\delta\mathbf{u}(t)$ to guarantee that $\mathbf{x}^*(t)$, $\mathbf{p}^*(t)$, $\mathbf{u}^*(t)$ are indeed an optimal solution of the OCP (5.44). As a consequence the following necessary optimality conditions can be derived [37, 77, 87]:

$$\mathbf{0} = \dot{\mathbf{x}}^* - \mathbf{f}(\mathbf{x}^*, \mathbf{u}^*, t) \tag{5.49a}$$

$$\dot{\mathbf{p}}^* = -\frac{\partial \mathcal{L}}{\partial \mathbf{x}}(\mathbf{x}^*, \mathbf{u}^*, t) - \left(\frac{\partial \mathbf{f}}{\partial \mathbf{x}}\right)^T (\mathbf{x}^*, \mathbf{u}^*, t)\,\mathbf{p}^* \tag{5.49b}$$

$$\mathbf{0} = \frac{\partial \mathcal{L}}{\partial \mathbf{u}}(\mathbf{x}^*, \mathbf{u}^*, t) + \left(\frac{\partial \mathbf{f}}{\partial \mathbf{u}}\right)^T (\mathbf{x}^*, \mathbf{u}^*, t)\,\mathbf{p}^* \tag{5.49c}$$

Furthermore the following terminal conditions can be derived from (5.48):

$$0 = \left(\frac{\partial \Phi}{\partial \mathbf{x}} - \mathbf{p}^*\right)^T\bigg|_{t_f} \delta\mathbf{x}(t_f) = \sum_{i=1}^n \left(\frac{\partial \Phi}{\partial x_i} - p_i^*\right)\bigg|_{t_f} \delta x_i(t_f) \tag{5.50}$$

This means that a Lagrange multiplier $\mathbf{p}^*(t)$ must exist for $\mathbf{x}^*(t), \mathbf{u}^*(t)$, such that the differential equations (5.49b), the algebraic constraint equations (5.49c) and the terminal conditions (5.50) hold for all admissible variations $\delta\mathbf{x}(t_f)$.

Due to the partial terminal conditions (5.44) it is meaningful to formulate the transversality condition (5.50) also as partial condition. The variations $\delta x_i(t_f) = 0$ must hold for the fixed states $x_i(t_f)$, $i \in \mathcal{I}_f$ at the terminal time t_f. The variations of the free states $x_i(t_f)$, $i \notin \mathcal{I}_f$ are arbitrary. As a consequence, the equation

$$p_i^*(t_f) = \frac{\partial \Phi}{\partial x_i}\bigg|_{t_f}, i \notin \mathcal{I}_f \tag{5.51}$$

has to be fulfilled for the free states in order that the transversality condition (5.50) is fulfilled for all admissible variations.

5.4. DYNAMIC OPTIMIZATION PROBLEMS

Derivation from the Hamiltonian

By using a scalar function H (the *Hamiltonian*), the optimality conditions to derive the optimal solution $\mathbf{x}^*(t)$, $\mathbf{u}^*(t)$ can be written in a compact form. The Hamiltonian is defined as

$$H(\mathbf{x},\mathbf{u},\mathbf{p},t) = \mathcal{L}(\mathbf{x},\mathbf{u},t) + \mathbf{p}^T \mathbf{f}(\mathbf{x},\mathbf{u},t) \tag{5.52}$$

Pontryagin's maximum principle implies that an optimal control must minimize the Hamiltonian [14, 37, 87]. As a consequence, the necessary optimality conditions of 1st order for the OCP (5.44) read as [37, 87]

$$\dot{\mathbf{x}}^* = \frac{\partial H}{\partial \mathbf{p}}(\mathbf{x}^*,\mathbf{u}^*,\mathbf{p}^*,t) = \mathbf{f}(\mathbf{x}^*,\mathbf{u}^*,t) \tag{5.53a}$$

$$\dot{\mathbf{p}}^* = -\frac{\partial H}{\partial \mathbf{x}}(\mathbf{x}^*,\mathbf{u}^*,\mathbf{p}^*,t) = -\frac{\partial \mathcal{L}}{\partial \mathbf{x}} - \left(\frac{\partial \mathbf{f}}{\partial \mathbf{x}}\right)^T \mathbf{p}^* \tag{5.53b}$$

$$0 = \frac{\partial H}{\partial \mathbf{u}}(\mathbf{x}^*,\mathbf{u}^*,\mathbf{p}^*,t) = \frac{\partial \mathcal{L}}{\partial \mathbf{u}} + \left(\frac{\partial \mathbf{f}}{\partial \mathbf{u}}\right)^T \mathbf{p}^* \tag{5.53c}$$

with $t \in [t_0, t_f]$. Eqs. (5.53b) are called *adjoint equations* or *co-state equations* and (5.53c) are called control equations. The differential equations (5.53a) and (5.53b) together are called *canonical equations* or *Hamilton-equations*. Eqs. (5.53c) are a simplified statement of *Pontryagin's maximum principle* [37].
Additionally to (5.53), the boundary conditions have to be defined.

$$\mathbf{x}^*(t_0) = \mathbf{x}_0^* \tag{5.54a}$$
$$x_i^*(t_f) = x_{f,i}^*, \quad i \in \mathcal{I}_f \tag{5.54b}$$

In (5.54b) partial terminal conditions are used, i.e. not all terms of $\mathbf{x}(t_f)$ have to be specified. Furthermore, *transversality conditions* are formulated

$$p_i^*(t_f) = \left.\frac{\partial \Phi}{\partial x_i}\right|_{t=t_f} \quad \text{for } i \notin \mathcal{I}_f \tag{5.55a}$$

$$H(\mathbf{x}^*,\mathbf{u}^*,\mathbf{p}^*,t)|_{t=t_f} = -\left.\frac{\partial \Phi}{\partial t}\right|_{t=t_f} \quad \text{if } t_f \text{ is free} \tag{5.55b}$$

The set of necessary conditions (5.53) consists of a DAE system with boundary conditions (5.54) and (5.55). This is referred to as a *two-point boundary value problem* (BVP) for the optimal states $\mathbf{x}^*(t)$, $\mathbf{p}^*(t)$ and the optimal control $\mathbf{u}^*(t)$ [87]. If t_f is fixed, then the BVP (5.53)-(5.55a) includes $2n$ equations with $2n$ boundary conditions. If t_f is free, the transversality condition (5.55b) must also be considered to determine the optimal terminal time t_f^*.
With the final conditions (5.54b) and (5.55a) following cases can occur [77, 87]:

- Fixed final states $\mathbf{x}(t_f) = \mathbf{x}_f$, i.e. $\mathcal{I}_f = \{1,\ldots,n\}$:
  All states $\mathbf{x}$ are fixed at the beginning and at the end of the interval $[t_0, t_f]$ while the adjoint final states $\mathbf{p}(t_f)$ are free.

- Free final states $\mathbf{x}(t_f)$, i.e. $\mathcal{I}_f = \{\}$:
  The complete adjoint final states are fixed $\mathbf{p}(t_f) = \frac{\partial \Phi}{\partial \mathbf{x}}\big|_{t_f}$. The boundary conditions are separated in initial conditions $\mathbf{x}(t_0)$ and final conditions $\mathbf{p}(t_f)$.

- Free terminal time t_f with Mayer-term $\Phi = 0$ or $\Phi = \Phi(\mathbf{x}(t_f))$:
  In this case the transversality condition (5.55b) is homogeneous, i.e. $H|_{t=t_f} = 0$.

In addition to the necessary optimality conditions (5.53) a Legendre-condition can be formulated [87].

$$\frac{\partial^2 H}{\partial \mathbf{u}^2} \geq 0 \quad (\text{positive semi} - \text{definite}) \quad \forall t \in [t_0, t_f] \tag{5.56}$$

This is a necessary optimality condition of 2nd order for an optimal solution $\mathbf{x}^*(t)$, $\mathbf{u}^*(t)$. It guarantees that $\mathbf{u}^*(t)$ causes H to be a local minimum. The Hamiltonian (5.52) shows a specific behavior along an optimal solution:

$$\frac{d}{dt}H = \frac{\partial H}{\partial t} + \underbrace{\left(\frac{\partial H}{\partial \mathbf{x}}\right)^T \dot{\mathbf{x}}}_{=\left(\frac{\partial H}{\partial \mathbf{x}}\right)^T \mathbf{f}} + \underbrace{\left(\frac{\partial H}{\partial \mathbf{u}}\right)^T \dot{\mathbf{u}}}_{=0} + \underbrace{\left(\frac{\partial H}{\partial \mathbf{p}}\right)^T \dot{\mathbf{p}}}_{=-\mathbf{f}^T \frac{\partial H}{\partial \mathbf{x}}} = \frac{\partial H}{\partial t} \tag{5.57}$$

For time-invariant problems (5.44) where neither the cost functional $\mathcal{L}$ nor the system equations $\mathbf{f}$ explicitly depend on the time t, the Hamiltonian H is constant along an optimal trajectory [77, 87].

Fig. 5.2 illustrates how the calculus of variations can be used in mechanical systems to derive the Euler-Lagrange equations (2.47) from Hamilton's principle and in optimal control theory to derive the necessary optimality conditions (5.53)-(5.55) from Pontryagin's maximum principle.

5.4.4 Solution Strategies

A general solution strategy for the OCP (5.44) with the optimality conditions (5.53)-(5.55) can be subdivided into the following steps [77, 87]:

1. Calculation of the Hamiltonian $H(\mathbf{x}, \mathbf{u}, \mathbf{p}, t) = \mathcal{L}(\mathbf{x}, \mathbf{u}, t) + \mathbf{p}^T \mathbf{f}(\mathbf{x}, \mathbf{u}, t)$

2. Calculation of $\mathbf{u}$ from (5.53c) $\frac{\partial H}{\partial \mathbf{u}} = \mathbf{0}$ in order that $\mathbf{u}$ can be expressed as a function of $\mathbf{x}$, $\mathbf{p}$, t:

$$\mathbf{u} = \boldsymbol{\psi}(\mathbf{x}, \mathbf{p}, t) \tag{5.58}$$

5.4. DYNAMIC OPTIMIZATION PROBLEMS

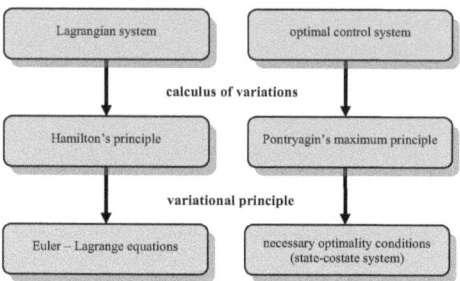

Figure 5.2: Calculus of variations in mechanics and optimal control theory [115]

3. Insertion of (5.58) into the canonical equations (5.53a) and (5.53b). This results in the BVP

$$\dot{\mathbf{x}} = \mathbf{f}(\mathbf{x}, \boldsymbol{\psi}(\mathbf{x}, \mathbf{p}, t), t) \qquad (5.59a)$$
$$\dot{\mathbf{p}} = -H_{\mathbf{x}}(\mathbf{x}, \boldsymbol{\psi}(\mathbf{x}, \mathbf{p}, t), t) \qquad (5.59b)$$

with the boundary conditions (5.54)-(5.55). This resulting BVP is independent of the control $\mathbf{u}$. If the final time t_f is free, the final condition (5.55b) has to be included as well.

4. The solution of the BVP are the states $\mathbf{x}^*(t)$, $\mathbf{p}^*(t)$, $t \in [t_0, t_f]$. Now $\mathbf{u}^*(t)$ can be calculated from (5.58).

In most applications the BVP has to be solved numerically. Therefore, standard routines like the *bvp4c*-solver in *Matlab* can be used.

5.4.5 Singular Case

A problem occurs, if one or more elements u_i from $\mathbf{u}$ cannot be calculated from the stationarity condition (5.53c). If the integral part $\mathcal{L}$ of the cost functional (5.44a) is independent from u and $\mathbf{f}(\mathbf{x}, \mathbf{u})$ is linear with respect to u, the Hamiltonian and its first derivate with respect to u reads as

$$H(\mathbf{x}, u, \mathbf{p}, t) = \mathcal{L}(\mathbf{x}, t) + \mathbf{p}^T \left[\mathbf{f}_0(\mathbf{x}) + \mathbf{f}_1(\mathbf{x}) u \right] \qquad (5.60a)$$
$$H_u(\mathbf{x}, \mathbf{p}) = \mathbf{p}^T \mathbf{f}_1(\mathbf{x}) = 0 \qquad (5.60b)$$

In this case a singular problem occurs, because u_i cannot be calculated from $\frac{\partial H}{\partial u} = 0$. However, the stationarity condition $H_u(\mathbf{x}^*(t), \mathbf{p}^*(t)) = 0$ has to be fulfilled, even if it does not provide any information about the optimal control. The order of singularity r can be determined with the minimum number of derivatives with respect to time such that the input u explicitly occurs.

$$\frac{d^r}{dt^r} H_u(\mathbf{x}^*(t), \mathbf{p}^*(t)) = 0 \quad \text{with} \quad \frac{\partial}{\partial u} \left(\frac{d^r}{dt^r} H_u(\cdot) \right) \neq 0 \qquad (5.61)$$

5.4. DYNAMIC OPTIMIZATION PROBLEMS

The equation $\frac{d^r}{dt^r}H_u(\cdot)$ can now be used to calculate $u^*(t)$.

In practical applications a *regularization term* $\sum_{i=1}^{m} r_i u_i^2$ can be used to avoid this singularity [77]. This term with a small weighting factor $r_i > 0$ can be added to the cost functional. As a consequence u_i does not vanish in the stationarity conditions $\frac{\partial H}{\partial u_i} = 0$.

5.4.6 Constrained Problems

In the previous considerations it was assumed that the OCP is an unconstrained problem (5.44). This section focuses on constraints in the control variables.

$$\mathbf{u}(t) \in U \subseteq \mathbb{R}^{m_c} \quad \forall t \in [t_0, t_f] \tag{5.62}$$

U is a subset of $\mathbb{R}^{m_c}$. The constraint has to be fulfilled at any time. In most practical purposes constraints in the control variables occur. Such constraints can be considered with *Pontryagin's maximum principle*. Constraints in the state variables are much more complicated. However, they are not as relevant as constraints in the control variables and should not be discussed further.

5.4.7 Pontryagin's Maximum Principle

Pontryagin's maximum principle is based on the Hamiltonian and the canonical equations (5.53a) and (5.53b). However, the necessary optimality condition $\frac{\partial H}{\partial \mathbf{u}} = 0$ is not valid any more. This is illustrated in Fig. 5.3, where the Hamiltonian $H(\mathbf{x}^*, u, \mathbf{p}^*, t)$ is plotted over a scalar input variable u at a fixed time t. Following

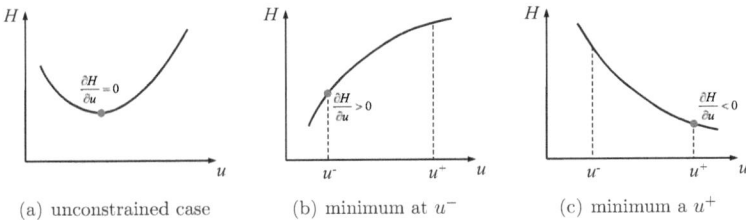

(a) unconstrained case (b) minimum at u^- (c) minimum a u^+

Figure 5.3: Boundary minima in a constrained case $u \in U = [u^-, u^+]$ [77]

cases are distinguished:

a) Unconstrained case:
   The necessary conditions (5.53c) and (5.56) guarantee the minimum of the Hamiltonian, i.e. $\frac{\partial H}{\partial u} = 0$ and $\frac{\partial^2 H}{\partial u^2} \geq 0$ (positive semi-definite).

b) Minimum at u^-:
   The control variable u is constrained in the interval $U = [u^-, u^+]$ and the minimum is located at u^-, where $\frac{\partial H}{\partial u} = 0$ is not fulfilled.

5.5. NUMERICAL METHODS FOR DYNAMIC OPTIMIZATION PROBLEMS

c) Minimum at u^+:
The control variable u is constrained in the interval $U = [u^-, u^+]$ and the minimum is located at u^+, where $\frac{\partial H}{\partial u} = 0$ is not fulfilled.

As a consequence, the condition (5.53c) $\frac{\partial H}{\partial \mathbf{u}} = \mathbf{0}$ is replaced by

$$H(\mathbf{x}^*, \mathbf{u}^*, \mathbf{p}^*, t) = \min_{\mathbf{u} \in U} H(\mathbf{x}^*, \mathbf{u}, \mathbf{p}^*, t) \qquad (5.63)$$

This condition is the basis of Pontryagin's maximum principle. To solve such a constrained OCP, the second task in the solution strategy 5.4.4 has to be replaced by:

$$\mathbf{u}^* = \arg\min_{\mathbf{u} \in U} H(\mathbf{x}, \mathbf{u}, \mathbf{p}, t) \qquad (5.64)$$

The minimization task (5.64) has to be performed for each possible combination of $(\mathbf{x}, \mathbf{p})$.

5.5 Numerical Methods for Dynamic Optimization Problems

In nearly all applications an OCP cannot be solved fully analytically. Rather numerical methods are used. Basically, two different classes are defined [14, 32, 33]:

- **Indirect Methods:**
  Indirect methods are based on the optimality conditions (5.53) - (5.55). The resulting two-point-BVP can be solved by collocation methods, shooting methods or by a gradient method [14, 115].

- **Direct Methods:**
  Direct methods discretize the input trajectory $\mathbf{u}(t)$ and transcribe the OCP to finite dimensional static optimization problems. This procedure is known as direct transcription method [14, 28, 29]. These resulting problems can be solved with methods described in section 5.1.

5.5.1 Indirect Methods: Solving the Optimality Conditions

Collocation Method:
The basis for this method is the two-point BVP, which results from the optimality conditions (5.53) - (5.55). Problems with a fixed terminal time t_f are considered. As already stated in section 5.4.4, the canonical equations can be expressed independently from $\mathbf{u}$ (5.59). Both equations can be summarized by defining a new state vector $\bar{\mathbf{x}} = [\mathbf{x}, \mathbf{p}]^T$ [77]:

$$\left.\begin{array}{l} \dot{\mathbf{x}} = \mathbf{f}(\mathbf{x}, \boldsymbol{\psi}(\mathbf{x}, \mathbf{p}), t) \\ \dot{\mathbf{p}} = -H_\mathbf{x}(\mathbf{x}, \boldsymbol{\psi}(\mathbf{x}, \mathbf{p}), \mathbf{p}, t) \end{array}\right\} \dot{\bar{\mathbf{x}}} = \mathbf{F}(\bar{\mathbf{x}}, t) \qquad (5.65)$$

5.5. NUMERICAL METHODS FOR DYNAMIC OPTIMIZATION PROBLEMS

Initial conditions and partial terminal conditions are summarized as well:

$$\mathbf{x}(t_0) = \mathbf{x}_0, \quad \mathbf{G}(\mathbf{x}(t_f)) = \begin{bmatrix} (x_i(t_f) - x_{f,i}) \; \forall i \in \mathcal{I}_f \\ \left(p_i(t_f) - \frac{\partial \Phi}{\partial x_i}\Big|_{t_f}\right) \; \forall i \notin \mathcal{I}_f \end{bmatrix} = \mathbf{0} \qquad (5.66)$$

The resulting system consists of $2n$ boundary conditions for the $2n$ differential equations. Now the two-point BVP is discretized in the time interval $[t_0, t_f]$

$$t_0 = t^0 < t^1 < \cdots < t^N = t_f$$

and the solution is approximated at the sampling points

$$\bar{\mathbf{x}}^k \approx \bar{\mathbf{x}}(t^k), \quad k = 0, 1, \ldots, N \qquad (5.67)$$

For discretization e.g. a trapezoidal rule can be used:

$$\frac{\bar{\mathbf{x}}^{k+1} - \bar{\mathbf{x}}^k}{t^{k+1} - t^k} = \frac{1}{2}\left[\mathbf{F}(\bar{\mathbf{x}}^k, t^k) + \mathbf{F}(\bar{\mathbf{x}}^{k+1}, t^{k+1})\right], \quad i = 0, 1, \ldots, N-1 \qquad (5.68)$$

Additionally, the $2n$ boundary conditions must be fulfilled.

$$\mathbf{x}^0 = \mathbf{x}_0, \quad \mathbf{G}(\bar{\mathbf{x}}^N) = \mathbf{0} \qquad (5.69)$$

Eqs. (5.68) and (5.69) describe a system of $2n(N+1)$ nonlinear equations for the $2n(N+1)$ unknown variables. Instead of the trapezoidal rule other basis functions like polynomials can be used.
Certain characteristics of collocation methods are listed below [14, 77]:

- Collocation methods are more robust than shooting methods

- The initial guess of the adjoint variables $\mathbf{p}(t)$ is of great importance and influences the convergence speed dramatically (cf. examples in sections 7.1, 7.2 with the *Matlab-solver bvp4c*)

- The implementation effort is relatively high

- The number of discretization points influence the accuracy of the solution as well as the convergence speed

- The exactness of the solution depends on the method of discretization. If e.g. a trapezoidal rule of 2nd order is used, the solution coincides with a Taylor series till the quadratic terms

Shooting Methods:
Instead of using a large system of equations like in the collocation method, shooting methods solve an IVP with unknown initial values for the adjoint variables $\mathbf{p}_0$.

$$\dot{\bar{\mathbf{x}}} = \mathbf{F}(\bar{\mathbf{x}}, t), \quad \bar{\mathbf{x}}(t_0) = \begin{bmatrix} \mathbf{x}_0 \\ \mathbf{p}_0 \end{bmatrix} \qquad (5.70)$$

5.5. NUMERICAL METHODS FOR DYNAMIC OPTIMIZATION PROBLEMS

Therefore, the solution which depends on $\mathbf{p}_0$ can be formulated as [14, 77]

$$\bar{\mathbf{x}}(t, \mathbf{p}_0) = \begin{bmatrix} \mathbf{x}_0 \\ \mathbf{p}_0 \end{bmatrix} + \int_{t_0}^{t_f} \mathbf{F}(\bar{\mathbf{x}}(\tau, \mathbf{p}_0), \tau) \, d\tau \qquad (5.71)$$

Normally the solution of the IVP with the initial guess for $\mathbf{p}_0$ does not fulfill the boundary condition

$$\mathbf{G}(\bar{\mathbf{x}}(t_f, \mathbf{p}_0)) = \mathbf{0} \qquad (5.72)$$

However, the deviation from $\mathbf{G}(\bar{\mathbf{x}}(t_f, \mathbf{p}_0))$ to $\mathbf{0}$ can be used to improve the guess for the adjoint variables $\mathbf{p}_0$. As a consequence, Eq. 5.72 formulates the equations to find the roots for the initial values $\mathbf{p}_0$. This has to be done numerically, e.g. by using Newton's method. The subordinated IVP (5.70) has to be integrated in every Newton-step to determine the trajectory (5.71). Fig. 5.4 illustrates the shooting method for a scalar function $\bar{x}(t, p_0)$.

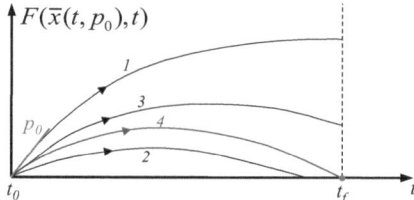

Figure 5.4: Shooting method [14]

Certain characteristics of shooting methods are listed below [14, 77]:

- The implementation effort is not as high as for collocation methods
- The initial values of the adjoint variables $\mathbf{p}_0$ are very important. Shooting methods are generally very sensitive with respect to initial guesses
- The integration of the IVP can be numerically critical
- Multiple shooting methods are better suited than single shooting methods [32]

Gradient Method, Kelley-Bryson-Method [143]:
The main application of the gradient method are OCPs with a fixed terminal time and a free final state. For this special case the optimality conditions (5.53) are formulated as

$$\dot{\mathbf{x}} = H_{\mathbf{p}}(\mathbf{x}, \mathbf{u}, \mathbf{p}, t) = \mathbf{f}(\mathbf{x}, \mathbf{u}, t), \quad \mathbf{x}(t_0) = \mathbf{x}_0 \qquad (5.73\text{a})$$
$$\dot{\mathbf{p}} = -H_{\mathbf{x}}(\mathbf{x}, \mathbf{u}, \mathbf{p}, t), \quad \mathbf{p}(t_f) = \Phi_{\mathbf{x}}(\mathbf{x}(t_f)) \qquad (5.73\text{b})$$
$$\mathbf{0} = H_{\mathbf{u}}(\mathbf{x}, \mathbf{u}, \mathbf{p}, t) \qquad (5.73\text{c})$$

5.5. NUMERICAL METHODS FOR DYNAMIC OPTIMIZATION PROBLEMS

The gradient method takes advantage of the decoupling of the boundary conditions. The canonical equations (5.73a) and (5.73b) can be integrated as a function of $\mathbf{u}$ sequentially forward and backward in time. In this way the trajectories for the states $\mathbf{x}^j(t)$ and costates $\mathbf{p}^j(t)$ are calculated. Generally the stationarity condition (5.73c) is not fulfilled and hence a residual gradient exists.

$$\mathbf{g}^j(t) = H_\mathbf{u}(\mathbf{x}^j(t), \mathbf{u}^j(t), \mathbf{p}^j(t), t), \quad t \in [t_0, t_f] \tag{5.74}$$

The negative gradient can now be used to minimize the cost functional [87].

In this thesis an algorithm is used which follows an idea of *Henry J. Kelly* [86] and *Arthur E. Bryson, Jr.* [37]. The basis of this method is the calculation of the gradient of the cost functional, i.e. the variation of the control variable $\mathbf{u}(t)$ that causes a maximum increase or decrease in J. For a given control $\mathbf{u}(t)$ the cost functional

$$J = \Phi(\mathbf{x}_f, t_f) + \int_{t_0}^{t_f} \mathcal{L}(\mathbf{x}, \mathbf{u}, t)\, dt \tag{5.75}$$

can be calculated straight forward. The state equations (5.73a) have to be integrated and the results have to be inserted into (5.75).
In trajectory tracking problems the general formulation (5.75) is specified to (5.39b).

$$J = \int_{t_0}^{t_f} \|\mathbf{y}(\mathbf{x}, t) - \tilde{\mathbf{y}}(t)\|^2\, dt \to \min \tag{5.76}$$

where the scrap function $\Phi(\mathbf{x}_f, t_f)$ is neglected.
Now the linear change of J has to be calculated, if $\mathbf{u}(t)$ is slightly modified by the variation $\delta\mathbf{u}(t)$. It is evident that a small variation of the control $\mathbf{u}(t)$ causes a small variation of the states $\mathbf{x}(t)$. Therefore, the linear change of the performance measure is calculated by

$$\delta J = \Phi_{\mathbf{x}_f}^T \delta\mathbf{x}_f + \int_{t_0}^{t_f} \mathcal{L}_\mathbf{x}^T \delta\mathbf{x}\, dt \tag{5.77}$$

$\mathcal{L}_\mathbf{x}$ and $\Phi_{\mathbf{x}_f}$ denote the partial derivatives of $\mathcal{L}$ and Φ with respect to $\mathbf{x}$ and $\mathbf{x}_f$. Now it must be determined how $\delta\mathbf{x}$ is related to $\delta\mathbf{u}$. For small variations the state equations (5.73a) can be linearized around a nominal trajectory $\mathbf{x}(t)$, if the control $\mathbf{u}(t)$ is given.

$$\begin{aligned}\delta\dot{\mathbf{x}} &= \mathbf{f}_\mathbf{x}(\mathbf{x}(t), \mathbf{u}(t), t)\delta\mathbf{x} + \mathbf{f}_\mathbf{u}(\mathbf{x}(t), \mathbf{u}(t), t)\delta\mathbf{u} \\ &= \mathbf{A}(t)\delta\mathbf{x} + \mathbf{B}(t)\delta\mathbf{u}\end{aligned} \tag{5.78}$$

A linearization around a trajectory results in time-variant matrices $\mathbf{A}(t)$ and $\mathbf{B}(t)$.

$$\mathbf{A}(t) = \mathbf{f}_\mathbf{x}(\mathbf{x}(t), \mathbf{u}(t), t), \quad \mathbf{B}(t) = \mathbf{f}_\mathbf{u}(\mathbf{x}(t), \mathbf{u}(t), t) \tag{5.79}$$

Instead of solving (5.78) for $\delta\mathbf{x}(t)$, a costate equation is defined:

$$\dot{\mathbf{p}} = -\mathbf{A}^T\mathbf{p} - \mathcal{L}_\mathbf{x} \tag{5.80}$$

5.5. NUMERICAL METHODS FOR DYNAMIC OPTIMIZATION PROBLEMS

This costate equation is used in the following auxiliary calculation:

$$\frac{d}{dt}\left(\mathbf{p}^T \delta \mathbf{x}\right) = \dot{\mathbf{p}}^T \delta \mathbf{x} + \mathbf{p}^T \delta \dot{\mathbf{x}}$$
$$= (-\mathbf{A}^T \mathbf{p} - \mathcal{L}_\mathbf{x})^T \delta \mathbf{x} + \mathbf{p}^T (\mathbf{A} \delta \mathbf{x} + \mathbf{B} \delta \mathbf{u})$$
$$= -\mathcal{L}_\mathbf{x}^T \delta \mathbf{x} + \mathbf{p}^T \mathbf{B} \delta \mathbf{u}$$

or

$$\mathcal{L}_\mathbf{x}^T \delta \mathbf{x} = \mathbf{p}^T \mathbf{B} \delta \mathbf{u} - \frac{d}{dt}\left(\mathbf{p}^T \delta \mathbf{x}\right) \tag{5.81}$$

As a consequence, the integral term in (5.77) results in

$$\int_{t_0}^{t_f} \mathcal{L}_\mathbf{x}^T \delta \mathbf{x}\, dt = \int_{t_0}^{t_f} \mathbf{p}^T \mathbf{B} \delta \mathbf{u}\, dt + \mathbf{p}(t_0)^T \delta \mathbf{x}(t_0) - \mathbf{p}(t_f)^T \delta \mathbf{x}(t_f) \tag{5.82}$$

The states at t_0 are given by the initial conditions $\mathbf{x}(t_0) = \mathbf{x}_0$ and therefore the variation $\delta \mathbf{x}(t_0)$ vanishes. Now Eq. (5.82) is inserted into (5.77) which results in

$$\delta J = \int_{t_0}^{t_f} \mathbf{p}^T \mathbf{B} \delta \mathbf{u}\, dt + \left[\Phi_{\mathbf{x}_f} - \mathbf{p}(t_f)\right]^T \delta \mathbf{x}_f \tag{5.83}$$

If the boundary condition (5.73b) $\mathbf{p}(t_f) = \Phi_{\mathbf{x}_f}$ is used, the second term vanishes and $\delta \mathbf{x}_f$ has not to be calculated to determine δJ. As a consequence, the cost functional is simplified to

$$\delta J = \int_{t_0}^{t_f} \mathbf{p}^T \mathbf{B} \delta \mathbf{u}\, dt \tag{5.84}$$

The costate equations (5.80) have to be solved backwards in time, beginning from the initial conditions $\mathbf{p}(t_f) = \Phi_{\mathbf{x}_f}$. By using the Hamiltonian (5.52), the variation (5.84) can also be written as

$$\delta J = \int_{t_0}^{t_f} H_\mathbf{u}^T \delta \mathbf{u}\, dt \tag{5.85}$$

For the gradient method the largest variation δJ, which can be obtained by all admissible control variations $\delta \mathbf{u}$, is of interest. This results in the variation

$$\delta \mathbf{u} = -\kappa H_\mathbf{u} = -\kappa \mathbf{B}^T \mathbf{p} \tag{5.86}$$

that results from the calculus of variations (see Appendix A).
If (5.86) is inserted into (5.85), the variation of the cost functional is always negative, assuming that $\kappa > 0$.

$$\delta J = -\kappa \int_{t_0}^{t_f} H_\mathbf{u}^T H_\mathbf{u}\, dt \tag{5.87}$$

5.5. NUMERICAL METHODS FOR DYNAMIC OPTIMIZATION PROBLEMS

Hence, the variation of the control variable $\delta \mathbf{u}$ (5.86) causes a decrease in the cost functional J. However, it should be mentioned that δJ is only the linear part of the cost functional J and therefore κ must be sufficiently small. On the other side, the step length κ should be as large as possible in order to increase the convergence speed of the algorithm. Therefore, a line search algorithm is used.

The main steps of the gradient method are summarized in Table 5.1, [87, 144]. The algorithm (5.1) is implemented in *Matlab* and is applied to specific academic

---

1. Initialization:
   - Set iteration counter to zero $j = 0$
   - Define a termination criterion ε
   - Choose an initial guess for the trajectory of the control variables $\mathbf{u}(t)$
   - Define initial values for the states $\mathbf{x}_0$
2. Integrate the state equations (5.73a) with respect to $\mathbf{x}(t)$
3. Compute the time-variant matrices $\mathbf{A}(t)$ and $\mathbf{B}(t)$ (5.79) and the vectors $\mathcal{L}_\mathbf{x}, \Phi_{\mathbf{x}_f}$
4. Integrate the costate equations (5.80) with respect to $\mathbf{p}(t)$ backwards in time, starting from t_f
5. Determine the step length κ and compute the variation of the controls $\delta \mathbf{u}$ by (5.86)
6. Compute the new control trajectory $\mathbf{u}^{j+1}(t) = \mathbf{u}^j(t) + \delta \mathbf{u}(t)$
7. Solve the state equations again and compute the cost functional
8. Repeat from step 5 to find the optimal step length κ which causes the maximum decrease of J
9. $j = j + 1$
10. Repeat from step 2 until $\|H_\mathbf{u}\| = \|\mathbf{B}^T\mathbf{p}\| \leq \varepsilon$.

Table 5.1: Algorithm of the gradient method

examples, cf. chapter 7.

It should be mentioned that if no scrap function $\Phi(\mathbf{x}_f, t_f)$ would be defined as in (5.75), i.e. the formulation (5.76) is used, it would always result in $\mathbf{p}(t_f) = \mathbf{0}$ and furthermore $\delta \mathbf{u}(t_f) = \mathbf{0}$, i.e. the control at the end point would never be updated in (5.86). This problem can be avoided by adding a weighted end point error to (5.76).

$$\Phi(\mathbf{x}_f, t_f) = \frac{1}{2}\alpha \|\mathbf{y}(\mathbf{x}_f, t_f) - \tilde{\mathbf{y}}(t_f)\|^2 \qquad (5.88)$$

The scalar weighting factor α can be chosen in a way that $\Phi(\mathbf{x}_f, t_f)$ and the integral (5.76) divided by $t_f - t_0$ are in the same range. The resulting cost functional reads as

$$J = \frac{1}{2}\alpha \|\mathbf{y}(\mathbf{x}_f, t_f) - \tilde{\mathbf{y}}(t_f)\|^2 + \int_{t_0}^{t_f} \frac{1}{2} \|\mathbf{y}(\mathbf{x}, t) - \tilde{\mathbf{y}}(t)\|^2 \, dt \qquad (5.89)$$

Certain characteristics of the gradient method are listed below [14, 77]:

5.5. NUMERICAL METHODS FOR DYNAMIC OPTIMIZATION PROBLEMS

- A big advantage of the gradient method is that **u** has not explicitly to be calculated from $H_\mathbf{u} = \mathbf{0}$ or $\min_{\mathbf{u} \in U} H$, respectively. Especially for nonlinear problems this benefit is of great importance.

- Constraints in the control variables $\mathbf{u} \in U$ can be considered by a projection at an admissible set U

- The gradient method is numerically more robust than shooting methods, because the adjoint system has not to be integrated in the unstable forward direction

- An initial guess of the adjoint variables is not needed

- The convergence speed is slow near the optimal solution

- Partial boundary conditions and a free terminal time diminish robustness and convergence behavior

- Conjugate gradient methods can be used to improve the convergence

Gradient Method for Constrained MBS [143]:
The basis are the DAEs (2.51) from chapter 2.5. In a general form they can be formulated as

$$\dot{\mathbf{x}} = \mathbf{f}(\mathbf{x}, \mathbf{u}, \boldsymbol{\lambda}, t) \tag{5.90a}$$
$$\mathbf{0} = \mathbf{g}(\mathbf{x}) \tag{5.90b}$$

where $\mathbf{g}(\mathbf{x})$ are the holonomic algebraic constraints and $\boldsymbol{\lambda}$ denotes the vector of Lagrange multipliers. As in the previous section the variation δJ is calculated by a linearization of the equations of motion (5.90) about a nominal trajectory.

$$\delta\dot{\mathbf{x}} = \mathbf{f}_\mathbf{x}\delta\mathbf{x} + \mathbf{f}_\mathbf{u}\delta\mathbf{u} + \mathbf{f}_\boldsymbol{\lambda}\delta\boldsymbol{\lambda} \tag{5.91a}$$
$$\mathbf{0} = \mathbf{g}_\mathbf{x}\delta\mathbf{x} \tag{5.91b}$$

If the time-variant matrices $\mathbf{A}(t)$, $\mathbf{B}(t)$, $\mathbf{C}(t)$ and $\mathbf{D}(t)$ are introduced,

$$\mathbf{A}(t) = \mathbf{f}_\mathbf{x}, \quad \mathbf{B}(t) = \mathbf{f}_\mathbf{u}, \quad \mathbf{C}(t) = \mathbf{f}_\boldsymbol{\lambda}, \quad \mathbf{D}(t) = \mathbf{g}_\mathbf{x} \tag{5.92}$$

the linearized differential-algebraic equations (5.91) read as

$$\delta\dot{\mathbf{x}} = \mathbf{A}(t)\delta\mathbf{x} + \mathbf{B}(t)\delta\mathbf{u} + \mathbf{C}(t)\delta\boldsymbol{\lambda} \tag{5.93a}$$
$$\mathbf{0} = \mathbf{D}(t)\delta\mathbf{x} \tag{5.93b}$$

The costate equations (5.80) for an ODE system are extended to

$$\dot{\mathbf{p}} = -\mathbf{A}^T\mathbf{p} - \mathbf{D}^T\boldsymbol{\mu} - \mathcal{L}_\mathbf{x} \tag{5.94a}$$
$$\mathbf{0} = \mathbf{C}^T\mathbf{p} \tag{5.94b}$$

5.5. NUMERICAL METHODS FOR DYNAMIC OPTIMIZATION PROBLEMS

for a DAE system. The vector $\boldsymbol{\mu}(t)$ includes algebraic variables. Again, an auxiliary calculation is done:

$$\begin{aligned}\frac{d}{dt}\left(\mathbf{p}^T\delta\mathbf{x}\right) &= \dot{\mathbf{p}}^T\delta\mathbf{x} + \mathbf{p}^T\delta\dot{\mathbf{x}} \\ &= \left(-\mathbf{A}^T\mathbf{p} - \mathbf{D}^T\boldsymbol{\mu} - \mathcal{L}_\mathbf{x}\right)^T\delta\mathbf{x} + \mathbf{p}^T\left(\mathbf{A}\delta\mathbf{x} + \mathbf{B}\delta\mathbf{u} + \mathbf{C}\delta\boldsymbol{\lambda}\right) \\ &= -\mathcal{L}_\mathbf{x}^T\delta\mathbf{x} + \mathbf{p}^T\mathbf{B}\delta\mathbf{u} - \boldsymbol{\mu}^T\mathbf{D}\delta\mathbf{x} + (\mathbf{C}^T\mathbf{p})^T\delta\boldsymbol{\lambda}\end{aligned}$$

The last two terms vanish as a result of (5.91b) and (5.94b). As a consequence, the same result as in (5.81) is achieved. Hence, the variation of the cost functional is also identical to (5.83).

It can be claimed that $\mathbf{p}(t_f) = \Phi_{\mathbf{x}_f}$ in order that the integrated term in (5.83) is identical to zero. However, if $\Phi \neq 0$, the costate variables will probably not fulfill the constraint $\mathbf{C}_f^T\mathbf{p}_f$.

To find compatible boundary conditions for the costate variables, the fact can be used that $\delta\mathbf{x}_f$ is not completely independent but rather subject to the constraint $\mathbf{D}(t_f)\delta\mathbf{x}_f = 0$. Then a new Lagrange multiplier $\boldsymbol{\xi}$ can be introduced and the term $\boldsymbol{\xi}^T\mathbf{D}(t_f)\delta\mathbf{x}_f = 0$ can be added to (5.83):

$$\begin{aligned}\delta J &= \int\limits_{t_0}^{t_f}\mathbf{p}^T\mathbf{B}\delta\mathbf{u}\,dt + (\Phi_{\mathbf{x}_f} - \mathbf{p}(t_f))^T\delta\mathbf{x}_f + \boldsymbol{\xi}^T\mathbf{D}(t_f)\delta\mathbf{x}_f \\ &= \int\limits_{t_0}^{t_f}\mathbf{p}^T\mathbf{B}\delta\mathbf{u}\,dt + (\Phi_{\mathbf{x}_f} - \mathbf{p}_f + \mathbf{D}_f^T\boldsymbol{\xi})^T\delta\mathbf{x}_f\end{aligned} \qquad (5.95)$$

Now it can be claimed that

$$\Phi_{\mathbf{x}_f} - \mathbf{p}_f + \mathbf{D}_f^T\boldsymbol{\xi} = 0 \qquad (5.96)$$

The Lagrange multiplier $\boldsymbol{\xi}$ must be chosen in a way that

$$\mathbf{C}_f^T\mathbf{p}_f = 0 \qquad (5.97)$$

Then the variation of the cost functional is identical to (5.84) and $\mathbf{p}_f$ fulfills the constraint equation. The system (5.96) and (5.97) can be solved for $\mathbf{p}_f$ and $\boldsymbol{\xi}$. Inserting $\mathbf{p}_f$ from (5.96) into (5.97) results in

$$\mathbf{C}_f^T(\Phi_{\mathbf{x}_f} + \mathbf{D}_f^T\boldsymbol{\xi}) = 0 \qquad (5.98)$$

Hence,

$$\boldsymbol{\xi} = -(\mathbf{C}_f^T\mathbf{D}_f^T)^{-1}\mathbf{C}_f^T\Phi_{\mathbf{x}_f}$$

Now the costate variables can be calculated from (5.96)

$$\mathbf{p}_f = \Phi_{\mathbf{x}_f} - \mathbf{D}_f^T(\mathbf{C}_f^T\mathbf{D}_f^T)^{-1}\mathbf{C}_f^T\Phi_{\mathbf{x}_f} \qquad (5.99)$$

This equation can only be calculated if the quadratic matrix $\mathbf{C}_f^T\mathbf{D}_f^T$ is nonsingular. Otherwise, the boundary term cannot be removed from δJ except for the case $\Phi_{\mathbf{x}_f} = \mathbf{0}$, i.e. $\mathbf{p}_f = \mathbf{0}$.

Finally, the costate equations (5.94) can be integrated backwards in time, using the

5.5. NUMERICAL METHODS FOR DYNAMIC OPTIMIZATION PROBLEMS

initial conditions (5.99).
The variation of the control trajectory can be computed in the same way as for ODE-systems (5.86) by

$$\delta \mathbf{u} = -\kappa \mathbf{B}^T \mathbf{p} \qquad (5.100)$$

In the case of constrained MBS a Hamiltonian can be defined as well:

$$H(\mathbf{x}, \mathbf{p}, \mathbf{u}, \boldsymbol{\lambda}, \boldsymbol{\mu}, t) = \mathcal{L}(\mathbf{x}, \mathbf{u}, t) + \mathbf{p}^T \mathbf{f}(\mathbf{x}, \mathbf{u}, \boldsymbol{\lambda}, t) + \boldsymbol{\mu}^T \mathbf{g}(\mathbf{x}) \qquad (5.101)$$

As a consequence, the state equations (5.90) can be formulated as

$$\dot{\mathbf{x}} = H_\mathbf{p} \qquad (5.102a)$$
$$0 = H_\boldsymbol{\mu} \qquad (5.102b)$$

Furthermore, the costate equations (5.94) read as

$$\dot{\mathbf{p}} = -H_\mathbf{x} \qquad (5.103a)$$
$$0 = H_\boldsymbol{\lambda} \qquad (5.103b)$$

The update of the control variables can also be calculated as for the unconstrained case (5.86).

$$\delta \mathbf{u} = -\kappa H_\mathbf{u} \qquad (5.104)$$

Application to Multibody Systems [143]:
If the equations of motions are formulated by the index 3 DAE system (2.51), (5.90) a problem occurs in the boundary condition of the adjoint equations (5.99). The term $(\mathbf{C}_f^T \mathbf{D}_f^T)$ is always singular due to zero-entries that result from the partial derivatives with respect to the states $\mathbf{x}$, i.e. with respect to $\mathbf{q}$ and $\mathbf{v}$, Eq. (5.92). As a consequence, this matrix cannot be inverted and the initial conditions cannot be formulated for the adjoint variables $\mathbf{p}_f$, which have to be integrated backwards in time.
However, it can be shown that the index 2 Gear-Gupta-Leimkuhler (GGL) formulation, cf. Eq. (2.76) in section 2.8.4, yields a regular matrix regarding to the boundary condition (5.99). The second advantage of the GGL-formulation is that the adjoint equations are also index 2 equations, which is beneficial for the numerical integration.
The GGL-formulation should be repeated and reads as

$$\mathbf{M}(\mathbf{q})\dot{\mathbf{q}} = \mathbf{M}(\mathbf{q})\mathbf{v} - \mathbf{G}^T(\mathbf{q})\boldsymbol{\nu} \qquad (5.105a)$$
$$\mathbf{M}(\mathbf{q})\dot{\mathbf{v}} = \mathbf{f}(\mathbf{q}, \mathbf{v}, \mathbf{u}) - \mathbf{G}^T(\mathbf{q})\boldsymbol{\lambda} \qquad (5.105b)$$
$$\mathbf{g}(\mathbf{q}) = \mathbf{0} \qquad (5.105c)$$
$$\mathbf{G}(\mathbf{q})\mathbf{v} = \mathbf{0} \qquad (5.105d)$$

The gradient method requires the time-variant matrices $\mathbf{A}$, $\mathbf{B}$, $\mathbf{C}$ and $\mathbf{D}$, Eq. (5.93) that result from a forward solution of the original DAEs. Hence, the GGL-formulation

5.5. NUMERICAL METHODS FOR DYNAMIC OPTIMIZATION PROBLEMS

(5.105) must be rewritten in the general state-space form (5.90). Therefore, the mass matrix has to be inverted and the right hand side of the equations must be differentiated with respect to $\mathbf{q}$ and $\mathbf{v}$. However, these computations can be avoided. If the differential equation

$$\mathbf{M}(\mathbf{x})\dot{\mathbf{x}} = \mathbf{f}(\mathbf{x}) \tag{5.106}$$

is differentiated with respect to x_i, it results in

$$\frac{\partial \mathbf{M}}{\partial x_i}\dot{\mathbf{x}} + \mathbf{M}\frac{\partial \dot{\mathbf{x}}}{\partial x_i} = \frac{\partial \mathbf{f}}{\partial x_i}$$

or

$$\frac{\partial \dot{\mathbf{x}}}{\partial x_i} = \mathbf{M}^{-1}\left(\frac{\partial \mathbf{f}}{\partial x_i} - \frac{\partial \mathbf{M}}{\partial x_i}\dot{\mathbf{x}}\right)$$

Now a new function can be introduced:

$$\mathcal{F}(\mathbf{x}, \dot{\mathbf{x}}) = \mathbf{f}(\mathbf{x}) - \mathbf{M}(\mathbf{x})\dot{\mathbf{x}} \tag{5.107}$$

As a result the Jacobian matrix of $\dot{\mathbf{x}}(\mathbf{x})$ is given by

$$\mathbf{A} = \frac{\partial \dot{\mathbf{x}}}{\partial \mathbf{x}} = \mathbf{M}^{-1}\left.\frac{\partial \mathcal{F}}{\partial \mathbf{x}}\right|_{\dot{\mathbf{x}} = \mathbf{M}^{-1}\mathbf{f}} \tag{5.108}$$

The Jacobian $\mathbf{A}$ can be derived from the Jacobian of $\mathcal{F}$ with respect to $\mathbf{x}$ where $\dot{\mathbf{x}}$ is kept constant and inserted from the forward simulation of Eq. (5.106). If this procedure is applied to the GGL-DAEs (5.105), it results in

$$\mathbf{A} = \begin{bmatrix} \mathbf{M}^{-1}\mathcal{Q}_\mathbf{q} & \mathbf{M}^{-1}\mathcal{Q}_\mathbf{v} \\ \mathbf{M}^{-1}\mathcal{V}_\mathbf{q} & \mathbf{M}^{-1}\mathcal{V}_\mathbf{v} \end{bmatrix} \tag{5.109}$$

The abbreviations $\mathcal{Q}$ and $\mathcal{V}$ read as

$$\mathcal{Q}(\mathbf{q}, \mathbf{v}, \dot{\mathbf{q}}) = \mathbf{M}(\mathbf{q})\mathbf{v} - \mathbf{G}^T(\mathbf{q})\boldsymbol{\nu} - \mathbf{M}(\mathbf{q})\dot{\mathbf{q}} \tag{5.110a}$$
$$\mathcal{V}(\mathbf{q}, \mathbf{v}, \dot{\mathbf{v}}) = \mathbf{f}(\mathbf{q}, \mathbf{v}, \mathbf{u}) - \mathbf{G}^T(\mathbf{q})\boldsymbol{\lambda} - \mathbf{M}(\mathbf{q})\dot{\mathbf{v}} \tag{5.110b}$$

The Lagrange multiplier $\boldsymbol{\nu}$ will be zero and therefore it is not necessary to compute the second term in (5.110a). Furthermore, the term $\mathbf{M}^{-1}\mathcal{Q}_\mathbf{v}$ is the identity matrix $\mathbf{I}_{(n\times n)}$ and $\mathbf{M}^{-1}\mathcal{Q}_\mathbf{q} = \mathbf{0}$ due to $\dot{\mathbf{q}} = \mathbf{v}$. As a consequence, the Jacobian (5.109) simplifies to

$$\mathbf{A} = \begin{bmatrix} \mathbf{0} & \mathbf{I} \\ \mathbf{M}^{-1}\mathcal{V}_\mathbf{q} & \mathbf{M}^{-1}\mathcal{V}_\mathbf{v} \end{bmatrix} \tag{5.111}$$

The remaining matrices $\mathbf{B}(t)$, $\mathbf{C}(t)$ and $\mathbf{D}(t)$ read as follows

$$\mathbf{B} = \begin{bmatrix} \mathbf{0} \\ \mathbf{f}_\mathbf{u} \end{bmatrix} \tag{5.112}$$

$$\mathbf{C} = \begin{bmatrix} \mathbf{0} & -\mathbf{M}^{-1}\mathbf{G}^T \\ -\mathbf{M}^{-1}\mathbf{G}^T & \mathbf{0} \end{bmatrix} \tag{5.113}$$

5.5. NUMERICAL METHODS FOR DYNAMIC OPTIMIZATION PROBLEMS

$$\mathbf{D} = \begin{bmatrix} \mathbf{G} & \mathbf{0} \\ \mathbf{R} & \mathbf{G} \end{bmatrix}, \quad \mathbf{R} = D_{\mathbf{q}}\left(\mathbf{G}(\mathbf{q})\mathbf{v}\right) \tag{5.114}$$

The costate variables $\mathbf{p}(t)$ and $\mathbf{w}(t)$, which are associated to $\mathbf{q}(t)$ and $\mathbf{v}(t)$, are introduced for the adjoint equations (5.94). Furthermore, the Lagrange multipliers $\boldsymbol{\lambda}$ and $\boldsymbol{\nu}$ in (5.105) correspond to the adjoint multipliers $\boldsymbol{\mu}$ and $\boldsymbol{\xi}$.
As a result the adjoint equations (5.94) are formulated by

$$\dot{\mathbf{p}} = -\boldsymbol{\mathcal{V}}_{\mathbf{q}}^T \mathbf{M}^{-1} \mathbf{w} - \mathbf{G}^T \boldsymbol{\mu} - \mathbf{R}^T \boldsymbol{\xi} - \mathcal{L}_{\mathbf{q}} \tag{5.115a}$$
$$\dot{\mathbf{w}} = -\mathbf{p} - \boldsymbol{\mathcal{V}}_{\mathbf{v}}^T \mathbf{M}^{-1} \mathbf{w} - \mathbf{G}^T \boldsymbol{\xi} - \mathcal{L}_{\mathbf{v}} \tag{5.115b}$$
$$0 = \mathbf{G}\mathbf{M}^{-1}\mathbf{w} \tag{5.115c}$$
$$0 = \mathbf{G}\mathbf{M}^{-1}\mathbf{p} \tag{5.115d}$$

With the matrix $\mathbf{B}$ from Eq. (5.112), the variation of the control $\mathbf{u}$ (5.100) reads as

$$\delta \mathbf{u} = -\kappa \mathbf{f}_{\mathbf{u}}^T \mathbf{w} \tag{5.116}$$

Finally, the boundary condition (5.99) is derived for the adjoint system (5.115). The matrix $\mathbf{C}^T \mathbf{D}^T$ reads as

$$\mathbf{C}^T\mathbf{D}^T = \begin{bmatrix} 0 & -\mathbf{GM}^{-1} \\ -\mathbf{GM}^{-1} & 0 \end{bmatrix} \begin{bmatrix} \mathbf{G}^T & \mathbf{R}^T \\ 0 & \mathbf{G}^T \end{bmatrix} = -\begin{bmatrix} 0 & \mathbf{GM}^{-1}\mathbf{G}^T \\ \mathbf{GM}^{-1}\mathbf{G}^T & \mathbf{GM}^{-1}\mathbf{R}^T \end{bmatrix} \tag{5.117}$$

This matrix is non-singular for the GGL-formulation (5.105), while it is always singular for the original index 3 formulation ((5.90)). In the GGL-system the boundary condition (5.99) of the adjoint system is given by

$$\begin{bmatrix} \mathbf{p}(t_f) \\ \mathbf{w}(t_f) \end{bmatrix} = \mathbf{P}_f \begin{bmatrix} \Phi_{\mathbf{q}}(t_f) \\ \Phi_{\mathbf{v}}(t_f) \end{bmatrix} \tag{5.118}$$

where the abbreviation $\mathbf{P}_f$ denotes the matrix

$$\mathbf{P}_f = \begin{bmatrix} \mathbf{I} & 0 \\ 0 & \mathbf{I} \end{bmatrix} - \underbrace{\begin{bmatrix} \mathbf{G}^T & \mathbf{R}^T \\ 0 & \mathbf{G}^T \end{bmatrix}}_{\mathbf{D}^T} \underbrace{\begin{bmatrix} 0 & \mathbf{GM}^{-1}\mathbf{G}^T \\ \mathbf{GM}^{-1}\mathbf{G}^T & \mathbf{GM}^{-1}\mathbf{R}^T \end{bmatrix}^{-1}}_{-(\mathbf{C}^T\mathbf{D}^T)^{-1}} \underbrace{\begin{bmatrix} 0 & \mathbf{GM}^{-1} \\ \mathbf{GM}^{-1} & 0 \end{bmatrix}}_{-\mathbf{C}^T} \Bigg|_{t=t_f} \tag{5.119}$$

The index 2 adjoint equations (5.115) are integrated backwards in time, starting from (5.118). The update of the control $\delta \mathbf{u}$ is computed by Eq. (5.116).

5.5.2 Direct Methods: Reduction to Static Optimization Problems

The idea of direct methods is to transcribe the infinite dimensional OCP (5.36) to a finite dimensional optimization problem, specifically a nonlinear programming

5.5. NUMERICAL METHODS FOR DYNAMIC OPTIMIZATION PROBLEMS

(NLP) problem [14, 115]. In this context the procedure is called *direct transcription method*. Applications are published e.g. in [28, 29, 30, 31, 32, 33, 34, 115].
Generally, it must be distinguished between a partly discretization, where the control trajectory is discretized over the time interval and a full discretization, where control trajectory and differential equations are discretized.
In the partly discretized procedure simulation and optimization are executed sequentially. The model equations are solved numerically for the current guess of control parameters. The control trajectory fulfills the differential equations in each iteration.
In the fully discretized method simulation and optimization are performed simultaneously. The discretized differential equations enter as nonlinear constraints into the transcribed optimization. During the iterations the constraints can be violated, but in the final solution they have to be fulfilled [115].

Partly discretization:
Generally, direct methods first subdivide the time interval $[t_0, t_f]$ into a certain number of sub-intervals $[t_0, t_1]$, $[t_1, t_2]$, $[t_2, t_3]$ etc. Then the control variables $\mathbf{u}(t)$ are defined at each sub-interval $[t_i, t_{i+1}]$. The simplest way would be to define constant controls during one interval. An improvement can be achieved with piecewise linear functions, as illustrated in Fig. 5.5. Therefore, the control variables can be

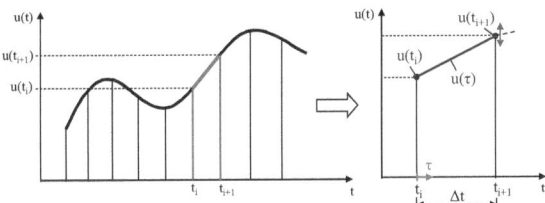

Figure 5.5: Piecewise linear control functions

formulated as

$$\mathbf{u}(\tau) = \mathbf{u}_i + \mathbf{v}_i \tau, \quad \mathbf{v}_i = \frac{\mathbf{u}_{i+1} - \mathbf{u}_i}{t_{i+1} - t_i}, \quad \tau \in [t_i, t_{i+1}] \qquad (5.120)$$

within one sufficiently small time interval. The term $\mathbf{v}_i$ describes the slope of the linear function. The controls at the end of an interval are calculated by

$$\mathbf{u}_{i+1} = \mathbf{u}_i + \mathbf{v}_i(t_{i+1} - t_i) \qquad (5.121)$$

Certainly, also polynomials of second order can be formulated.

$$\mathbf{u}(\tau) = \mathbf{a} + \mathbf{b}\tau + \mathbf{c}\tau^2 \qquad (5.122\text{a})$$
$$\dot{\mathbf{u}}(\tau) = \mathbf{b} + 2\mathbf{c}\tau \qquad (5.122\text{b})$$

5.5. NUMERICAL METHODS FOR DYNAMIC OPTIMIZATION PROBLEMS

The vectors **a**, **b** and **c** include unknown constants. These constants can be calculated by inserting the boundary conditions

$$\mathbf{u}(0) = \mathbf{u}_i = \mathbf{a} \tag{5.123a}$$
$$\dot{\mathbf{u}}(0) = \mathbf{v}_i = \mathbf{b} \tag{5.123b}$$
$$\mathbf{u}_{i+1} = \mathbf{a} + \mathbf{b}\Delta t + \mathbf{c}\Delta t^2 \tag{5.123c}$$

Eqs. (5.123a) and (5.123b) formulate the initial values for a single time interval $[t_i, t_{i+1}]$ and (5.123c) describes the final value at t_{i+1}. From (5.123c) the constant **c** can be calculated and furthermore the polynomial of second order (5.122a) can be specified as

$$\mathbf{u}(\tau) = \mathbf{u}_i + \mathbf{v}_i \tau + \frac{\mathbf{u}_{i+1} - \mathbf{u}_i - \mathbf{v}_i \Delta t}{\Delta t^2} \tau^2 \tag{5.124}$$

The derivative of the control variable at the end time of an interval $\mathbf{v}_{i+1}$ can be calculated from (5.122b) and results in

$$\mathbf{v}_{i+1} = \mathbf{b} + 2\mathbf{c}\Delta t = \frac{2\mathbf{u}_{i+1} - 2\mathbf{u}_i - \mathbf{v}_i \Delta t}{\Delta t} \tag{5.125}$$

If N denotes the number of discretized time points for each control variable $\mathbf{u}(t) \in \mathbb{R}^{m_c}$, an optimization problem with $(N \times m_c)$ dimensions is achieved. In a general form, a partly discretized OCP can be formulated by the system (5.126) [14, 77].

$$\min_{\mathbf{u}} \; J(\mathbf{u}) = \Phi(\mathbf{x}(t_f)) + \sum_{i=0}^{N-1} \int_{t_i}^{t_{i+1}} \mathcal{L}(\mathbf{x}(t), \mathbf{u}(t), t) \, dt \tag{5.126a}$$

$$\text{s.t.} : \; \dot{\mathbf{x}} = \mathbf{f}(\mathbf{x}, \mathbf{u}, t), \; t \in [t_i, t_{i+1}], \quad i = 0, \ldots, N-1 \tag{5.126b}$$

$$\mathbf{x}(t_0) = \mathbf{x}_0, \quad \mathbf{x}(t_i) = \mathbf{x}(t_{i-}), \quad i = 0, \ldots, N-1 \tag{5.126c}$$

$$\mathbf{g}(\mathbf{x}(t_N)) = \mathbf{0} \tag{5.126d}$$

$$\mathbf{h}(\mathbf{x}, \mathbf{u}, t) \leq \mathbf{0} \; \forall t \in [t_i, t_{i+1}], \quad i = 0, \ldots, N-1 \tag{5.126e}$$

For the inverse problem the system (5.126) can be treated as black box [28], cf. Fig 5.6, which is excited by the control variables $\mathbf{u}(t)$. By measuring the system outputs $\mathbf{y}(t)$ and comparing them with the target signals $\tilde{\mathbf{y}}(t)$, the cost functional (5.126a) can be evaluated in each time interval $[t_i, t_{i+1}]$. Therefore, the form (5.39b) is used:

$$J(\mathbf{u}) = \int_{t_i}^{t_{i+1}} \|\mathbf{y}(t) - \tilde{\mathbf{y}}(t)\|^2 \, dt \tag{5.127}$$

The idea of our approach is to discretize $\mathbf{u}(t)$, perform a numerical integration of the system (5.126b) (e.g.: *Matlab-solver ode45, ode15s, ode23s, ...*) and compute the cost functional (5.127). The numerical integration can be done without going

5.5. NUMERICAL METHODS FOR DYNAMIC OPTIMIZATION PROBLEMS

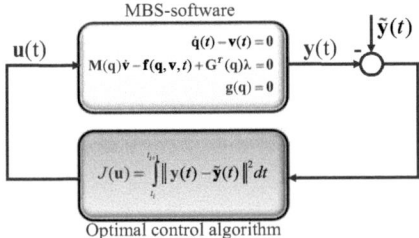

Figure 5.6: Optimal control approach which treats the MBS as black box

into the structure of the equations (5.126b).
If the time intervals are chosen be be sufficiently small, the cost functional can be written in the form

$$J_i(\mathbf{v}_i) \approx \frac{1}{2} \left[\|\mathbf{y}(t_i) - \tilde{\mathbf{y}}(t_i)\|^2 + \|\mathbf{y}(t_{i+1}) - \tilde{\mathbf{y}}(t_{i+1})\|^2 \right] (t_{i+1} - t_i) \qquad (5.128)$$

Eq. (5.128) approximates the integral (5.127) by using the trapezoidal rule. Furthermore, the slope $\mathbf{v}_i$ of the linear function $\mathbf{u}(\tau)$ does not affect the states $\mathbf{x}(t_i)$ and hence only the second part of (5.128) has to be minimized.

$$\hat{J}(\mathbf{v}_i) = \frac{1}{2} \|\mathbf{y}(t_{i+1}) - \tilde{\mathbf{y}}(t_{i+1})\|^2 + \frac{\varepsilon}{2} \|\mathbf{u}(t_{i+1})\|^2 \to \min \qquad (5.129)$$

This means that the cost functional has to be calculated at the end of each interval. Due to the quadratic form of (5.129), the functional has an elliptical stationary point. This fact is a big advantage for the minimization purpose, because local minima and the global minimum coincide. However, this is only valid if the time interval is sufficiently small. This effect is illustrated in Fig. 5.7, which shows the performance measure for different step sizes. The performance measure of Fig. 5.7 are taken from the example of a nonlinear oscillator, cf. section 7.1. In Fig. 5.7a and 5.7c a nominal step length of $\Delta t = 0.01s$ was chosen to be sufficiently small. It can be seen that the performance measure has a parabolic form and the minimum point has an elliptical structure. If the step size in increased to $\Delta t = 0.05s$ (see Fig. 5.7b and 5.7d), the elliptical minimum point is not distinctive and thus it can be problematic for the minimization task.
The term $\frac{\varepsilon}{2} \|\mathbf{u}(t_{i+1})\|^2$ in (5.129)is a so-called *Tikhonov regularization* term [7, 102]. This regularization term causes the minimization of the variations of the control $\mathbf{u}$. Hence, the zigzagging trend of the input variables can be minimized depending on the scalar weighting factor ε. However, ε should not be chosen too large, because then the optimization problem would diverge from the inverse dynamics problem. The influence of the Tikhonov-regularization is illustrated in Fig. 5.8 where the optimal control algorithm is applied to the nonlinear oscillator from section 7.1. In this numerical example a step size of $\Delta t = 0.005\,s$ and a weighting factor of $\varepsilon = 10^{-4}$

5.5. NUMERICAL METHODS FOR DYNAMIC OPTIMIZATION PROBLEMS

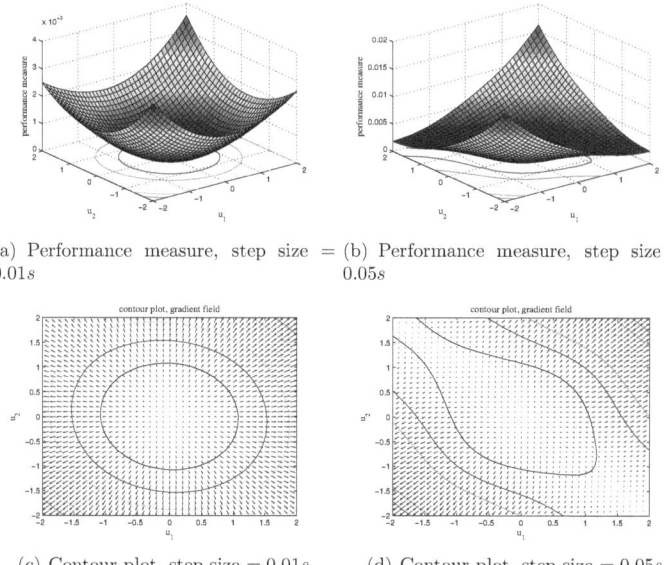

(a) Performance measure, step size = 0.01s

(b) Performance measure, step size = 0.05s

(c) Contour plot, step size = 0.01s

(d) Contour plot, step size = 0.05s

Figure 5.7: Performance measure with contour plot and gradient field for different step sizes

are used. In [40] weighting factors in the range of $\varepsilon = 10^{-4} \cdots 10^{-8}$ are suggested. For the minimization of the cost functional $J(\mathbf{u})$ a steepest descent method (cf.

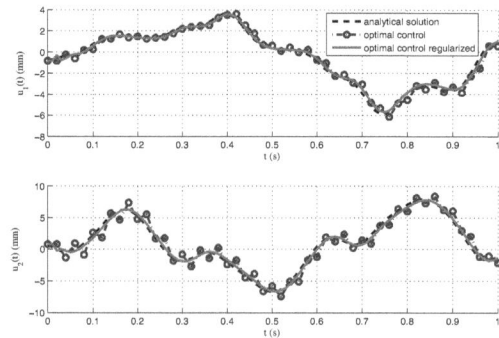

Figure 5.8: Influence of the Tikhonov regularization in the computed control variables

5.5. NUMERICAL METHODS FOR DYNAMIC OPTIMIZATION PROBLEMS

section 5.2.1) can be used:

$$\mathbf{u}_{i+1}^{k+1} = \mathbf{u}_{i+1}^{k} + \alpha^k \mathbf{d}_{i+1}^{k} \qquad (5.130)$$

The descent direction in the steepest descent method is the negative gradient of the cost functional $\mathbf{d}_{i+1}^{k} = -\nabla \hat{J}_i^k$. The step length α^k is computed by a line search algorithm. Due to the fact that only input and output variables are known in this black-box approach, the gradient has to be calculated numerically. Therefore, the forward difference quotient is used.

$$\frac{\partial \hat{J}_i^k}{\partial u_{i,j}^k} = \frac{\hat{J}_i^k(u_j^k + \varepsilon_j^k, t_{i+1}) - \hat{J}_i^k(u_j^k, t_{i+1})}{\varepsilon_j^k}, \quad j = 1, \ldots, m_c \qquad (5.131)$$

The gradient is assembled from the individual changes of the performance measures:

$$\nabla \hat{J}_i^k = \left[\frac{\partial \hat{J}_i^k}{\partial u_{i,1}^k}, \frac{\partial \hat{J}_i^k}{\partial u_{i,2}^k}, \ldots, \frac{\partial \hat{J}_i^k}{\partial u_{i,m_c}^k} \right]^T \qquad (5.132)$$

To improve the convergence speed, a conjugate gradient method (see section 5.2.2) is implemented instead of the steepest descent method. Hence, the descent direction is calculated as

$$\mathbf{d}_{i+1}^{k} = -\nabla \hat{J}_i^k + \beta^k \mathbf{d}_{i+1}^{k-1} \qquad (5.133)$$

The minimization by the conjugate gradient method and the line search algorithm are basically illustrated in Fig. 5.9. The algorithm is implemented in *Matlab* and applied to specific academic examples, cf. chapter 7. The main steps of the al-

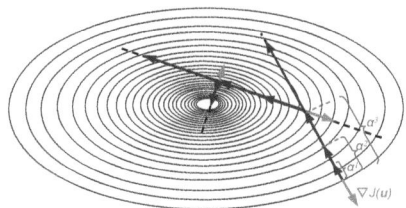

Figure 5.9: Conjugate gradient method with a line search algorithm

gorithm, which is implemented in *Matlab*, can be seen in the Nassi-Shneiderman diagram 5.10.

Full discretization:
In a full discretized version of the OCP the control trajectories and the differential equations (5.126b) are discretized as well. The discretization of the differential equations can be done e.g. by a trapezoidal rule:

$$\frac{\mathbf{x}_{i+1} - \mathbf{x}_i}{t_{i+1} - t_i} = \frac{1}{2} \left[\mathbf{f}(\mathbf{x}_i, \mathbf{u}_i, t_i) + \mathbf{f}(\mathbf{x}_{i+1}, \mathbf{u}_{i+1}, t_{i+1}) \right], \quad i = 0, 1, \ldots, N-1 \qquad (5.134)$$

5.5. NUMERICAL METHODS FOR DYNAMIC OPTIMIZATION PROBLEMS

Figure 5.10: Nassi-Shneiderman diagram of the optimal control approach with a partly discretization

5.5. NUMERICAL METHODS FOR DYNAMIC OPTIMIZATION PROBLEMS

As a consequence, the OCP can be formulated as [14, 77]

$$\min_{\mathbf{u}} \; J(\mathbf{x}, \mathbf{u}) = \Phi(\mathbf{x}_N) + \sum_{i=0}^{N-1} \frac{t_{i+1} - t_i}{2} \left[\mathcal{L}(\mathbf{x}_i, \mathbf{u}_i, t_i) + \mathcal{L}(\mathbf{x}_{i+1}, \mathbf{u}_{i+1}, t_{i+1}) \right]$$

$$s.t. : \; \frac{\mathbf{x}_{i+1} - \mathbf{x}_i}{t_{i+1} - t_i} = \frac{1}{2} \left[\mathbf{f}(\mathbf{x}_i, \mathbf{u}_i, t_i) + \mathbf{f}(\mathbf{x}_{i+1}, \mathbf{u}_{i+1}, t_{i+1}) \right], \quad i = 0, \ldots, N-1$$

$$\mathbf{x}_{i=0} = \mathbf{x}_0$$

$$\mathbf{g}(\mathbf{x}_N) = \mathbf{0}$$

$$\mathbf{h}(\mathbf{x}_i, \mathbf{u}_i, t_i) \leq \mathbf{0}, \qquad i = 0, \ldots, N$$

(5.135a)

(5.135b)

(5.135c)
(5.135d)
(5.135e)

In contrast to a partially discretized scheme the number of optimization variables is drastically higher $(m+n)(N+1)$, but the differential equations (5.126) are already discretized and hence they do not have to be integrated.

A piecewise discretization of inputs and the DAEs of a multibody system is presented in [54, 65, 66, 67, 68, 69]. An application of the direct transcription method to flexible MBS can be found in [5]. Direct optimal control algorithms which are applied to MBS in *Adams* are published in [28].

5.5.3 Comparison of Direct and Indirect Methods

Indirect methods allow more insight into the structure of the optimal control problem. Solutions are accurate and the adjoint variables can be used for sensitivity analysis and controller design [77]. In direct methods the canonical equations do not have to be derived. Constraints in the state variables can be treated easier and sometimes the convergence area is larger compared to indirect methods. Another advantage is that initial guesses for the costates are not required. Direct methods are most commonly used in practical applications because of their applicability and robustness [14]. A classification of solution strategies for OCPs is given in Fig. 5.11.

5.5.4 Software

The following list presents either free or commercially available direct optimal control software that are based on partly or fully discretization [77, 115]:

5.5. NUMERICAL METHODS FOR DYNAMIC OPTIMIZATION PROBLEMS

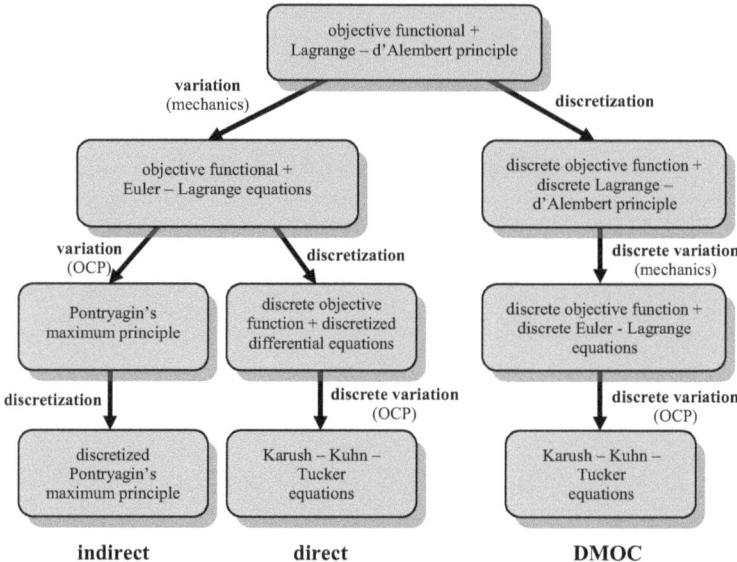

Figure 5.11: Classification of solution strategies for optimal control problems [115]

- HQP (huge quadratic programming, direct multiple shooting)
  http://hqp.sourceforge.net

- MUSCOD II (multiple shooting code for optimization)
  http://www.iwr.uni-heidelberg.de/~agbock/RESEARCH/muscod.php

- NTG (nonlinear trajectory generation, direct collocation)
  http://www.cds.caltech.edu/~murray/software/2002a_ntg.html

- TOMLAB/PROPT (direct collocation)
  http://tomdyn.com

- gOPT (direct single shooting)
  http://www.psenterprise.com/gproms/index.html

- DYNOPT (direct collocation)
  http://www.kirp.chtf.stuba.sk/moodle/course/view.php?id=187

- SOCS (sparse optimal control software, direct collocation) [14]
  http://www.boeing.com/phantom/socs/

- OCPRSQP (direct collocation)

5.5. NUMERICAL METHODS FOR DYNAMIC OPTIMIZATION PROBLEMS

- DIRCOL (direct collocation)
  http://www.sim.tu-darmstadt.de/sw/dircol.html
- IPOPT (interior point optimizer, direct collocation)
  https://projects.coin-or.org/Ipopt

> *A theory is something nobody believes, except the person who made it. An experiment is something everybody believes, except the person who made it.*
>
> Albert Einstein

Chapter 6

Flatness-Based Trajectory Tracking

In this section explicit control laws are derived for the control inputs $\mathbf{u}(t) \in \mathbb{R}^{m_c}$ in order that the system outputs $\mathbf{y}(t) \in \mathbb{R}^k$ follow a predefined desired trajectory $\mathbf{y}_d(t) \in \mathbb{R}^k$. As a basis, the nonlinear equations of motion are written in the state space form of the following type [83]:

$$\dot{\mathbf{x}} = \mathbf{f}(\mathbf{x}) + \sum_{j=1}^{m_c} \mathbf{g}_j(\mathbf{x}) u_j \tag{6.1a}$$

$$y_i = h_i(\mathbf{x}) \quad i = 1, ..., k \tag{6.1b}$$

The state vector consists of the generalized coordinates and velocities $\mathbf{x} = [\mathbf{q}, \dot{\mathbf{q}}]^T \in \mathbb{R}^{2n}$. The following considerations are restricted to fully determined systems where the number of inputs is identical to the number of outputs m_c. The theory of exact linearization is taken from [83, 91, 112, 135, 136, 137, 146, 165].

The considered examples in chapter 7 are all MIMO-systems. However, for reasons of clarity the basic theory of exact linearization is firstly illustrated for SISO-systems.

6.1 Nonlinear Feedback for SISO-systems

The basic idea is to calculate the derivatives of the output $y \in \mathbb{R}$ up to a certain order until the input $u \in \mathbb{R}$ explicitly appears.

6.1.1 Exact Input-Output Linearization

The subsequent methods are based on nonlinear systems that are given as affine input systems (AI systems) of the form

$$\dot{\mathbf{x}} = \mathbf{f}(\mathbf{x}) + \mathbf{g}(\mathbf{x}) u \tag{6.2a}$$
$$y = h(\mathbf{x}) \tag{6.2b}$$

6.1. NONLINEAR FEEDBACK FOR SISO-SYSTEMS

The state vector is denoted as $\mathbf{x} \in \mathbb{R}^{2n}$, the input as $u \in \mathbb{R}$ and the output as $y \in \mathbb{R}$. Furthermore, it is assumed that the vector fields $\mathbf{f}(\mathbf{x})$ and $\mathbf{g}(\mathbf{x})$ and the function $h(\mathbf{x})$ are sufficiently smooth, i.e. continuously differentiable.

The derivatives of y with respect to time are calculated from Eq. (6.2b) and the derivation $\dot{\mathbf{x}}$ is inserted from (6.2a). This procedure is repeated until the input u appears in the rth derivative of the output y.

$$\begin{aligned} y &= h(\mathbf{x}) \\ \dot{y} &= L_\mathbf{f} h(\mathbf{x}) + \underbrace{L_\mathbf{g} h(\mathbf{x})}_{=0} u \\ \ddot{y} &= L_\mathbf{f}^2 h(\mathbf{x}) + \underbrace{L_\mathbf{g} L_\mathbf{f} h(\mathbf{x})}_{=0} u \\ &\vdots \quad \vdots \quad \vdots \\ y^{(r-1)} &= L_\mathbf{f}^{r-1} h(\mathbf{x}) + \underbrace{L_\mathbf{g} L_\mathbf{f}^{r-2} h(\mathbf{x})}_{=0} u \\ y^{(r)} &= L_\mathbf{f}^r h(\mathbf{x}) + \underbrace{L_\mathbf{g} L_\mathbf{f}^{r-1} h(\mathbf{x})}_{\neq 0} u \end{aligned} \qquad (6.3)$$

The terms $L_\mathbf{f} h(\mathbf{x})$ and $L_\mathbf{g} h(\mathbf{x})$ in Eq. (6.3) are called *Lie-derivatives* of the scalar function $h(\mathbf{x})$ along the vector fields $\mathbf{f}(\mathbf{x})$ and $\mathbf{g}(\mathbf{x})$. They are calculated by

$$L_\mathbf{f} h(\mathbf{x}) = \frac{\partial h}{\partial \mathbf{x}} \mathbf{f}(\mathbf{x}), \quad L_\mathbf{g} h(\mathbf{x}) = \frac{\partial h}{\partial \mathbf{x}} \mathbf{g}(\mathbf{x}) \qquad (6.4)$$

Based on these derivatives, the relative degree r of the SISO-system can be defined: The system (6.2) has the relative degree r at the point $\mathbf{x}^\circ$, if the conditions

$$L_\mathbf{g} L_\mathbf{f}^k h(\mathbf{x}) = 0, \quad k = 0, ..., (r-2) \quad \forall \mathbf{x} \text{ in the neighborhood of } \mathbf{x}^\circ \qquad (6.5a)$$
$$L_\mathbf{g} L_\mathbf{f}^{r-1} h(\mathbf{x}^\circ) \neq 0 \qquad (6.5b)$$

are fulfilled [83, 165]. This means that the relative degree r is equal to the number of time derivatives that have to be calculated from the output y until the input u explicitly arises.

From Eq. (6.3) the state control law can be formulated for the input u.

$$u = \frac{1}{L_\mathbf{g} L_\mathbf{f}^{r-1} h(\mathbf{x})} (-L_\mathbf{f}^r h(\mathbf{x}) + v) \qquad (6.6)$$

The control law (6.6) results in a linear input-output behavior in the form of an integrator chain

$$y^{(r)} = v \qquad (6.7)$$

6.1.2 Transformation to the Byrnes-Isidori Normal Form

The system (6.2) can be transformed to the so-called *Byrnes-Isidori normal form* by using a local invertible diffeomorphism $\mathbf{z} = \mathbf{\Phi}(\mathbf{x})$. The nonlinear state-space

6.1. NONLINEAR FEEDBACK FOR SISO-SYSTEMS

transformation can be written in the form [83]

$$\mathbf{z} = \mathbf{\Phi}(\mathbf{x}) = \begin{bmatrix} z_1 \\ \vdots \\ z_{2n} \end{bmatrix} = \begin{bmatrix} h(\mathbf{x}) \\ L_\mathbf{f} h(\mathbf{x}) \\ \vdots \\ L_\mathbf{f}^{r-1} h(\mathbf{x}) \\ \phi_{r+1}(\mathbf{x}) \\ \vdots \\ \phi_{2n}(\mathbf{x}) \end{bmatrix} \qquad (6.8)$$

If r is strictly less than $2n$, it is always possible to find $(2n-r)$ functions $\phi_{r+1}(\mathbf{x}), \ldots, \phi_{2n}(\mathbf{x})$ such that a local diffeomorphism is given in the neighborhood of $\mathbf{x}°$. The functions $\phi_{r+1}(\mathbf{x}), \ldots, \phi_{2n}(\mathbf{x})$ can be chosen in order that $L_\mathbf{g} \phi_k(\mathbf{x}) = 0$, $k = (r+1), \ldots, 2n$ for all $\mathbf{x}$ in the neighborhood of $\mathbf{x}°$.

If the nonlinear state-space transformation is applied to the system (6.2), the transformed system in Byrnes-Isidori normal form reads as [83, 91, 146, 165]

$$\begin{aligned} \dot{z}_1 &= z_2 \\ \dot{z}_2 &= z_3 \\ &\vdots \\ \dot{z}_r &= L_\mathbf{f}^r h(\mathbf{\Phi}^{-1}(\mathbf{z})) + L_\mathbf{g} L_\mathbf{f}^{r-1} h(\mathbf{\Phi}^{-1}(\mathbf{z})) u = b(\mathbf{z}) + a(\mathbf{z}) u \\ \dot{z}_{r+1} &= L_\mathbf{f} \phi_{r+1}(\mathbf{\Phi}^{-1}(\mathbf{z})) = q_{r+1}(\mathbf{z}) \\ &\vdots \\ \dot{z}_{2n} &= L_\mathbf{f} \phi_{2n}(\mathbf{\Phi}^{-1}(\mathbf{z})) = q_{2n}(\mathbf{z}) \end{aligned} \qquad (6.9)$$

$$y = z_1$$

Fig. 6.1 illustrates the equations of the Byrnes-Isidori normal form. The input u

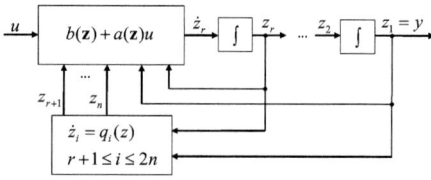

Figure 6.1: Block diagram of the Byrnes-Isidori normal form [83]

can be calculated from Eq. (6.9) in the same way as from Eq. (6.3), which is not given in the new coordinates.

$$u = \frac{1}{a(\mathbf{z})}(-b(\mathbf{z}) + v) = \frac{1}{L_\mathbf{g} L_\mathbf{f}^{r-1} h(\mathbf{\Phi}^{-1}(\mathbf{z}))}(-L_\mathbf{f}^r h(\mathbf{\Phi}^{-1}(\mathbf{z})) + v) \qquad (6.10)$$

The control law (6.10) transforms the system (6.2) or (6.9) into a system with an exact linear input-output behavior from the new input v to the output y. Hence, the transfer matrix can be written as

$$G(s) = \frac{\hat{Y}}{\hat{V}} = \frac{1}{s^r} \qquad (6.11)$$

The resulting Byrnes-Isidori normal form splits the transformed system in a reachable and observable system with the states $\boldsymbol{\xi}$, $\dim(\boldsymbol{\xi}) = r$ and a system with the states $\boldsymbol{\eta}$, $\dim(\boldsymbol{\eta}) = (2n - r)$, which is not observable. The observable system consists of a chain of integrators and presents an exact input-output-linearization. The non-observable system is called internal dynamics of the system [135].

6.1.3 Zero Dynamics

The *output-zeroing problem* discusses how the initial states $\mathbf{x}_0$ and the control inputs $u(t)$ can be calculated in order that the output $y(t)$ is equal to zero at any time [146]. This problem can be solved by considering the Byrnes-Isidori normal form (6.9). The states of the observable system are denoted by $\boldsymbol{\xi} = [z_1, ..., z_r]^T$ and the states of the non-observable system by $\boldsymbol{\eta} = [z_{r+1}, ..., z_{2n}]^T$.

$$\begin{aligned}
\dot{z}_1 &= z_2 \\
\dot{z}_2 &= z_3 \\
&\vdots \\
\dot{z}_r &= b(\boldsymbol{\xi}, \boldsymbol{\eta}) + a(\boldsymbol{\xi}, \boldsymbol{\eta})u \\
\dot{\boldsymbol{\eta}} &= \mathbf{q}(\boldsymbol{\xi}, \boldsymbol{\eta}) \\
y &= z_1
\end{aligned} \qquad (6.12)$$

From the condition $y(t) = z_1 = 0$ it follows that all states of the observable system are equal to zero $z_2 = z_3 = \cdots = z_r = 0$. Hence, the input $u(t)$ can be calculated for the output-zeroing problem [83, 112].

$$b(\mathbf{0}, \boldsymbol{\eta}) + a(\mathbf{0}, \boldsymbol{\eta})u = 0 \Rightarrow u(t) = -\frac{b(\mathbf{0}, \boldsymbol{\eta}(t))}{a(\mathbf{0}, \boldsymbol{\eta}(t))} \qquad (6.13)$$

The states $\boldsymbol{\eta}(t)$ of the non-observable system are a solution of the differential equation

$$\dot{\boldsymbol{\eta}} = \mathbf{q}(\mathbf{0}, \boldsymbol{\eta}) \qquad (6.14)$$

with the initial conditions $\boldsymbol{\xi}(0) = \mathbf{0}$ and an arbitrary $\boldsymbol{\eta}(0) = \boldsymbol{\eta}_0$. The differential equation (6.14) describes the so-called *internal dynamics* of the system. If the outputs of the internal dynamics are restricted to be zero, it is called *zero-dynamics*. The zero dynamics and hence the dynamics of the non-observable system is generally nonlinear. The zero dynamics is crucial regarding to the stability of the closed-loop system. The closed-circle-system can only be stabilized with a stable zero dynamics, i.e. the system is a minimum phase system [137]. A nonlinear minimum phase system is characterized by a locally asymptotically stable equilibrium point $\boldsymbol{\eta}_s$ of

the zero dynamics (6.14).
If the new input v is chosen as [91, 112]

$$v = -\sum_{j=1}^{r} a_{j-1}\xi_j = -\sum_{j=1}^{r} a_{j-1} L_\mathbf{f}^{j-1} h(\mathbf{x}) \qquad (6.15)$$

and inserted in the control law (6.10)

$$u = \frac{1}{a(\boldsymbol{\xi},\boldsymbol{\eta})}\left(-b(\boldsymbol{\xi},\boldsymbol{\eta}) - \sum_{j=1}^{r} a_{j-1}\xi_j\right) = \frac{1}{L_\mathbf{g} L_\mathbf{f}^{r-1} h(\mathbf{x})}\left(-L_\mathbf{f}^r h(\mathbf{x}) - \sum_{j=1}^{r} a_{j-1} L_\mathbf{f}^{j-1} h(\mathbf{x})\right)$$

$$(6.16)$$

the closed loop reads as

$$\dot{\boldsymbol{\xi}} = \mathbf{A}_r \boldsymbol{\xi} \qquad (6.17a)$$
$$\dot{\boldsymbol{\eta}} = \mathbf{q}(\boldsymbol{\xi},\boldsymbol{\eta}) \qquad (6.17b)$$
$$y = \xi_1 \qquad (6.17c)$$

with the dynamic matrix

$$\mathbf{A}_r = \begin{bmatrix} 0 & 1 & \cdots & 0 \\ \vdots & \vdots & \ddots & \vdots \\ 0 & 0 & \cdots & 1 \\ -a_0 & -a_1 & \cdots & -a_{r-1} \end{bmatrix} \qquad (6.18)$$

The coefficients a_j, $j = 0, ..., (r-1)$ can be chosen in a way that the matrix $\mathbf{A}_r$ is a Hurwitz matrix. If (6.18) is a Hurwitz matrix and if the system (6.2) is locally exponentially minimum phase at $\mathbf{x}_s = \mathbf{0}$ (i.e. $\boldsymbol{\xi}_s = \mathbf{0}$, $\boldsymbol{\eta}_s = \mathbf{0}$), the dynamic matrix of the linearized closed circle (6.17) is also a Hurwitz matrix. The system is locally exponentially minimum phase, if all eigenvalues of $\partial \mathbf{q}(\mathbf{0},\boldsymbol{\eta}_s)/\partial \boldsymbol{\eta}$ have a negative real part [91, 137]. The dynamic matrix of the linearized closed loop reads as

$$\frac{d}{dt}\begin{bmatrix} \Delta\boldsymbol{\xi} \\ \Delta\boldsymbol{\eta} \end{bmatrix} = \begin{bmatrix} \mathbf{A}_r & \mathbf{0} \\ \frac{\partial \mathbf{q}(\mathbf{0},\boldsymbol{\eta}_s)}{\partial \boldsymbol{\xi}} & \frac{\partial \mathbf{q}(\mathbf{0},\boldsymbol{\eta}_s)}{\partial \boldsymbol{\eta}} \end{bmatrix}\begin{bmatrix} \Delta\boldsymbol{\xi} \\ \Delta\boldsymbol{\eta} \end{bmatrix} \qquad (6.19)$$

If the dynamic matrix (6.19) is a Hurwitz matrix, the equilibrium point $\mathbf{x}_s = \mathbf{0}$ (i.e. $\boldsymbol{\xi}_s = \mathbf{0}$, $\boldsymbol{\eta}_s = \mathbf{0}$) of the closed circle is locally asymptotically stable.
This means that the exact input-output linearization results in a stable closed circle, if the system is asymptotically minimum phase.

6.1.4 Exact Input-State Linearization

The zero dynamics vanishes, if the relative degree is equal to the number of states $r = 2n$. Then the control law (6.10) reads as

$$u = \frac{1}{a(\mathbf{z})}(-b(\mathbf{z}) + v) = \frac{1}{L_\mathbf{g} L_\mathbf{f}^{n-1} h(\boldsymbol{\Phi}^{-1}(\mathbf{z}))}(-L_\mathbf{f}^n h(\boldsymbol{\Phi}^{-1}(\mathbf{z})) + v) \qquad (6.20)$$

6.1. NONLINEAR FEEDBACK FOR SISO-SYSTEMS

If this control law is inserted in the Byrnes-Isidori normal form (6.9), it simplifies to

$$\begin{aligned} \dot{z}_1 &= z_2 \\ \dot{z}_2 &= z_3 \\ &\vdots \\ \dot{z}_{2n} &= v \end{aligned} \quad (6.21)$$

This system can also be written in the form

$$\dot{\mathbf{z}} = \begin{bmatrix} 0 & 1 & \cdots & 0 \\ \vdots & \vdots & \ddots & \vdots \\ 0 & 0 & \cdots & 1 \\ 0 & 0 & 0 & 0 \end{bmatrix} \mathbf{z} + \begin{bmatrix} 0 \\ 0 \\ 0 \\ 1 \end{bmatrix} v \quad (6.22)$$

The system (6.22) is known as *Brunovsky canonical form* [83]. It can be compared with the normal form of a linear system. The equations of the closed loop system (6.22) describe a system that is linear and controllable. Hence, it can be concluded that any nonlinear system with a relative degree $r = 2n$ can be transformed into a linear and controllable system. Fig. 6.2 illustrates the Brunovsky canonical form. A parameterization of a fictitious output $y = \lambda(\mathbf{x})$ with a relative degree $r = 2n$ can

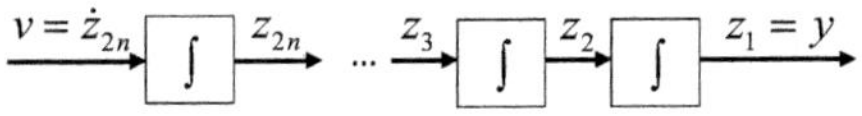

Figure 6.2: Block diagram of the Brunovsky canonical form [83]

always be found, if the conditions

$$L_\mathbf{g}\lambda(\mathbf{x}) = L_\mathbf{g}L_\mathbf{f}\lambda(\mathbf{x}) = \cdots = L_\mathbf{g}L_\mathbf{f}^{2n-2}\lambda(\mathbf{x}) = 0 \quad \forall\, \mathbf{x} \quad (6.23a)$$
$$L_\mathbf{g}L_\mathbf{f}^{2n-1}\lambda(\mathbf{x}^\circ) \neq 0 \quad (6.23b)$$

are fulfilled [83]. The function $\lambda(\mathbf{x})$ is involved in a system of partial differential equations (PDEs) (6.23a). However, this system of higher order is equivalent to a system of first order partial differential equations of the co-called *Frobenius-type*.

$$L_\mathbf{g}\lambda(\mathbf{x}) = L_{ad_\mathbf{f}\mathbf{g}(\mathbf{x})}\lambda(\mathbf{x}) = \cdots = L_{ad_\mathbf{f}^{2n-2}\mathbf{g}(\mathbf{x})}\lambda(\mathbf{x}) = 0 \quad (6.24a)$$
$$L_{ad_\mathbf{f}^{2n-1}\mathbf{g}(\mathbf{x})}\lambda(\mathbf{x}^\circ) \neq 0 \quad (6.24b)$$

For the computations in Eq. (6.24) the definitions of a *Lie-bracket* should be mentioned. The Lie-bracket $[\mathbf{f},\mathbf{g}](\mathbf{x})$ is defined as [91, 95]

$$[\mathbf{f},\mathbf{g}](\mathbf{x}) = \frac{\partial \mathbf{g}}{\partial \mathbf{x}}\mathbf{f}(\mathbf{x}) - \frac{\partial \mathbf{f}}{\partial \mathbf{x}}\mathbf{g}(\mathbf{x}) \quad (6.25)$$

The kth Lie bracket can be calculated recursively by

$$ad_\mathbf{f}^k\mathbf{g}(\mathbf{x}) = \left[\mathbf{f}, ad_\mathbf{f}^{k-1}\mathbf{g}\right](\mathbf{x}), \quad ad_\mathbf{f}^0\mathbf{g}(\mathbf{x}) = \mathbf{g}(\mathbf{x}) \quad (6.26)$$

The existence of a solution $\lambda(\mathbf{x})$ of the system of partial differential equations of first order (6.24a) is strongly related to the conditions [83]

the matrix $\left[\mathbf{g}, ad_\mathbf{f}\mathbf{g}, \ldots, ad_\mathbf{f}^{2n-1}\mathbf{g}\right](\mathbf{x}°)$ has rank $2n$ (6.27a)

the distribution $D = \text{span}\left\{\mathbf{g}, ad_\mathbf{f}\mathbf{g}, \ldots, ad_\mathbf{f}^{2n-2}\mathbf{g}\right\}$
is involutive in a neighborhood of $\mathbf{x}°$ (6.27b)

If the conditions (6.27) are fulfilled, the system is exactly input-state linearizable in a neighborhood of $\mathbf{x}°$. It is assumed that the output $y = h(\mathbf{x})$ has a relative degree $r = 2n$. Then the derivatives (6.3) can be written in the form

$$
\begin{aligned}
y &= h(\mathbf{x}) \\
\dot{y} &= L_\mathbf{f} h(\mathbf{x}) \\
\ddot{y} &= L_\mathbf{f}^2 h(\mathbf{x}) \\
&\vdots \quad \vdots \quad \vdots \\
y^{(2n-1)} &= L_\mathbf{f}^{2n-1} h(\mathbf{x}) \\
y^{(2n)} &= L_\mathbf{f}^{2n} h(\mathbf{x}) + L_\mathbf{g} L_\mathbf{f}^{2n-1} h(\mathbf{x}) u
\end{aligned}
\qquad (6.28)
$$

Due to the regular transformation (6.8) in the case $r = 2n$, all states $\mathbf{x}$ can be parameterized by the output y and its time derivatives up to order $2n - 1$.

$$\mathbf{x} = \psi_1\left(y, \dot{y}, \ldots, y^{(2n-1)}\right) = \Phi^{-1}(\mathbf{z}) \qquad (6.29)$$

The states $\mathbf{z}$ are formed by the outputs and its time derivatives $\mathbf{z} = \left[y, \dot{y}, \ldots, y^{(2n-1)}\right]^T$. Furthermore, the input u can be parameterized by the output y and its time derivatives up to order $2n$, cf. the last equation in (6.28).

$$u = \psi_2\left(y, \dot{y}, \ldots, y^{(2n)}\right) = \frac{y^{(2n)} - L_\mathbf{f}^{2n} h(\Phi^{-1}(\mathbf{z}))}{L_\mathbf{g} L_\mathbf{f}^{2n-1} h(\Phi^{-1}(\mathbf{z}))} \qquad (6.30)$$

By using the parameterizations (6.29) and (6.30) the term *differential flatness* can be explained. A dynamical system of the form (6.2) is called *differentially flat*, if all state variables and input variables can be expressed by the output y and its derivatives with respect to time up to a certain order. The output of such a system is called *flat output*.

In a SISO-system the characteristics of differential flatness and exact input-state linearization are coherent. A system that is input-state linearizable, is differentially flat and each output with a relative degree of $r = 2n$ is a flat output.

6.1.5 Trajectory Tracking Control

The goal of the trajectory tracking control is to design a controller in order that the output $y(t)$ follows a predefined sufficiently often differentiable desired trajectory $y_d(t)$. The parameterization (6.30) of the input u results in a feedforward control

6.1. NONLINEAR FEEDBACK FOR SISO-SYSTEMS

that does not take the error between $y(t)$ and $y_d(t)$ into account.
In the following considerations it is assumed that all state variables **x** can physically be measured. If this is not the case, the control law has to be extended by an observer which estimates the non-measurable states $\hat{\mathbf{x}}$. Furthermore, the trajectory tracking control can be applied to an output which is not differentially flat. In such a case another parametrization of the non-flat output can be found. The implementation of an observer and a treatment of non-flat outputs are not considered here but can be found e.g. in [91].

The basis of the trajectory tracking control is the system given in Brunovsky canonical form (6.22). If the error $e_1 = z_1 - y_d = y - y_d$ is introduced, the derivatives of the error with respect to time $e_2 = \dot{e}_1$, $e_3 = \dot{e}_2$ read as

$$\begin{aligned}
\dot{e}_1 &= \dot{z}_1 - \dot{y}_d = \dot{y} - \dot{y}_d \\
\dot{e}_2 &= \dot{z}_2 - \ddot{y}_d = \ddot{y} - \ddot{y}_d \\
&\vdots \\
\dot{e}_{2n} &= \dot{z}_{2n} - y_d^{(2n)} = v - y_d^{(2n)} = y^{(2n)} - y_d^{(2n)}
\end{aligned} \tag{6.31}$$

This results in a control law with parameters a_j, $j = 0, ..., (2n-1)$ that can be chosen arbitrarily. If the parameters are chosen properly (*pole setting*), an asymptotically stable error dynamic is obtained [135].

$$\begin{bmatrix} \dot{e}_1 \\ \dot{e}_2 \\ \vdots \\ \dot{e}_{2n} \end{bmatrix} = \underbrace{\begin{bmatrix} 0 & 1 & 0 & 0 & 0 \\ 0 & 0 & 1 & 0 & 0 \\ \vdots & \vdots & \cdots & \ddots & \vdots \\ 0 & 0 & \cdots & 0 & 1 \\ -a_0 & -a_1 & \cdots & -a_{2n-2} & -a_{2n-1} \end{bmatrix}}_{\mathbf{A}_e} \begin{bmatrix} e_1 \\ e_2 \\ \vdots \\ e_{2n} \end{bmatrix} \tag{6.32}$$

The matrix $\mathbf{A}_e$ in (6.32) is called error dynamic matrix, whose eigenvalues can be influenced by the parameters a_j. As a consequence, the control law (6.30) for the input u is extended by the trajectory error $e_1 = y - y_d$ and its time derivatives up to order $(2n-1)$.

$$u = \frac{1}{L_g L_f^{2n-1} h(\mathbf{x})} \left(y_d^{(2n)}(t) - L_f^{2n} h(\mathbf{x}) - \sum_{j=1}^{2n} a_{j-1} \left(y^{(j-1)} - y_d^{(j-1)}(t) \right) \right) \tag{6.33}$$

The $(j-1)^{th}$ derivative of the output y is equal to the Lie-derivative $y^{(j-1)} = L_f^{j-1} h(\mathbf{x})$. Again, it should be mentioned that the control law (6.33) for the trajectory tracking problem can only be used, if all states **x** can be measured.

Exact Feedforward Linearization with Output Stabilization:
In the case where no measurements of the outputs exist, the flatness-based control

can be designed by the parameterizations (6.29) and (6.30).

$$\mathbf{x}_d = \boldsymbol{\psi}_1\left(y_d, \dot{y}_d, ..., y_d^{(2n-1)}\right) = \boldsymbol{\Phi}^{-1}(\mathbf{z}_d) \tag{6.34a}$$

$$u_d = \psi_2\left(y_d, \dot{y}_d, ..., y_d^{(2n)}\right) = \frac{y_d^{(2n)}(t) - L_\mathbf{f}^{2n}h(\mathbf{x}_d)}{L_\mathbf{g}L_\mathbf{f}^{2n-1}h(\mathbf{x}_d)} \tag{6.34b}$$

The states $\mathbf{z}_d$ in the new coordinates include the desired outputs and its time derivatives $\mathbf{z}_d = \begin{bmatrix} y_d, \dot{y}_d, ..., y_d^{(2n-1)} \end{bmatrix}^T$. The flatness-based control $u_d(t)$ (6.34b) is also known as *exact feedforward linearization* [136].
If the desired outputs $y_d(t)$ are consistent with the initial conditions $\mathbf{x}_0$ of the system (6.2), the mathematical model is exact and no parameter variations and disturbances occur, the flatness-based control $u_d(t)$ applied to the system (6.2) results exactly in the desired outputs $y_d(t)$ [83].
However, if the initial conditions are not consistent or parameters variations occur, the solution will drift apart from the desired solution. For sufficiently small disturbances a linear controller can be used to counterbalance the disturbances (feedback control, cf. Fig. 6.3). Hence, the flatness-based control is extended by a control law u_c. If a proportional-integral controller (PI controller) is used, the control algorithm reads as

$$u_c = k_p w_c + \int_0^t w_c\, dt, \quad w_c = w - w_d \tag{6.35}$$

The term $w = l(\mathbf{x})$ denotes the measurable variables. The control law of the PI-controller (6.35) is added to the flatness-based control u_d.

$$u = u_d + u_c \tag{6.36}$$

The procedure (6.36) is called *two-degree-of-freedom design* and is illustrated in Fig. 6.3. The linear PI-controller can be justified by the fact that the flatness-based control $u_d(t)$ still results in system trajectories $\mathbf{x}(t)$ that are sufficiently near the desired trajectories $\mathbf{x}_d(t)$.

6.2 Nonlinear Feedback for MIMO-systems

6.2.1 Exact Input-Output Linearization

In this section the equations of motion of a multibody system with multiple inputs and multiple outputs (MIMO) are given as system with affine inputs. The state vector $\mathbf{x} \in \mathbb{R}^{2n}$ is summarized by $\mathbf{x} = [\mathbf{q}, \mathbf{v}]^T$, the inputs by $\mathbf{u} = [u_1, \ldots u_{m_c}]^T \in \mathbb{R}^{m_c}$ and the outputs by $\mathbf{y} = [y_1, \ldots y_{m_c}]^T \in \mathbb{R}^{m_c}$, i.e. the number of inputs is identical

6.2. NONLINEAR FEEDBACK FOR MIMO-SYSTEMS

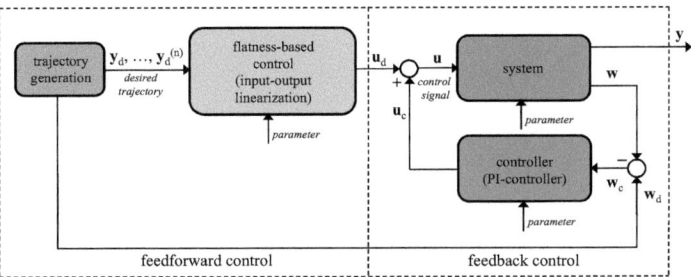

Figure 6.3: Two-degree-of-freedom design of a trajectory tracking control [91, 112]

to the number of outputs. This yields a system of the form [83]:

$$\dot{\mathbf{x}} = \mathbf{f}(\mathbf{x}) + \sum_{j=1}^{m_c} \mathbf{g}_j(\mathbf{x}) u_j$$
$$y_1 = h_1(\mathbf{x})$$
$$\vdots$$
$$y_{m_c} = h_{m_c}(\mathbf{x})$$
(6.37)

Furthermore, it is assumed that the vector fields $\mathbf{f}(\mathbf{x})$ and $\mathbf{g}_j(\mathbf{x})$ as well as the functions $h_j(\mathbf{x})$ are sufficiently smooth, i.e. continuously differentiable.
The (vector) relative degree $\{r_1, \ldots r_{m_c}\}$, $r = \sum_{j=1}^{m} r_j \leq 2n$ can be defined at the point $\mathbf{x}^\circ$, if the following conditions are fulfilled [83]:

$$L_{\mathbf{g}_j} L_{\mathbf{f}}^k h_i(\mathbf{x}) = 0, \quad j = 1, \ldots, m_c, \ i = 1, \ldots, m_c, \ k = 0, \ldots, (r_i - 2)$$
$$\forall \ \mathbf{x} \text{ in the neighborhood of } \mathbf{x}^\circ$$
(6.38a)

Furthermore, the $(m_c \times m_c)$ decoupling matrix

$$\mathbf{A}(\mathbf{x}) = \begin{bmatrix} L_{\mathbf{g}_1} L_{\mathbf{f}}^{r_1-1} h_1(\mathbf{x}) & \cdots & L_{\mathbf{g}_{m_c}} L_{\mathbf{f}}^{r_1-1} h_1(\mathbf{x}) \\ L_{\mathbf{g}_1} L_{\mathbf{f}}^{r_2-1} h_2(\mathbf{x}) & \cdots & L_{\mathbf{g}_{m_c}} L_{\mathbf{f}}^{r_2-1} h_2(\mathbf{x}) \\ \vdots & \ddots & \vdots \\ L_{\mathbf{g}_1} L_{\mathbf{f}}^{r_{m_c}-1} h_{m_c}(\mathbf{x}) & \cdots & L_{\mathbf{g}_{m_c}} L_{\mathbf{f}}^{r_{m_c}-1} h_{m_c}(\mathbf{x}) \end{bmatrix}$$
(6.38b)

must be regular.

6.2. NONLINEAR FEEDBACK FOR MIMO-SYSTEMS

If the system (6.37) has the (vector) relative degree $\{r_1, \ldots r_{m_c}\}$, the derivatives of the output $y_j = h_j(\mathbf{x})$ in the neighborhood $\mathbf{x}^\circ$ read as

$$\begin{aligned}
y_j &= h_j(\mathbf{x}) \\
\dot{y}_j &= L_\mathbf{f} h_j(\mathbf{x}) + \underbrace{L_{\mathbf{g}_1} h_j(\mathbf{x})}_{=0} u_1 + \cdots + \underbrace{L_{\mathbf{g}_{m_c}} h_j(\mathbf{x})}_{=0} u_{m_c} \\
\ddot{y}_j &= L_\mathbf{f}^2 h_j(\mathbf{x}) + \underbrace{L_{\mathbf{g}_1} L_\mathbf{f} h_j(\mathbf{x})}_{=0} u_1 + \cdots + \underbrace{L_{\mathbf{g}_{m_c}} L_\mathbf{f} h_j(\mathbf{x})}_{=0} u_{m_c} \\
&\vdots \quad \vdots \quad \vdots \\
y_j^{(r_j-1)} &= L_\mathbf{f}^{r_j-1} h_j(\mathbf{x}) + \underbrace{L_{\mathbf{g}_1} L_\mathbf{f}^{r_j-2} h_j(\mathbf{x})}_{=0} u_1 + \cdots + \underbrace{L_{\mathbf{g}_{m_c}} L_\mathbf{f}^{r_j-2} h_j(\mathbf{x})}_{=0} u_{m_c} \\
y_j^{(r_j)} &= L_\mathbf{f}^{r_j} h_j(\mathbf{x}) + L_{\mathbf{g}_1} L_\mathbf{f}^{r_j-1} h_j(\mathbf{x}) u_1 + \cdots + L_{\mathbf{g}_{m_c}} L_\mathbf{f}^{r_j-1} h_j(\mathbf{x}) u_{m_c}
\end{aligned} \tag{6.39}$$

If these derivatives are applied to all outputs $y_j = h_j(\mathbf{x})$, $j = 1, \ldots, m_c$, the last equation in (6.39) is extended to

$$\underbrace{\begin{bmatrix} y_1^{(r_1)} \\ \vdots \\ y_{m_c-1}^{(r_{m_c-1})} \\ y_{m_c}^{(r_{m_c})} \end{bmatrix}}_{\mathbf{v}} = \underbrace{\begin{bmatrix} L_\mathbf{f}^{r_1} h_1(\mathbf{x}) \\ \vdots \\ L_\mathbf{f}^{r_{m_c-1}} h_{m_c-1}(\mathbf{x}) \\ L_\mathbf{f}^{r_{m_c}} h_{m_c}(\mathbf{x}) \end{bmatrix}}_{\mathbf{b}(\mathbf{x})} + \mathbf{A}(\mathbf{x}) \underbrace{\begin{bmatrix} u_1 \\ \vdots \\ u_{m_c-1} \\ u_{m_c} \end{bmatrix}}_{\mathbf{u}} \tag{6.40}$$

This equation can be rearranged to find a state-space control law for the inputs $\mathbf{u}$ in a neighborhood of $\mathbf{x}^\circ$ [83, 91]:

$$\boxed{\mathbf{u} = \mathbf{A}^{-1}(\mathbf{x}) \left(\mathbf{v} - \mathbf{b}(\mathbf{x}) \right)} \tag{6.41}$$

It can be seen that the control law (6.41) for MIMO-systems is of the same type as the control law for SISO-systems (6.10). Eq. (6.41) results in an exact linear input-output behavior from the new inputs $\mathbf{v} = [v_1, \ldots, v_{m_c}]^T$ to the outputs $\mathbf{y} = [y_1, \ldots, y_{m_c}]^T$ in the form of m_c integrator chains, which read as

$$\begin{bmatrix} y_1 \\ \vdots \\ y_{m_c-1}^{(r_{m_c-1})} \\ y_{m_c}^{(r_{m_c})} \end{bmatrix} = \begin{bmatrix} v_1 \\ \vdots \\ v_{m_c-1} \\ v_{m_c} \end{bmatrix} \tag{6.42}$$

6.2.2 Transformation to the Byrnes-Isidori Normal Form

The nonlinear state-space transformation, which transforms the system (6.37) to the Byrnes-Isidori normal form, can also be applied to MIMO-systems [83].

$$\mathbf{z} = \Phi(\mathbf{x}) = \begin{bmatrix} z_1 \\ \vdots \\ z_{2n} \end{bmatrix} = \begin{bmatrix} \boldsymbol{\xi} \\ \boldsymbol{\eta} \end{bmatrix} = \begin{bmatrix} h_1(\mathbf{x}) \\ \vdots \\ L_\mathbf{f}^{r_1-1} h_1(\mathbf{x}) \\ \vdots \\ h_{m_c}(\mathbf{x}) \\ \vdots \\ L_\mathbf{f}^{r_{m_c}-1} h_{m_c}(\mathbf{x}) \\ \phi_{r+1}(\mathbf{x}) \\ \vdots \\ \phi_{2n}(\mathbf{x}) \end{bmatrix} \qquad (6.43)$$

In contrast to SISO-systems, the functions $\Phi_{r+1}(\mathbf{x}), ..., \Phi_{2n}(\mathbf{x})$ cannot be chosen such that $L_{\mathbf{g}_j}\phi_k(\mathbf{x}) = 0$, $j = 1, ..., m_c$, $k = (r+1), ..., 2n$ is fulfilled, except the distribution $G_0 = \text{span}\{\mathbf{g}_1, ..., \mathbf{g}_{m_c}\}$ is involutive in a neighborhood of $\mathbf{x}^\circ$ [83].
If the state-space transformation (6.43) is applied to the MIMO-system (6.37), the resulting Byrnes-Isidori normal form reads as [83, 91]

$$\begin{aligned}
\dot{\xi}_{1,1} &= \xi_{1,2} \\
\dot{\xi}_{1,2} &= \xi_{1,3} \\
&\vdots \\
\dot{\xi}_{1,r_1} &= \tilde{b}_1(\boldsymbol{\xi}, \boldsymbol{\eta}) + \sum_{j=1}^{m_c} \tilde{A}_{1,j}(\boldsymbol{\xi}, \boldsymbol{\eta}) u_j \\
&\vdots\vdots\vdots \\
\dot{\xi}_{m_c,1} &= \xi_{m_c,2} \\
\dot{\xi}_{m_c,2} &= \xi_{m_c,3} \\
&\vdots \\
\dot{\xi}_{m_c,r_{m_c}} &= \tilde{b}_{m_c}(\boldsymbol{\xi}, \boldsymbol{\eta}) + \sum_{j=1}^{m_c} \tilde{A}_{m_c,j}(\boldsymbol{\xi}, \boldsymbol{\eta}) u_j \\
\dot{\eta}_1 &= q_1(\boldsymbol{\xi}, \boldsymbol{\eta}) + \sum_{j=1}^{m_c} P_{1,j}(\boldsymbol{\xi}, \boldsymbol{\eta}) u_j \\
&\vdots\vdots\vdots \\
\dot{\eta}_{2n-r} &= q_{2n-r}(\boldsymbol{\xi}, \boldsymbol{\eta}) + \sum_{j=1}^{m_c} P_{2n-r,j}(\xi, \eta) u_j \\
\mathbf{y} &= [\xi_{1,1}, \xi_{2,1}, ..., \xi_{m_c,1}]^T
\end{aligned} \qquad (6.44)$$

The abbreviations $\tilde{b}_j(\boldsymbol{\xi},\boldsymbol{\eta})$, $\tilde{A}_{l,j}(\boldsymbol{\xi},\boldsymbol{\eta})$, $q_i(\boldsymbol{\xi},\boldsymbol{\eta})$ and $P_{i,i}(\boldsymbol{\xi},\boldsymbol{\eta})$ stand for

$$\tilde{b}_j(\boldsymbol{\xi},\boldsymbol{\eta}) = b_j(\boldsymbol{\Phi}^{-1}(\boldsymbol{\xi},\boldsymbol{\eta})) = L_{\mathbf{f}}^{r_j} h_j(\boldsymbol{\Phi}^{-1}(\boldsymbol{\xi},\boldsymbol{\eta})), \quad j = 1,...,m_c \quad (6.45a)$$

$$\tilde{A}_{l,j}(\boldsymbol{\xi},\boldsymbol{\eta}) = A_{l,j}(\boldsymbol{\Phi}^{-1}(\boldsymbol{\xi},\boldsymbol{\eta})) = L_{\mathbf{g}_j} L_{\mathbf{f}}^{r_l-1} h_l(\boldsymbol{\Phi}^{-1}(\boldsymbol{\xi},\boldsymbol{\eta})) \quad j,l = 1,...,m_c \quad (6.45b)$$

$$q_i(\boldsymbol{\xi},\boldsymbol{\eta}) = L_{\mathbf{f}}\phi_{r+i}(\boldsymbol{\Phi}^{-1}(\boldsymbol{\xi},\boldsymbol{\eta})), \quad i = 1,...,(2n-r) \quad (6.45c)$$

$$P_{i,i}(\boldsymbol{\xi},\boldsymbol{\eta}) = L_{\mathbf{g}_j}\phi_{r+i}(\boldsymbol{\Phi}^{-1}(\boldsymbol{\xi},\boldsymbol{\eta})), \quad i = 1,...,(2n-r),\ l = 1,...,m_c \quad (6.45d)$$

Hence, the control law (6.41) in the new coordinates reads as

$$\mathbf{u} = \tilde{\mathbf{A}}^{-1}(\boldsymbol{\xi},\boldsymbol{\eta})\left(\mathbf{v} - \mathbf{b}(\boldsymbol{\xi},\boldsymbol{\eta})\right) \quad (6.46)$$

6.2.3 Zero Dynamics

The stability of a closed MIMO-system is analogous to that of SISO-systems. The method of exact input-output linearization yields only in a stable closed circuit, if the zero dynamics is asymptotically stable and hence that the system is minimum phase. The zero dynamics of a MIMO-system reads as

$$\dot{\boldsymbol{\eta}} = \mathbf{q}(\mathbf{0},\boldsymbol{\eta}) + \mathbf{P}(\mathbf{0},\boldsymbol{\eta})\tilde{\mathbf{A}}^{-1}(\mathbf{0},\boldsymbol{\eta})\left(-\tilde{\mathbf{b}}(\mathbf{0},\boldsymbol{\eta})\right) \quad (6.47)$$

6.2.4 Exact Input-State Linearization

If the (vector) relative degree $\{r_1, r_2, ..., r_{m_c}\}$ is equal to the number of states $r = \sum_{j=1}^{m_c} r_j = 2n$, the zero dynamics vanishes. Fictitious output variables $\lambda_1(\mathbf{x}), ..., \lambda_{m_c}(\mathbf{x})$ with the (vector) relative degree $r = \sum_{j=1}^{m_c} r_j = 2n$ can be found as a solution of the PDEs [83]

$$L_{\mathbf{g}_j} L_{\mathbf{f}}^k \lambda_i(\mathbf{x}) = 0, \quad j = 1,...,m_c,\ i = 1,...,m_c,\ k = 0,...,(r_i - 2) \quad (6.48)$$

Furthermore, the decoupling matrix $\mathbf{A}(\mathbf{x})$ must be regular as in the case of the input-output linearization. The PDEs (6.48) can also be written as PDEs of first order in the Frobenius-form [91].

$$L_{ad_{\mathbf{f}}^k \mathbf{g}_j(\mathbf{x})}\lambda_i(\mathbf{x}) = 0, \quad j = 1,...,m_c,\ i = 1,...,m_c,\ k = 0,...,(r_i - 2) \quad (6.49)$$

The existence of solutions $\lambda_1(\mathbf{x}), ..., \lambda_{m_c}(\mathbf{x})$ is strongly related to the fact that the system of PDEs (6.49) fulfills following conditions: (i) the decoupling matrix $\mathbf{A}(\mathbf{x})$ is regular and (ii) the distribution

$$G_i(\mathbf{x}) = \mathrm{span}\left\{ad_{\mathbf{f}}^k \mathbf{g}_j(\mathbf{x}) : 0 \le k \le i,\ 1 \le j \le m_c\right\} \quad (6.50)$$

fulfills the conditions

$G_0(\mathbf{x}^\circ)$ has rank m_c (6.51a)

$G_i(\mathbf{x})$ has constant rank in the neighborhood $\mathbf{x}^\circ\ \forall\ i = 1,...,(2n-1)$ (6.51b)

$G_{2n-1}(\mathbf{x}^\circ)$ has rank $2n$ (6.51c)

$G_i(\mathbf{x})$ is involutive in the neighborhood of $\mathbf{x}^\circ\ \forall\ i = 0,...,(2n-2)$ (6.51d)

for the (vector) relative degree $\{r_1, r_2, ..., r_{m_c}\}$, $r = \sum_{j=1}^{m_c} r_j = 2n$. If these conditions are fulfilled, the system (6.37) is exactly input-state linearizable in the neighborhood of $\mathbf{x}°$ [83].
The (vector) relative degree $\{r_1, r_2, ..., r_{m_c}\}$ can be calculated by the auxiliary variables

$$\delta_i = \text{rank}\,(G_i(\mathbf{x}°)) - \text{rank}\,(G_{i-1}(\mathbf{x}°))\,, \quad i = 1, ..., (2n-1) \quad (6.52)$$

The component r_j, $j = 1, ..., m_c$ of the (vector) relative degree is always by one greater than the number of δ_i's, $i = 1, ..., (2n-1)$, which are greater or equal to j [91].
By using the state-space transformation (6.43) and the control law (6.46) the system (6.37) is transformed in an exact linear system with the new states $\mathbf{z}$ and the new inputs $\mathbf{v}$. The linear system consists of m_c integrator chains of the length $\{r_1, ..., r_{m_c}\}$. This form is also known as Brunovsky canonical form, Eq. (6.22).

An important difference between the linearization of SISO-systems and MIMO-systems is the connection between *differential flatness* and *input-state linearization*. In SISO-systems these two characteristics are strongly related. A differentially flat SISO-system is input-state linearizable and vice versa. In a MIMO-system a flatness based parametrization of state- and input variables can be found, even if the system is not input-state-linearizable. On the other hand, an input-state linearizable system is always differentially flat [91, 95, 112].

6.2.5 Trajectory Tracking Control

The control laws (6.41) or (6.46) can furthermore be extended by a stabilization term as it was performed for SISO-systems, Eq. (6.33).

$$\mathbf{u} = \mathbf{A}^{-1}(\mathbf{x}) \left(\mathbf{v} - \mathbf{b}(\mathbf{x}) - \begin{bmatrix} \sum_{j=1}^{r_1} a_{1,j-1} \left(L_\mathbf{f}^{j-1} h_1(\mathbf{x}) - y_{1,d}^{(j-1)}(t) \right) \\ \vdots \\ \sum_{j=1}^{r_{m_c}} a_{m_c,j-1} \left(L_\mathbf{f}^{j-1} h_{m_c}(\mathbf{x}) - y_{m_c,d}^{(j-1)}(t) \right) \end{bmatrix} \right) \quad (6.53)$$

6.3 Flatness Based Trajectory Tracking

As already mentioned, a flatness-based parameterization of all input- and state variables can be found for a MIMO-system, even if the conditions (6.38) and (6.51) are not fulfilled [131]. As a consequence, the large symbolic computations for the nonlinear coordinate transformation (6.8), (6.43) do not have to be performed.
Differential flatness was introduced by [57, 58]. The following section gives a definition of differential flatness as it can also be found e.g. in [82, 95, 131].
A general nonlinear MIMO-system with the states $\mathbf{x} \in \mathbb{R}^{2n}$ and the inputs $\mathbf{u} \in \mathbb{R}^{m_c}$

6.3. FLATNESS BASED TRAJECTORY TRACKING

of the form
$$\dot{\mathbf{x}} = \mathbf{f}(\mathbf{x}, \mathbf{u}), \quad \mathbf{x}(0) = \mathbf{x}_0 \qquad (6.54)$$
is considered. The nonlinear system (6.54) is called differentially flat, if a fictitious output $\mathbf{y} = [y_1, ..., y_{m_c}]^T$ with $m_c = \dim(\mathbf{u})$ exists, that fulfills the following conditions:

(i) The variables y_i, $i = 1, ..., m_c$ can be parameterized by the system variables x_j, $j = 1, ..., 2n$ and u_i, $i = 1, ..., m_c$ and a finite number of derivatives with respect to time $u_i^{(k)}$, $k = 1, ..., \alpha_i$. This means that the fictitious output of the nonlinear system (6.54) can be parameterized by

$$\begin{aligned}\mathbf{y} &= \boldsymbol{\phi}\left(\mathbf{x}, u_1, ..., u_1^{(\alpha_1)}, ..., u_{m_c}, ..., u_{m_c}^{(\alpha_{m_c})}\right) \\ &= \boldsymbol{\phi}\left(\mathbf{x}, \mathbf{u}, \dot{\mathbf{u}}, ..., \mathbf{u}^{(\alpha)}\right)\end{aligned} \qquad (6.55)$$

(ii) The system variables x_i, $i = 1, ..., 2n$ and u_i, $i = 1, ..., m_c$ can be parameterized by functions of y_i, $i = 1, ..., m_c$ and a finite number of derivatives with respect to time $y_i^{(k)}$, $k = 1, ..., \beta_i$, i.e.

$$\begin{aligned}\mathbf{x} &= \boldsymbol{\psi}_1\left(y_1, ..., y_1^{(\beta_1 - 1)}, ..., y_{m_c}, ..., y_{m_c}^{(\beta_{m_c} - 1)}\right) \\ &= \boldsymbol{\psi}_1\left(\mathbf{y}, \dot{\mathbf{y}}, ..., \mathbf{y}^{(\beta - 1)}\right)\end{aligned} \qquad (6.56a)$$

$$\begin{aligned}\mathbf{u} &= \boldsymbol{\psi}_2\left(y_1, ..., y_1^{(\beta_1)}, ..., y_{m_c}, ..., y_{m_c}^{(\beta_{m_c})}\right) \\ &= \boldsymbol{\psi}_2\left(\mathbf{y}, \dot{\mathbf{y}}, ..., \mathbf{y}^{(\beta)}\right)\end{aligned} \qquad (6.56b)$$

(iii) The components of $\mathbf{y}$ are differentially independent, i.e. they do not fulfill differential equations of the form

$$\boldsymbol{\phi}\left(\mathbf{y}, \dot{\mathbf{y}}, ..., \mathbf{y}^{(\gamma)}\right) = \mathbf{0} \qquad (6.57)$$

If these conditions are fulfilled at least locally, the fictitious output (6.55) is called flat output and the system (6.54) is called flat [131].
A flatness-based parametrization can be performed directly based on the equations of motion. The state-space representation (ODEs of first order) is not required. As a consequence, the nonlinear control law is drastically simplified.

In [21] it is shown that the number of necessary time derivatives β of the output (6.56) is related to the index i of the DAEs with control constraints. The value of β is smaller by one than the index i of the DAEs (4.2).

$$\beta = i - 1 \qquad (6.58)$$

If the flatness-based trajectory tracking is compared to the DAE-approach with control constraints (cf. section 4), it can be stated that the DAE-approach can be applied easier to arbitrary outputs and it requires less pre-computations [137].

However, the DAE-approach is numerically less efficient. On the other side it should be mentioned that an analytical solution based on the flatness-based control can only be found for small multibody systems. It can be very complicated and even impossible to find an analytical solution for larger multibody systems [21, 136].

> *Make things as simple as possible, but not simpler.*
>
> Albert Einstein

Chapter 7
Academic Examples

Three academic examples are considered in this section. The first example represents a nonlinear oscillator which is fully actuated. The second example is an underactuated planar overhead crane and the third example an underactuated 3D-rotary crane. All these examples illustrate an inverse problem where a mass (or two masses) should follow a pre-defined sufficiently smooth trajectory. The control inputs are computed by using the different methods described in the previous sections, namely the DAE-approach with control constraints, the optimal control and the flatness-based trajectory tracking control. The basis of all these nonlinear models are the equations of motion, which can be derived in a symbolic form, either as ODEs with minimal coordinates or as DAEs with redundant coordinates.

It should be mentioned that such problems are typical examples in nonlinear control and hence similar models can be found e.g. in [12, 13, 21, 25, 26, 52]. A flatness-based parameterization and a solution of the DAE-approach of a planar overhead crane are given in [12, 13]. The DAE-approach applied to an independent coordinates formulation is presented in [21]. In [25] a slightly different DAE-method is shown for the planar crane with redundant coordinates. In [26] the equations of motion, which are based on redundant coordinates, are derived for the 3D rotary crane. A flatness-based parameterization, which is based on independent coordinates of a similar rotating tower crane, can be found in [52]. In contrast to these references, the flatness-based parameterization and the DAE-approach of the examples in section 7.1, 7.2 and 7.3 are related to redundant coordinates. From the knowledge of the author it is the first time that the Kelley-Bryson method is applied to such examples. Especially the implementation for redundant coordinates can be seen as scientific novelty.

7.1 Nonlinear Oscillator

7.1.1 Problem Description

A typical example of a fully actuated system in the form of a nonlinear two-mass-oscillator should be considered, Fig. 7.1. The generalized coordinates $\mathbf{q} = [y_1, y_2]^T$ represent the positions of the two masses as well as the outputs $\mathbf{y}$ of the system.

7.1. NONLINEAR OSCILLATOR

The inputs $\mathbf{u} = [u_1, u_2]^T$ are displacements, which should be determined in a way that $\mathbf{y}$ is identical to a predefined desired motion $\mathbf{y}_d$. Hence, it can be seen that the number of DOFs is identical to the number of inputs and that the system is fully actuated.

The springs c_1 and c_3, as well as the damper k, are linear elements. The spring c_2 is modeled as nonlinear spring with the force $F_{c_2} = \alpha(y_2 - y_1)^3$.

The target signals are defined as harmonic sinusoidal curves (7.1), Fig. 7.2.

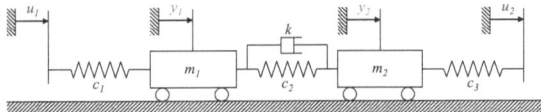

Figure 7.1: Nonlinear Oscillator

$$\begin{aligned} y_{1,d}(t) &= \hat{Y}_1 \sin(2\pi f_1 t), \quad \hat{Y}_1 = 2\,mm, \; f_1 = 1\,Hz \\ y_{2,d}(t) &= \hat{Y}_2 \sin(2\pi f_2 t) + \frac{\hat{Y}_2}{3}\sin(3 \cdot 2\pi f_2 t), \quad \hat{Y}_2 = 5\,mm, \; f_1 = 1.5\,Hz \end{aligned} \quad (7.1)$$

For numerical computations the following parameters are used:

$$m_1 = m_2 = 0.5\,kg, \; c_1 = c_3 = 150\,N/m, \; k = 1.5\,Ns/m, \; \alpha = 2 \cdot 10^6\,N/m^3$$

7.1.2 Equations of Motion

The equations of motion can be derived by applying Newton's second law. Due to the formulation with generalized coordinates, the system is given in the form (4.4b). Geometric constraints are not required and hence the equations of motion are given by the ODEs

$$\begin{bmatrix} m_1 & 0 \\ 0 & m_2 \end{bmatrix} \cdot \begin{bmatrix} \ddot{y}_1 \\ \ddot{y}_2 \end{bmatrix} = \begin{bmatrix} -c_1 y_1 + \alpha(y_2-y_1)^3 + k(\dot{y}_2 - \dot{y}_1) \\ -c_3 y_2 - \alpha(y_2-y_1)^3 - k(\dot{y}_2 - \dot{y}_1) \end{bmatrix} - \begin{bmatrix} -c_1 & 0 \\ 0 & -c_3 \end{bmatrix} \cdot \begin{bmatrix} u_1 \\ u_2 \end{bmatrix} \quad (7.2)$$

$$\Leftrightarrow \mathbf{M}\dot{\mathbf{v}} = \mathbf{f} - \mathbf{B}^T \mathbf{u}$$

7.1.3 DAE Approach with Control Constraints

The control constraints (4.4c) are formulated by

$$\mathbf{c}(\mathbf{q},t) = \mathbf{\Phi}(\mathbf{q}) - \boldsymbol{\gamma}(t) = \begin{bmatrix} y_1 \\ y_2 \end{bmatrix} - \begin{bmatrix} y_{1,d} \\ y_{2,d} \end{bmatrix} = \begin{bmatrix} 0 \\ 0 \end{bmatrix} \quad (7.3)$$

The equations of motion (7.2) in combination with the control constraints (7.3) yield an index 3 DAE. Two differentiations with respect to time yield the control constraints at acceleration level (4.7):

$$\mathbf{C}\dot{\mathbf{v}} + \boldsymbol{\xi} = \mathbf{C}\dot{\mathbf{v}} + \underbrace{\dot{\mathbf{C}}\mathbf{v}}_{=0} - \ddot{\boldsymbol{\gamma}} = \begin{bmatrix} 1 & 0 \\ 0 & 1 \end{bmatrix} \cdot \begin{bmatrix} \ddot{y}_1 \\ \ddot{y}_2 \end{bmatrix} - \begin{bmatrix} \ddot{y}_{1,d} \\ \ddot{y}_{2,d} \end{bmatrix} = \begin{bmatrix} 0 \\ 0 \end{bmatrix} \quad (7.4)$$

7.1. NONLINEAR OSCILLATOR

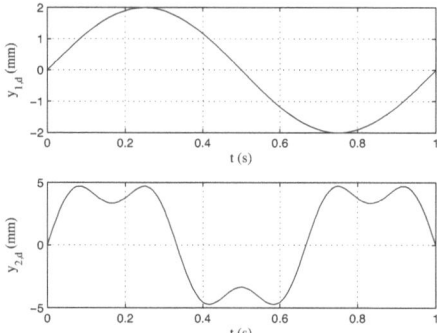

Figure 7.2: Target signals $y_{1,d}(t)$, $y_{2,d}(t)$

Hence, the projection matrix $\mathbf{C}$ is simply the (2×2) identity matrix $\mathbf{C} = \text{eye}(2)$. By using the notations of [21], it can be shown that the control constraints (7.2) of the system (7.3) are characterized by an orthogonal realization. Eq. (4.12) yields

$$p = \text{rank}(\mathbf{CM}^{-1}\mathbf{B}^T) = 2 = m_c \qquad (7.5)$$

and hence the first row in Table 4.1 is fulfilled.
The projection method (4.10) and the numerical discretization by the implicit Euler method (4.13) are applied to the system (7.2), (7.3). A step size of $\Delta t = 0.01\,s$ is used for time discretization. The results are shown in Fig. 7.4.

7.1.4 Flatness-Based Trajectory Tracking

The analytical solution for the control inputs $u_1(t)$ and $u_2(t)$ can directly be derived from the equations of motion (7.2).

$$u_1 = \frac{m_1}{c_1}\ddot{y}_{1,d} + y_{1,d} - \frac{\alpha}{c_1}(y_{2,d} - y_{1,d})^3 - \frac{k}{c_1}(\dot{y}_{2,d} - \dot{y}_{1,d}) \qquad (7.6a)$$

$$u_2 = \frac{m_2}{c_3}\ddot{y}_{2,d} + y_{2,d} + \frac{\alpha}{c_3}(y_{2,d} - y_{1,d})^3 + \frac{k}{c_3}(\dot{y}_{2,d} - \dot{y}_{1,d}) \qquad (7.6b)$$

Due to the possible parameterizations of the inputs u_1 and u_2 it can be stated that the system is differentially flat and the outputs y_1 and y_2 are flat outputs. In the parameterizations (7.6) it can be seen that the derivatives of the outputs up to second order are required.

$$u_1 = u_1\left(y_{1,d}, \dot{y}_{1,d}, \ddot{y}_{1,d}, y_{2,d}, \dot{y}_{2,d}\right) \qquad (7.7a)$$

$$u_2 = u_2\left(y_{1,d}, \dot{y}_{1,d}, y_{2,d}, \dot{y}_{2,d}, \ddot{y}_{2,d}\right) \qquad (7.7b)$$

7.1. NONLINEAR OSCILLATOR

By using the definition (6.58), it can be verified that the equations of motion are of index 3.

$$i = \beta + 1 = 3 \tag{7.8}$$

The highest number of derivation applied to the output **y** is denoted as β the index of the governing DAEs is denoted as i.
If the equations of motion (7.2) are written in the form of an affine input system (6.37), it results in

$$\begin{bmatrix} \dot{x}_1 \\ \dot{x}_2 \\ \dot{x}_3 \\ \dot{x}_4 \end{bmatrix} = \underbrace{\begin{bmatrix} x_3 \\ x_4 \\ -\frac{c_1}{m_1}x_1 + \frac{\alpha}{m_1}(x_2 - x_1)^3 + \frac{k}{m_1}(x_4 - x_3) \\ -\frac{c_3}{m_2}x_2 - \frac{\alpha}{m_2}(x_2 - x_1)^3 - \frac{k}{m_2}(x_4 - x_3) \end{bmatrix}}_{\mathbf{f(x)}} + \underbrace{\begin{bmatrix} 0 \\ 0 \\ \frac{c_1}{m_1} \\ 0 \end{bmatrix}}_{\mathbf{g_1(x)}} u_1 + \underbrace{\begin{bmatrix} 0 \\ 0 \\ 0 \\ \frac{c_3}{m_2} \end{bmatrix}}_{\mathbf{g_2(x)}} u_2$$

$y_1 = x_1$
$y_2 = x_2$

(7.9a)

(7.9b)
(7.9c)

If the necessary Lie-derivatives are computed, the definitions (6.38) and (6.51) can be verified. The (vector) relative degree results in

$$\{r_1, r_2\} = \{2, 2\}, \quad r = r_1 + r_2 = 4 = 2n \tag{7.10}$$

The decoupling matrix (6.38b) results in the regular matrix

$$\mathbf{A(x)} = \begin{bmatrix} \frac{c_1}{m_1} & 0 \\ 0 & \frac{c_3}{m_2} \end{bmatrix} \tag{7.11}$$

Furthermore, all conditions (6.51) are fulfilled and hence it can be stated that the system is exactly input-state linearizable.

7.1.5 Optimal Control

The direct optimal control algorithm from section 5.5.2 is applied to the system (7.2). A step size of $\Delta t = 0.005\,s$ is used for time discretization. To reduce the zig-zagging effect, a Tikhonov-regularization term was introduced with a weighting factor $\varepsilon = 10^{-4}$. Fig. 7.3 illustrates the convergence behavior within one time interval. Contour plot and gradient field from the performance measure $\hat{J}(\mathbf{v}_i)$ are shown. It can be seen that the initial point at the beginning of the time interval

7.1. NONLINEAR OSCILLATOR

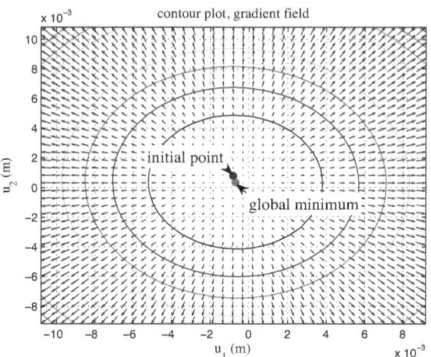

Figure 7.3: Contour plot and gradient field within one time interval

and the global minimum are close together. This results from the short step size Δt.

Furthermore, the indirect optimal control from section 5.5.1 is applied to the system (7.2). The optimality conditions (5.53) are formulated in combination with the boundary conditions (5.54) and the transversality conditions (5.55). Then the solution strategy from section 5.4.4 is applied to the resulting system of equations. For that reason the equations of motion (7.2) are formulated in state-space form. The state variables $x_1 = y_1$, $x_2 = y_2$, $x_3 = \dot{y}_1$, $x_4 = \dot{y}_2$ are introduced and the equations of motion result in

$$\begin{bmatrix} 1 & 0 & 0 & 0 \\ 0 & 1 & 0 & 0 \\ 0 & 0 & m_1 & 0 \\ 0 & 0 & 0 & m_2 \end{bmatrix} \cdot \begin{bmatrix} \dot{x}_1 \\ \dot{x}_2 \\ \dot{x}_3 \\ \dot{x}_4 \end{bmatrix} = \begin{bmatrix} x_3 \\ x_4 \\ -c_1 x_1 + \alpha(x_2 - x_1)^3 + k(x_4 - x_3) + c_1 u_1 \\ -c_3 x_2 - \alpha(x_2 - x_1)^3 - k(x_4 - x_3) + c_3 u_2 \end{bmatrix} \tag{7.12}$$

The cost functional for the inverse problem is formulated by

$$J(\mathbf{u}) = \int_{t_0}^{t_f} \left\{ \frac{1}{2}\left(x_1 - \hat{Y}_1 \sin(2\pi f_1 t)\right)^2 + \frac{1}{2}\left(x_2 - (\hat{Y}_2 \sin(2\pi f_2 t) + \frac{\hat{Y}_2}{3} \sin(3 \cdot 2\pi f_2 t))\right)^2 \right. \\ \left. + \frac{\varepsilon_1}{2} u_1^2 + \frac{\varepsilon_2}{2} u_2^2 \right\} dt$$

(7.13)

Cost functional (7.13) and equations of motion in state-space form (7.12) are used to calculate the Hamiltonian (5.52). The Hamiltonian is used to formulate the the state- and costate equations (5.53). By using the optimality condition $H_{\mathbf{u}} = 0$ (5.53c), the inputs u_1 and u_2 are calculated. These variables can be formulated as function of the states $\mathbf{x}$ and the costates $\mathbf{p}$. The resulting control input is inserted in the conditions (5.53a) and (5.53b) which yields a two-point BVP. The following

7.1. NONLINEAR OSCILLATOR

boundary conditions are used:

$$\mathbf{x}(t_0) = \begin{bmatrix} 0\,m \\ 0\,m \\ 0.0126\,m/s \\ 0.0942\,m/s \end{bmatrix}, \quad \mathbf{x}(t_f) = \begin{bmatrix} 0\,m \\ 0\,m \\ 0.0126\,m/s \\ -0.0942\,m/s \end{bmatrix} \tag{7.14}$$

The symbolic computations are performed in the software *Mathematica*. They are not explicitly presented because of the lengthly expressions. The BVP is numerically solved by using the *Matlab-solver bvp4c*.

7.1.6 Results

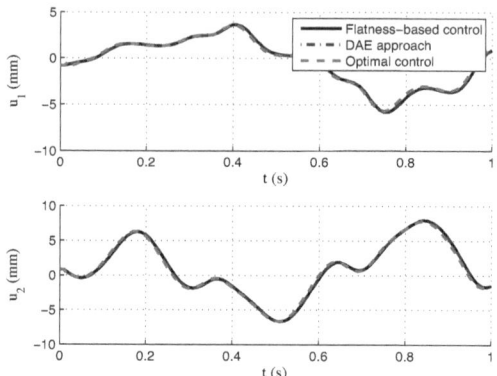

Figure 7.4: Inputs of the nonlinear oscillator

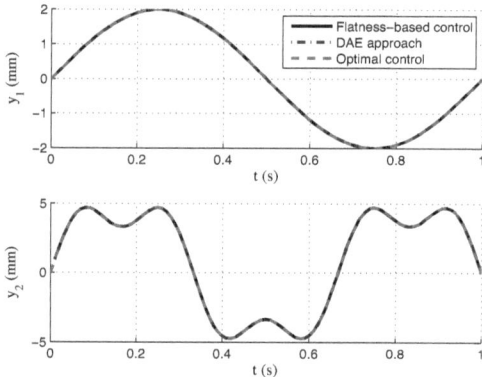

Figure 7.5: Outputs of the nonlinear oscillator

Fig. 7.4 shows the computed inputs of the nonlinear oscillator. The black curves present the flatness-based solution (7.6), which is taken as reference. The blue dotdashed lines show the solution from the DAE approach with control constraints. The red dashed curves present the solution from the direct optimal control algorithm. The solution of the optimal control is smoothed due to the Tikhonov regularization term. Without this regularization the solution would tend to be unsteady, as it was already presented in Fig. 5.8.

7.1. NONLINEAR OSCILLATOR

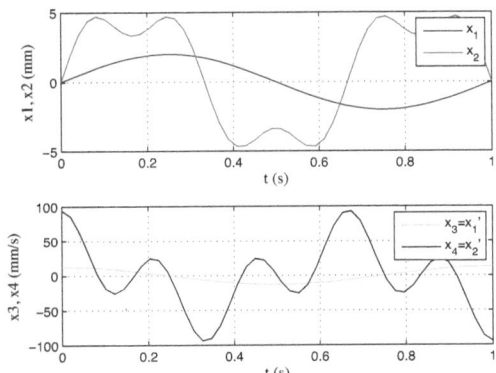

Figure 7.6: State variables of the nonlinear oscillator

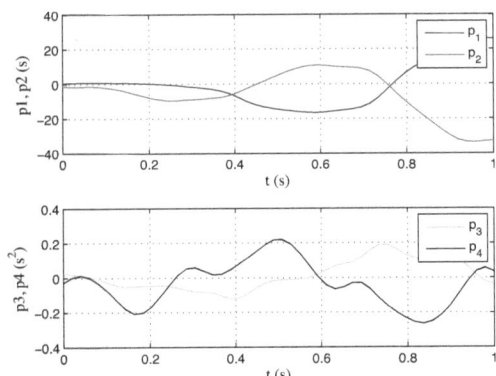

Figure 7.7: Costate variables of the nonlinear oscillator

7.1. NONLINEAR OSCILLATOR

The solutions of the different approaches are taken as inputs in a forward simulation. The resulting outputs are shown in Fig. 7.5. It can be seen that all inputs yield the same outputs which are congruent with the predefined target signals. In Fig. 7.6 and 7.7 the results of the indirect optimal control approach are presented. Fig. 7.6 shows the state variables $\mathbf{x}(t)$, i.e. the positions and velocities of the two masses. In Fig. 7.7 the costate variables $\mathbf{p}(t)$ are presented. The indirect optimal control approach results in the same inputs as the other methods. These results are already illustrated in Fig. 7.4 and therefore they are not presented again. Parts of the solutions are also published in [122, 124, 125].

7.1.7 Discussion

The different methods of (i) DAE approach with control constraints, (ii) flatness-based trajectory control, (iii) direct optimal control and (iv) indirect optimal control result in the same control inputs $\mathbf{u}(t)$, Fig. 7.4. The flatness-based trajectory control can be derived easily from the equations of motion. This results in an analytical solution, which does not require much computational effort. The DAE approach with control constraints can also be applied in a straightforward way. Due to the formulation with generalized coordinates y_1 and y_2, the governing DAEs are of index 3. As a consequence no projection method, i.e. index reduction procedure is required. The solution is numerically efficient and a relatively large step size of $\Delta t = 0.01s$ can be used. The direct optimal control approach requires much more computational effort. A shorter step size of $\Delta t = 0.005s$ is needed due to the optimization task in each subinterval. It was shown that a regularization could smooth the noisy control inputs drastically. However, the weighting factor ε must be chosen carefully. The indirect optimal control approach requires much more symbolic computations to formulate the necessary optimality conditions (5.53)-(5.55). The resulting two-point-BVP can only be solved numerically. The solver *bvp4c* is very sensitive regarding to the initial guesses of the adjoint variables. The initial guesses have a great influence on the convergence speed. All considered methods result in the same control inputs, except of small numerical differences.

7.2 Planar Overhead Crane

7.2.1 Problem Description

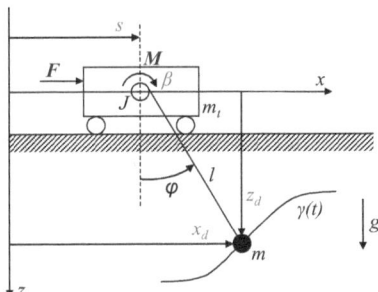

Figure 7.8: Planar overhead crane

Fig. 7.8 shows a planar overhead crane. The trolley with its mass m_t can move along the x-axis. Its absolute position is described by the coordinate $s(t)$. A winch with a moment of inertia J and a radius r is connected to the trolley via a revolute joint at the trolley's center of mass. It is assumed that J and r are constant and do not change due to the spooling of the rope. Furthermore it is assumed that the rope with its length $l(t)$ is massless and longitudinally stiff, i.e. the connection between winch and load is always a straight line. The load m is assumed to be a point mass. The multibody system is excited by two inputs $\mathbf{u} = [F(t), M(t)]^T$. The goal of the trajectory tracking problem is to calculate the input variables in order that the load follows a trajectory given by the Cartesian coordinates $x_d(t)$ and $z_d(t)$.

The equations of motion can either be formulated by using three independent coordinates $\mathbf{q} = [s, l, \varphi]^T$ or by four dependent coordinates $\mathbf{q} = [s, \beta, x, z]^T$. Both formulations result in an underactuated system ($m_c < n$).

This inverse dynamics problem is published e.g. in [21] for minimal coordinates and in [13] for redundant coordinates. In [25] it is shown that an augmented formulation with dependent variables is advantageous for the inverse dynamics problem. A flatness-based control approach of such a planar crane can be found in [95].

For numerical computations the parameters
$m_t = 10\,kg$, $m = 100\,kg$, $J = 0.1\,kgm^2$, $r = 0.1\,m$, $g = 9.81\,m/s^2$ are used.

7.2.2 Equations of Motion

Formulation with generalized (independent) coordinates:
The equations of motion with the minimal coordinates $\mathbf{q} = [s, l, \varphi]^T$ are derived

7.2. PLANAR OVERHEAD CRANE

from Lagrange's equations of the second kind (2.55).
The first step is the calculation of the position and the velocity of the mass m.

$$\mathbf{r_m} = \begin{bmatrix} s + l\sin\varphi \\ l\cos\varphi \end{bmatrix}, \quad \mathbf{v_m} = \begin{bmatrix} \dot{s} + \dot{l}\sin\varphi + l\dot{\varphi}\cos\varphi \\ \dot{l}\cos\varphi - l\dot{\varphi}\sin\varphi \end{bmatrix} \quad (7.15)$$

$$v_m^2 = \dot{s}^2 + \dot{s}\dot{l}\sin\varphi + \dot{s}l\dot{\varphi}\cos\varphi + \dot{s}\dot{l}\sin\varphi + \dot{l}^2\sin^2\varphi + l\dot{l}\dot{\varphi}\sin\varphi\cos\varphi +$$
$$\dot{s}l\dot{\varphi}\cos\varphi + l\dot{l}\dot{\varphi}\sin\varphi\cos\varphi + l^2\dot{\varphi}^2\cos^2\varphi + \dot{l}^2\cos^2\varphi - 2l\dot{l}\dot{\varphi}\sin\varphi\cos\varphi + l^2\dot{\varphi}^2\sin^2\varphi$$

$$v_m^2 = \dot{s}^2 + l^2\dot{\varphi}^2 + \dot{l}^2 + 2\dot{s}\dot{l}\sin\varphi + 2\dot{s}l\dot{\varphi}\cos\varphi \quad (7.16)$$

Thus, kinetic and potential energy can be calculated. Therefore, the rotation angle of the winch β has to be expressed by the cable length l and the winch radius r.

$$T = \frac{1}{2}m_t\dot{s}^2 + \frac{1}{2}m\left(\dot{s}^2 + l^2\dot{\varphi}^2 + \dot{l}^2 + 2\dot{s}\dot{l}\sin\varphi + 2\dot{s}l\dot{\varphi}\cos\varphi\right) + \frac{1}{2}\frac{J}{r^2}\dot{l}^2 \quad (7.17)$$

$$V = -mgl\cos\varphi \quad (7.18)$$

In the next step the partial derivatives with respect to the generalized coordinates s, l, φ have to be calculated.

$$\frac{\partial T}{\partial \dot{s}} = m_t\dot{s} + m\dot{s} + m\dot{l}\sin\varphi + m\dot{\varphi}l\cos\varphi$$

$$\frac{\partial T}{\partial s} = 0$$

$$\frac{d}{dt}\left(\frac{\partial T}{\partial \dot{s}}\right) = m_t\ddot{s} + m\ddot{s} + m\ddot{l}\sin\varphi + m\dot{l}\dot{\varphi}\cos\varphi + m\ddot{\varphi}l\cos\varphi + m\dot{\varphi}\dot{l}\cos\varphi - m\dot{\varphi}^2l\sin\varphi$$

$$\frac{\partial V}{\partial s} = 0$$

$$\frac{\partial T}{\partial \dot{l}} = \frac{J}{r^2}\dot{l} + m\dot{l} + m\dot{s}\sin\varphi$$

$$\frac{\partial T}{\partial l} = ml\dot{\varphi}^2 + m\dot{s}\dot{\varphi}\cos\varphi$$

$$\frac{d}{dt}\left(\frac{\partial T}{\partial \dot{l}}\right) = \frac{J}{r^2}\ddot{l} + m\ddot{l} + m\ddot{s}\sin\varphi + m\dot{s}\dot{\varphi}\cos\varphi$$

$$\frac{\partial V}{\partial l} = -mg\cos\varphi$$

$$\frac{\partial T}{\partial \dot{\varphi}} = ml^2\dot{\varphi} + m\dot{s}l\cos\varphi$$

$$\frac{\partial T}{\partial \varphi} = m\dot{s}\dot{l}\cos\varphi - m\dot{s}\dot{\varphi}l\sin\varphi$$

$$\frac{d}{dt}\left(\frac{\partial T}{\partial \dot{\varphi}}\right) = 2ml\dot{l}\dot{\varphi} + ml^2\ddot{\varphi} + m\ddot{s}l\cos\varphi + m\dot{s}\dot{l}\cos\varphi - m\dot{s}l\dot{\varphi}\sin\varphi$$

$$\frac{\partial V}{\partial \varphi} = mgl\sin\varphi$$

7.2. PLANAR OVERHEAD CRANE

Consequently, Lagrange's equations of the second kind (2.56) can be formulated.

$$\frac{d}{dt}\left(\frac{\partial T}{\partial \dot{q}_i}\right) - \frac{\partial T}{\partial q_i} + \frac{\partial V}{\partial q_i} = Q_i$$

$$\begin{bmatrix} m_t + m & m\sin\varphi & ml\cos\varphi \\ m\sin\varphi & \frac{J}{r^2} + m & 0 \\ ml\cos\varphi & 0 & ml^2 \end{bmatrix} \cdot \begin{bmatrix} \ddot{s} \\ \ddot{l} \\ \ddot{\varphi} \end{bmatrix} + \begin{bmatrix} 2ml\dot{\varphi}\cos\varphi - ml\dot{\varphi}^2\sin\varphi \\ -ml\dot{\varphi}^2 \\ 2ml\dot{l}\dot{\varphi} \end{bmatrix} -$$

$$\begin{bmatrix} 0 \\ mg\cos\varphi \\ -mgl\sin\varphi \end{bmatrix} - \begin{bmatrix} 1 & 0 \\ 0 & \frac{1}{r} \\ 0 & 0 \end{bmatrix} \cdot \begin{bmatrix} F \\ M \end{bmatrix} = \begin{bmatrix} 0 \\ 0 \\ 0 \end{bmatrix}$$

$$\Leftrightarrow \mathbf{M}\dot{\mathbf{v}} - \mathbf{f} + \mathbf{B}^T \mathbf{u} = 0 \tag{7.19}$$

In Eq. (7.19) the vector $\mathbf{f}(\mathbf{q}, \dot{\mathbf{q}}, t)$ is split into a vector, which includes Coriolis and centrifugal forces and a vector, which includes conservative forces that can be derived from the potential energy.

Formulation with redundant (dependent) coordinates:
The equations of motion for a redundant coordinates formulation with $\mathbf{q} = [s, \beta, x, z]^T$ results in the form (4.2b)

$$\begin{bmatrix} m_t & 0 & 0 & 0 \\ 0 & J & 0 & 0 \\ 0 & 0 & m & 0 \\ 0 & 0 & 0 & m \end{bmatrix} \cdot \begin{bmatrix} \ddot{s} \\ \ddot{\beta} \\ \ddot{x} \\ \ddot{z} \end{bmatrix} - \begin{bmatrix} 0 \\ 0 \\ 0 \\ mg \end{bmatrix} + 2\lambda \begin{bmatrix} s - x \\ -r^2\beta \\ x - s \\ z \end{bmatrix} +$$

$$\begin{bmatrix} -1 & 0 \\ 0 & -1 \\ 0 & 0 \\ 0 & 0 \end{bmatrix} \cdot \begin{bmatrix} F \\ M \end{bmatrix} = \begin{bmatrix} 0 \\ 0 \\ 0 \\ 0 \end{bmatrix} \tag{7.20}$$

$$\Leftrightarrow \mathbf{M}\dot{\mathbf{v}} - \mathbf{f} + \mathbf{G}^T \lambda + \mathbf{B}^T \mathbf{u} = 0$$

The redundancy of the coordinates requires the formulation of geometric (holonomic) constraints (4.2d) to link the position of the mass and the rotation of the winch.

$$\mathbf{g}(\mathbf{q}) = (x - s)^2 + z^2 - r^2 \beta^2 = 0 \tag{7.21}$$

By using (7.21), the constraint Jacobian $\mathbf{G} = D\mathbf{g}(\mathbf{q})$ can be calculated.

$$\mathbf{G} = \left[\frac{\partial \mathbf{g}(\mathbf{q})}{\partial s}, \frac{\partial \mathbf{g}(\mathbf{q})}{\partial \beta}, \frac{\partial \mathbf{g}(\mathbf{q})}{\partial x}, \frac{\partial \mathbf{g}(\mathbf{q})}{\partial z}\right] = 2\left[s - x, -r^2\beta, x - s, z\right] \tag{7.22}$$

The constraint Jacobian $\mathbf{G}$ (7.22) with its corresponding Lagrange multiplier λ ($m = 1$) is introduced in (7.20). The geometric constraints can also be formulated

7.2. PLANAR OVERHEAD CRANE

by $\mathbf{g}(\mathbf{q}) = \sqrt{(x-s)^2 + z^2} - r\beta = 0$. However, the constraint Jacobian $\mathbf{G} = D\mathbf{g}(\mathbf{q})$ and therefore the whole algorithm (4.10) would be more complex [25].
The input transformation matrix $\mathbf{B}$ is directly derived from the equations of motion (7.20).

$$\mathbf{B} = \begin{bmatrix} -1 & 0 & 0 & 0 \\ 0 & -1 & 0 & 0 \end{bmatrix} \quad (7.23)$$

Comparison of the Different Formulations:
If the equations of motion for the independent and the dependent variable formulation are compared, some remarkable differences can be seen. The mass matrix in Eq. (7.19) is nearly fully occupied and not constant. However, it is symmetric and positive definite. The mass matrix in the formulation with redundant coordinates (7.20) is a constant diagonal matrix. This is very advantageous for the numerical treatment. Furthermore, the inconvenient Coriolis and centrifugal terms that appear in (7.19) do not exist in (7.20).

The DAE approach with control constraints and the flatness-based trajectory tracking are applied to the formulation with redundant coordinates. This results from the advantages that are described in section 4.2. However, the DAE approach with control constraints can also be applied to the system that is formulated by independent coordinates and can be found e.g. in [21].

7.2.3 DAE Approach with Control Constraints

Control constraints are formulated in the form (4.2c). The aim of the planar overhead crane is that the load follows a predefined trajectory. Therefore, the positional coordinates of the load $[x, z]^T$ must be identical to desired coordinates $[x_d, z_d]^T$. With $\tau = t/(t_f - t_0)$ the control constraints read as

$$\mathbf{c}(\mathbf{q}, t) = \mathbf{\Phi}(\mathbf{q}) - \boldsymbol{\gamma}(t) = \mathbf{0} :$$

$$= \begin{bmatrix} x_d \\ z_d \end{bmatrix} - \left\{ \begin{bmatrix} 0 \\ 4 \end{bmatrix} + \begin{bmatrix} 5 \\ -3 \end{bmatrix} \cdot (70\tau^9 - 315\tau^8 + 540\tau^7 - 420\tau^6 + 126\tau^5) \right\} \quad (7.24)$$

Eq. (7.24) describes a trajectory of the mass, which is a straight line from the initial coordinates $[x_d, z_d] = [0, 4]$ at time $t_0 = 0s$ to the final coordinates $[x_d, z_d] = [5, 1]$ at time $t_f = 3s$. The polynomial in (7.24) is sufficiently smooth so that it can be continuously differentiated up to 4th order. In the DAE approach with control constraints the targets have to be differentiated only twice, Eq. (4.6). However, the inverse computation is compared with the flatness-based trajectory tracking control which needs a sufficiently smooth trajectory. The desired outputs $\boldsymbol{\gamma}(t)$ are taken from [13, 21]. In the next section 7.2.4 the derivation of this trajectory will be explained.
For the index reduction procedure the projection matrices $\mathbf{C}$ and $\mathbf{D}$ have to be determined. The matrix $\mathbf{C}$ is defined in (4.5) and results in

$$\mathbf{C} = D\mathbf{\Phi}(\mathbf{q}) = \begin{bmatrix} 0 & 0 & 1 & 0 \\ 0 & 0 & 0 & 1 \end{bmatrix} \quad (7.25)$$

7.2. PLANAR OVERHEAD CRANE

The projection matrix $\mathbf{D}$ has to be determined in a way that the complementarity condition (4.8) $\mathbf{CD} = \mathbf{0}$ is fulfilled. Hence, it results in

$$\mathbf{D} = \begin{bmatrix} 1 & 0 \\ 0 & 1 \\ 0 & 0 \\ 0 & 0 \end{bmatrix} \qquad (7.26)$$

It can be seen that the projection matrices (7.25) and (7.26) are simple sparse matrices of Boolean type, which is beneficial for the numerical computation.
The control constraint realization in this example is tangent which can be shown by

$$p = \operatorname{rank}(\mathbf{CM}^{-1}\mathbf{B}^T) = 0 \qquad (7.27)$$

Hence, the third row in Table 4.1 is fulfilled.
By using $\mathbf{C}$ and $\mathbf{D}$ the projection method (4.10) can be applied to the system. The implicit Euler algorithm with a step size of $\Delta t = 0.1\,s$ is used to solve the resulting equations.

7.2.4 Flatness-Based Trajectory Tracking

For a flatness-based parametrization of the state variables and the control inputs the differentially flat outputs of the system have to be identified. In [25, 26] it is shown that the load coordinates $\mathbf{r} = [x_d, y_d, z_d]^T$ of a crane represent flat outputs. This statement holds for planar as well as for different three-dimensional cranes [1, 22, 23, 26, 57, 58, 82, 90, 95].
By considering the system (4.10), all relevant variables can be parameterized by the flat outputs. Such a parametrization can be found e.g. in [13]. Eq. (4.10c) results in

$$\begin{aligned} 2(x_d - s)\lambda + m\ddot{x}_d &= 0 \\ 2z_d\lambda + m(\ddot{z}_d - g) &= 0 \end{aligned} \qquad (7.28)$$

From these equations the variables $\lambda(t)$ and $s(t)$ can be calculated.

$$\begin{aligned} \lambda &= \frac{m}{2z_d}(g - \ddot{z}_d) \\ s &= x_d + \frac{z_d \ddot{x}_d}{g - \ddot{z}_d} \end{aligned} \qquad (7.29)$$

In (7.29) it can be seen that derivatives of the flat outputs up to second order are needed to parameterize $\lambda(t)$ and $s(t)$. The state variable $\beta(t)$ can be calculated from the geometric constraints (4.10e), (7.21).

$$\beta = \pm \frac{z_d}{r(g - \ddot{z}_d)} \sqrt{\ddot{x}_d^2 + (g - \ddot{z}_d)^2} \qquad (7.30)$$

Finally, the control inputs $\mathbf{u}(t)$ can be derived from (4.10b). By considering that $\mathbf{D}^T\mathbf{B}^T = -\mathbf{I}_{2\times 2}$ (the negative (2×2) identity matrix), $\mathbf{u}$ can be calculated by

$$\mathbf{u} = \mathbf{D}^T\left\{\mathbf{M}\dot{\mathbf{v}} - \mathbf{f} + \mathbf{G}^T\lambda\right\} \qquad (7.31)$$

$$F = m_t \ddot{s} + 2(s - x_d)\lambda \tag{7.32a}$$
$$M = J\ddot{\beta} - 2r^2\beta\lambda \tag{7.32b}$$

This feedforward control law can furthermore be extended by a linear feedback control in order to compensate disturbances. In the flatness-based control (7.32) it can be seen that derivatives of the flat outputs $x_d(t)$, $z_d(t)$ up to fourth order are needed to parameterize the control inputs $F(t)$, $M(t)$.

$$F = F(x_d, \dot{x}_d, \ldots, x_d^{(4)}, z_d, \dot{z}_d, \ldots, z_d^{(4)}) \tag{7.33a}$$
$$M = M(x_d, \dot{x}_d, \ldots, x_d^{(4)}, z_d, \dot{z}_d, \ldots, z_d^{(4)}) \tag{7.33b}$$

Consequently, the index of the DAEs can be verified by using the definition (6.58).

$$i = \beta + 1 = 5 \tag{7.34}$$

In this way it can be shown that the DAEs with control constraints (7.20), (7.21), (7.24) are of index 5.

Due to (7.33), trajectories have to be defined for the load coordinates, which are continuously differentiable up to fourth order. Hence, five initial conditions and five final conditions have to be fulfilled. This results in a polynomial function of 9th order for the flat outputs $x_d(t)$ and $z_d(t)$. Therefore, the general polynomial function

$$s(t) = s_0 + (s_f - s_0) \sum_{i=\beta+1}^{2\beta+1} a_i \tau^i, \quad \tau = \frac{t}{t_f - t_0}, \ t \in [t_0, t_f] \tag{7.35}$$

can be used, where β denotes the highest number of time derivative. The following boundary conditions have to be fulfilled:

$$\begin{array}{ll} s(t_0) = s_0 & s(t_f) = s_f \\ \dot{s}(t_0) = 0 & \dot{s}(t_f) = 0 \\ \ddot{s}(t_0) = 0 & \ddot{s}(t_f) = 0 \\ s^{(3)}(t_0) = 0 & s^{(3)}(t_f) = 0 \\ s^{(4)}(t_0) = 0 & s^{(4)}(t_f) = 0 \end{array} \tag{7.36}$$

The coefficients a_i of the function (7.35) can be calculated by considering the boundary condition (7.36).

$$a_5 = 126, \ a_6 = -420, \ a_7 = 540, \ a_8 = -315, \ a_9 = 70 \tag{7.37}$$

Hence, the polynomial function (7.35) results in the previously presented trajectory (7.24).

The equations of motion (7.19) or (7.20) can also be written in the form of an affine input system (6.37). The resulting system is not explicitly written down here because of the lengthly expressions. If the Lie-derivatives are computed, the definitions (6.38) and (6.51) can be verified. The (vector) relative degree results in

7.2. PLANAR OVERHEAD CRANE

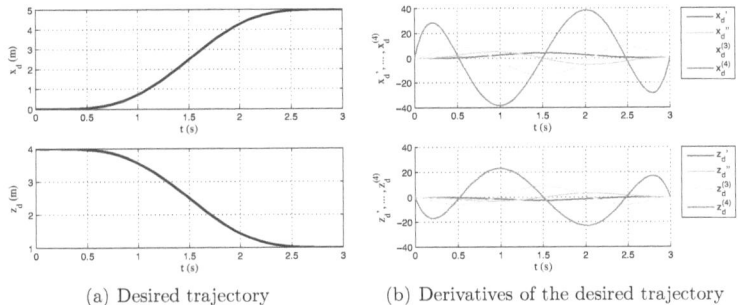

(a) Desired trajectory (b) Derivatives of the desired trajectory

Figure 7.9: Desired trajectory with time derivatives

$$\{r_1, r_2\} = \{4, 2\}, \quad r = r_1 + r_2 = 6 = 2n \tag{7.38}$$

if $2n$ denotes the number of state variables corresponding to minimal coordinates. In the formulation with independent coordinates the conditions (6.51) are fulfilled as long as $\varphi \neq 90°$. The decoupling matrix (6.38b) yields

$$\mathbf{A}(\mathbf{x}) = \begin{bmatrix} 1 & \frac{\sin(\varphi)}{r} \\ 0 & \frac{\cos(\varphi)}{r} \end{bmatrix} \tag{7.39}$$

Hence, the system is exactly input-state linearizable. However, the redundant coordinates formulation results in a singular decoupling matrix and therefore the system is not input-state linearizable. As already stated in section 6.2, a flatness-based parameterization can even be found if the MIMO-system is not input-state linearizable. This is shown in the parameterization (7.32), which is based on a redundant coordinates formulation.

7.2.5 Optimal Control

The direct optimal control algorithm from section 5.5.2 is applied to the planar overhead crane. To simplify the numerical integration in *Matlab* the formulation with independent coordinates (7.19) is used. As a consequence, the standard solver *ode45* can be applied. Numerical integration is required in each iteration, i.e. several times in each time interval. A step size of $\Delta t = 0.005\,s$ is used. The Tikhonov regularization term is weighted with a factor of $\varepsilon = 5 \cdot 10^{-5}$. Contour plot and gradient field within one time interval are shown in Fig. 7.10.
Additionally, the indirect optimal control algorithm from section 5.4.4 is applied to the system, which is formulated with independent coordinates (7.19). The equations of motion are rewritten in state-space form, where the states $x_1 = s$, $x_2 = l$, $x_3 =$

7.2. PLANAR OVERHEAD CRANE

φ, $x_4 = \dot{s}$, $x_5 = \dot{l}$, $x_6 = \dot{\varphi}$ are introduced.

$$\begin{bmatrix} 1 & 0 & 0 & 0 & 0 & 0 \\ 0 & 1 & 0 & 0 & 0 & 0 \\ 0 & 0 & 1 & 0 & 0 & 0 \\ 0 & 0 & 0 & m_t + m & m \sin x_3 & m x_2 \cos x_3 \\ 0 & 0 & 0 & m \sin x_3 & \frac{J}{r^2} + m & 0 \\ 0 & 0 & 0 & m x_2 \cos x_3 & 0 & m x_2^2 \end{bmatrix} \begin{bmatrix} \dot{x}_1 \\ \dot{x}_2 \\ \dot{x}_3 \\ \dot{x}_4 \\ \dot{x}_5 \\ \dot{x}_6 \end{bmatrix} +$$

$$\begin{bmatrix} 0 \\ 0 \\ 0 \\ 2 m x_5 x_6 \cos x_3 - m x_6^2 x_2 \sin x_3 \\ -m x_2 x_6^2 \\ 2 m x_2 x_5 x_6 \end{bmatrix} = \begin{bmatrix} x_4 \\ x_5 \\ x_6 \\ F(t) \\ \frac{M(t)}{r} + mg \cos x_3 \\ -mg x_2 \sin x_3 \end{bmatrix} \qquad (7.40)$$

The cost functional is given by

$$\begin{aligned} J(\mathbf{u}) &= \int_{t_0}^{t_f} \left\{ \frac{1}{2}(s + l \sin \varphi - x_d)^2 + \frac{1}{2}(l \cos \varphi - z_d)^2 + \frac{\varepsilon_1}{2} F^2 + \frac{\varepsilon_2}{2} M^2 \right\} dt \\ &= \int_{t_0}^{t_f} \left\{ \frac{1}{2}(x_1 + x_2 \sin x_3 - x_d)^2 + \frac{1}{2}(x_2 \cos x_3 - z_d)^2 + \frac{\varepsilon_1}{2} F^2 + \frac{\varepsilon_2}{2} M^2 \right\} dt \end{aligned}$$
(7.41)

By using (7.40) and (7.41) the Hamiltonian is derived and the necessary optimality

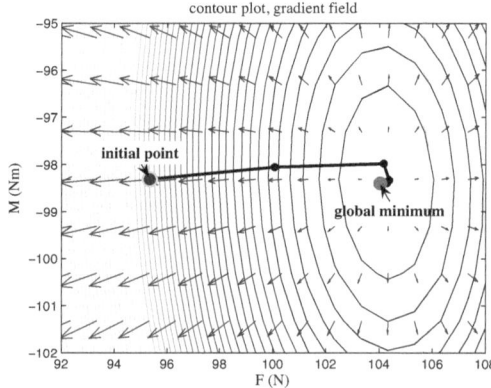

Figure 7.10: Contour plot and gradient field within one time interval

conditions (5.53), (5.54), (5.55) are formulated. Furthermore, the two-point BVP is derived in *Mathematica*.

The gradient method (Kelley-Bryson-method) from Table 5.1 from section 5.5.1

7.2. PLANAR OVERHEAD CRANE

is also applied to the planar overhead crane. If the ODE-system (7.19) is used in the procedure, the integral part of the cost functional reads as

$$\mathcal{L}(\mathbf{x},t) = \frac{1}{2}(s + l \sin\varphi - x_d)^2 + \frac{1}{2}(l \cos\varphi - z_d)^2 + \\ + \frac{\chi}{2}(\dot{s} + \dot{l}\sin\varphi + l\dot{\varphi}\cos\varphi - \dot{x}_d)^2 + \frac{\chi}{2}(\dot{l}\cos\varphi - l\dot{\varphi}\sin\varphi - \dot{z}_d)^2 \qquad (7.42)$$

Furthermore, the Mayer-term at the final time t_f is considered.

$$\Phi(\mathbf{x}_f, t_f) = \alpha\,\mathcal{L}(\mathbf{x},t)|_{t=t_f} \qquad (7.43)$$

As a consequence, the cost functional can be written in the form of Eq. (5.36a). Hence, the velocity deviations from the targets are considered as well. If the states $x_1 = s$, $x_2 = l$, $x_3 = \varphi$, $x_4 = \dot{s}$, $x_5 = \dot{l}$, $x_6 = \dot{\varphi}$ and the inputs $u_1 = F$, $u_2 = M/r$ are introduced, the Hamiltonian (5.52) can be formulated by

$$\begin{aligned}H &= \frac{1}{2}[(x_1 + x_2 \sin x_3 - x_d)^2 + (x_2 \cos x_3 - z_d)^2] \\ &+ \frac{\chi}{2}[(x_4 + x_5 \sin x_3 + x_2 x_6 \cos x_3 - \dot{x}_d)^2 + (x_5 \cos x_3 - x_2 x_6 \sin x_3 - \dot{z}_d)^2] \\ &+ p_1 x_4 + p_2 x_5 + p_3 x_6 + p_4 \left(\frac{u_1 + u_2 \sin x_3}{m_t}\right) \\ &+ p_5 \left(x_2 x_6^2 + g \cos x_3 + \frac{m u_2 \cos^2 x_3 - (m_t + m) u_2 - m u_1 \sin x_3}{m_t m}\right) \\ &- p_6 \left(\frac{m_t(2 x_5 x_6 + g \sin x_3) + (u_1 + u_2 \sin x_3)\cos x_3}{m_t x_2}\right)\end{aligned}$$

$$(7.44)$$

It should be mentioned, that the moment of inertia J is neglected in this formulation. As initial guess for the inputs the static solution $s = 0$, $l = l_0$, $\varphi = 0$

$$u_1(t) = F(t) = 0, \quad u_2(t) = \frac{M(t)}{r} = mg$$

is chosen [143]. This assumption is far away from the real drive signals. The weighting factors $\alpha = 0.1$, $\chi = 5$ and $\chi = 0$ are used in the cost functional.

The gradient method can also be applied to the DAE system (7.20). Therefore, the index 2 Gear-Gupta-Leimkuhler (5.105) DAEs have to be computed. By using the notations $q_1 = s$, $q_2 = \beta$, $q_3 = x$, $q_4 = z$ for the degrees of freedom, the

7.2. PLANAR OVERHEAD CRANE

GGL-system reads as

$$m_t \dot{q}_1 = m_t v_1 - 2\nu(q_1 - q_3) \quad (7.45\text{a})$$
$$J\dot{q}_2 = Jv_2 + 2\nu r^2 q_2 \quad (7.45\text{b})$$
$$m\dot{q}_3 = mv_3 - 2\nu(q_3 - q_1) \quad (7.45\text{c})$$
$$m\dot{q}_4 = mv_4 - 2\nu q_4 \quad (7.45\text{d})$$
$$m_t \dot{v}_1 = F - 2\lambda(q_1 - q_3) \quad (7.45\text{e})$$
$$J\dot{v}_2 = M + 2\lambda r^2 q_2 \quad (7.45\text{f})$$
$$m\dot{v}_3 = -2\lambda(q_3 - q_1) \quad (7.45\text{g})$$
$$m\dot{v}_4 = mg - 2\lambda q_4 \quad (7.45\text{h})$$
$$0 = (q_3 - q_1)^2 - r^2 q_2^2 + q_4^2 \quad (7.45\text{i})$$
$$0 = 2(q_3 - q_1)(v_3 - v_1) - 2r^2 q_2 v_2 + 2q_4 v_4 \quad (7.45\text{j})$$

This DAE-system can be integrated forwards in time by using an appropriate index 2 - solver. As a result $\mathbf{q}(t)$, $\mathbf{v}(t)$, $\lambda(t)$ and $\nu(t)$ are obtained.

The goal of the inverse problem is that the mass m follows the desired trajectory (7.24). For that reason the integral part of the cost functional (5.75) is formulated as

$$\mathcal{L}(\mathbf{x}, t) = \frac{1}{2}(q_3 - x_d)^2 + \frac{1}{2}(q_4 - z_d)^2 + \frac{\chi}{2}(v_3 - \dot{x}_d)^2 + \frac{\chi}{2}(v_4 - \dot{z}_d)^2 \quad (7.46)$$

This results in the gradients of $\mathcal{L}$

$$\mathcal{L}_{\mathbf{q}} = \begin{bmatrix} 0 \\ 0 \\ q_3 - x_d \\ q_4 - z_d \end{bmatrix}, \quad \mathcal{L}_{\mathbf{v}} = \begin{bmatrix} 0 \\ 0 \\ \chi(v_3 - \dot{x}_d) \\ \chi(v_4 - \dot{z}_d) \end{bmatrix} \quad (7.47)$$

To derive the adjoint equations (5.115), the matrices $\mathbf{R} = D_{\mathbf{q}}(\mathbf{G}(\mathbf{q})\mathbf{v})$ and $\mathcal{V}(\mathbf{q}, \mathbf{v}, \dot{\mathbf{v}}) = \mathbf{f} - \mathbf{G}^T \boldsymbol{\lambda} - \mathbf{M}\dot{\mathbf{v}}$ are needed.

$$\mathbf{R} = 2 \begin{bmatrix} v_1 - v_3, & -r^2 v_2, & v_3 - v_1, & v_4 \end{bmatrix} \quad (7.48)$$

$$\mathcal{V} = \begin{bmatrix} F - 2\lambda(q_1 - q_3) - m_t \dot{v}_1 \\ M + 2\lambda r^2 q_2 - J\dot{v}_2 \\ -2\lambda(q_3 - q_1) - m\dot{v}_3 \\ mg - 2\lambda q_4 - m\dot{v}_4 \end{bmatrix} \quad (7.49)$$

The Jacobian matrices of $\mathcal{V}$ yield

$$\mathcal{V}_{\mathbf{q}} = -2\lambda \begin{bmatrix} 1 & 0 & -1 & 0 \\ 0 & -1 & 0 & 0 \\ -1 & 0 & 1 & 0 \\ 0 & 0 & 0 & 1 \end{bmatrix}, \quad \mathcal{V}_{\mathbf{v}} = \mathbf{0} \quad (7.50)$$

Hence, the adjoint equations (5.115) can be formulated.

$$\dot{p}_1 = \frac{2\lambda}{m_t}w_1 - \frac{2\lambda}{m}w_3 - 2\mu(q_1 - q_3) - 2\xi(v_1 - v_3) \tag{7.51a}$$

$$\dot{p}_2 = -\frac{2\lambda}{J}w_2 + 2\mu q_2 + 2\xi v_2 \tag{7.51b}$$

$$\dot{p}_3 = \frac{2\lambda}{m}w_3 - \frac{2\lambda}{m_t}w_1 - 2\mu(q_3 - q_1) - 2\xi(v_3 - v_1) - (q_3 - x_d) \tag{7.51c}$$

$$\dot{p}_4 = \frac{2\lambda}{m}w_4 - 2\mu q_4 - 2\xi v_4 - (q_4 - z_d) \tag{7.51d}$$

$$\dot{w}_1 = -p_1 - 2\xi(q_1 - q_3) \tag{7.51e}$$

$$\dot{w}_2 = -p_2 + 2\xi q_2 \tag{7.51f}$$

$$\dot{w}_3 = -p_3 - 2\xi(q_3 - q_1) - \chi(v_3 - \dot{x}_d) \tag{7.51g}$$

$$\dot{w}_4 = -p_4 - 2\xi q_4 - \chi(v_4 - \dot{z}_d) \tag{7.51h}$$

$$0 = 2\frac{q_1 - q_3}{m_t}w_1 - 2\frac{q_2}{J}w_2 + 2\frac{q_3 - q_1}{m}w_3 + 2\frac{q_4}{m}w_4 \tag{7.51i}$$

$$0 = 2\frac{q_1 - q_3}{m_t}p_1 - 2\frac{q_2}{J}p_2 + 2\frac{q_3 - q_1}{m}p_3 + 2\frac{q_4}{m}p_4 \tag{7.51j}$$

This is also an index 2 DAE-system which can be solved for $\mathbf{p}(t)$, $\mathbf{w}(t)$, $\mu(t)$ and $\xi(t)$, once $\mathbf{q}(t)$, $\mathbf{v}(t)$, $\lambda(t)$ and $\nu(t)$ has been computed from a forward simulation of the system (7.45). The boundary conditions (5.118) for the adjoint equations are not written down here in detail because of the lengthly expressions. However, it should be remarked that the (1×1) matrix

$$\mathbf{G}\mathbf{M}^{-1}\mathbf{G}^T = 4\frac{(q_1 - q_3)^2}{m_t} + 4\frac{q_2^2}{J} + 4\frac{(q_1 - q_3)^2}{m} + 4\frac{q_4^2}{m}$$

is non-singular unless $l = 0$ and can therefore be inverted. The control update (5.116) follows from the solution of the adjoint equations (7.51) $\delta\mathbf{u} = -\kappa\mathbf{f}_\mathbf{u}^T\mathbf{w}$. Since

$$\mathbf{f}_\mathbf{u}^T = \begin{bmatrix} 1 & 0 & 0 & 0 \\ 0 & 1 & 0 & 0 \end{bmatrix}$$

the update reads as

$$\delta u_1 = \delta F = -\kappa w_1 \tag{7.52a}$$

$$\delta u_2 = \delta M = -\kappa w_2 \tag{7.52b}$$

7.2.6 Results

In Fig. 7.11 the inversely calculated inputs are presented. The analytical solution of the flatness-based trajectory tracking is presented by the black curves and serves as reference. The dot-dashed blue lines are the solutions of the DAE approach with control constraints. The red curves illustrate the smoothed solution of the optimal control algorithm. As already mentioned, the optimal control procedure is applied to the system with independent coordinates and hence the Lagrange multiplier does not exit in this computation.

In the curve of the winch torque M it can be seen that it starts and ends at $-98.1\,Nm$, which is the required static torque $M_s = -mgr$ to hold the load. The initial and final trolley force F is equal to zero due to the desired trajectory where initial and final velocity as well as acceleration are equal to zero, Fig. 7.9(b). The

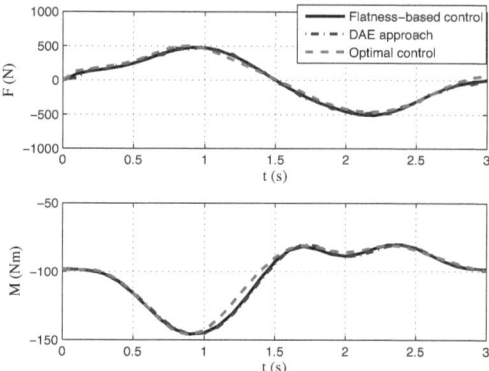

Figure 7.11: Inputs of the planar overhead crane

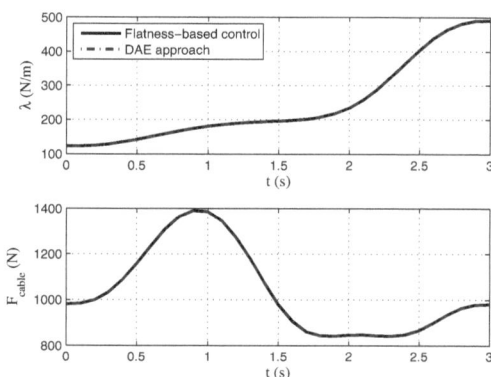

Figure 7.12: Lagrange multiplier and cable tension force of the planar overhead crane

Lagrange multiplier $\lambda(t)$ is presented in Fig. 7.12. It should be mentioned that $\lambda(t)$ has the physical unit N/m due to the formulation of the geometric constraints (7.21). If the constraints would be formulated as $\mathbf{g}(\mathbf{q}) = \sqrt{(x-s)^2 + z^2} - r\beta = 0$,

7.2. PLANAR OVERHEAD CRANE

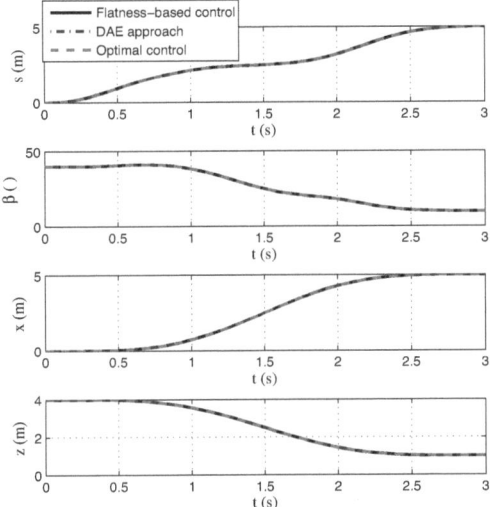

Figure 7.13: Outputs, redundant coordinates of the planar overhead crane

the Lagrange multiplier $\lambda(t)$ would be the constraint force in the rope. However, the constraint Jacobian and therefore the whole algorithm (4.10) would be more complex [25].

Nevertheless, the cable tension force can be computed from the Lagrange multiplier by $F_{cable} = 2l\lambda = 2r\beta\lambda$. This force is plotted in Fig. 7.12. It can be seen that the cable tension force is equal to $F_{cable,s} = mg = 981\,N$ at the initial time and the final time.

If the differently computed inputs from Fig. 7.11 are used in a forward dynamics simulation, the independent and furthermore the redundant coordinates in Fig. 7.13 are obtained. It can be seen that all input variables yield exactly in the desired targets x_d and z_d.

The results of the gradient method (Kelley-Bryson-method) from Table 5.1 are shown in Fig. 7.14. The formulation (7.45) - (7.51) is not used because of the numerical integration of the index 2 system. Rather the formulation based on the ODEs (7.19) is used in combination with the cost functional (7.42)-(7.43). In addition, the state and costate equations (7.45) and (7.51) are formulated as index 1 equations. The resulting system is solved by the DAE solver *ode15s*, which is able to integrate index 1 systems. ODE-formulation and DAE-formulation in conjunction with the Kelley-Bryson method yield exactly the same results.

Fig. 7.14 shows the calculated inputs from the Kelley-Bryson-method after 200 iterations. The weighting factors $\alpha = 0.1$ and $\chi = 5$ are used in the cost functional. Slight differences between Fig. 7.11 and Fig. 7.14 can be seen. This results from the fact that the moment of inertia J is neglected in the Kelley-Bryson-method.

7.2. PLANAR OVERHEAD CRANE

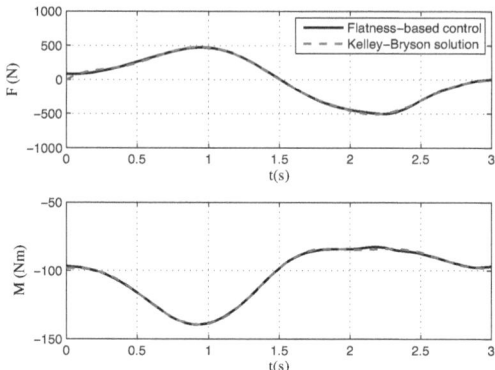

Figure 7.14: Inputs of the planar overhead crane computed by the Kelley-Bryson method

Fig. 7.15 presents the convergence characteristics of the Kelley-Bryson-method for the planar overhead crane. The motion of the planar overhead crane, which results

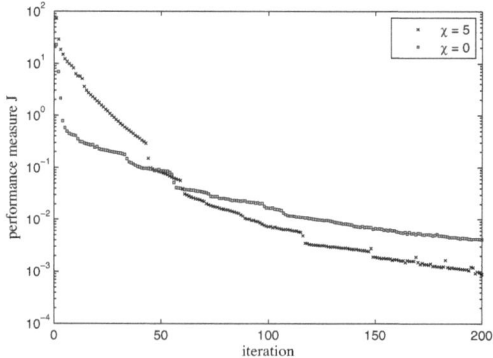

Figure 7.15: Convergence of the performance measure for $\chi = 5$ and $\chi = 0$

from the computed inputs, is captured by the snapshots presented in Fig. 7.16. The animation is performed in *Adams*.

7.2.7 Discussion

The formulation of the equations of motion is done for independent coordinates (7.19) as well as for dependent coordinates (7.20). The formulation with dependent coordinates, which results in an index 5 DAE-system, is shown as beneficial regarding to the inverse dynamics problem. The projection matrices, which reduce the DAE-system to an index 3 system, can be found in a straightforward way. The

7.2. PLANAR OVERHEAD CRANE

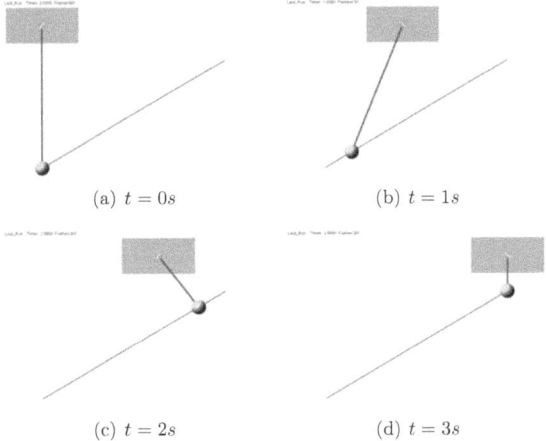

Figure 7.16: Snapshots of the motion of the planar overhead crane

projected DAEs are also the basis for the flatness-based trajectory control. Hence, an analytical solution can be found for the inputs $F(t)$ and $M(t)$, which are used as reference for the other numerical methods. The DAE approach with control constraints is quite efficient in this example. A large step size of $\Delta t = 0.1s$ could be used. The direct optimal control procedure requires a much shorter step size of $\Delta t = 0.005s$. The Tikhonov regularization term could smooth the inversely calculated inputs. However, this approach is seen as numerically inefficient. In contrast to that, the gradient method (Kelley-Bryson-method) is shown as the appropriate method for this specific inverse problem. After 200 iterations the outputs, which result from a subsequent forward simulation, are in a very good agreement with the desired targets. The weighting factor χ has a great influence in the computed control inputs. This weighting factor considers the velocity error. If $\chi = 0$, only the positional error is included in the cost functional, cf. Eq. (7.42). But due to the inertia of the multibody system, the position error may already be very small, even if the control inputs differ from the analytical solution [143]. Furthermore, the convergence speed can be improved by including the velocity error, Fig. 7.15. For $\chi = 5$ the drive signals coincide with the solution from the flatness-based parameterization. Details are also published in [122, 121, 143] as well as in previous work from other authors [13, 21].

7.3 3D Rotary Crane

7.3.1 Problem Description

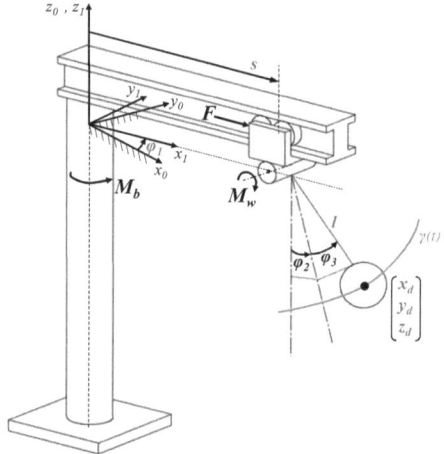

Figure 7.17: Rotary crane

In this section a three-dimensional rotary crane is considered, Fig. 7.17. The tower with its cantilever bridge can be rotated along the z_0-axis of the inertial frame. This rotation is described by the angle φ_1. The moment of inertia J_b is given in the rotation axis. The trolley with its mass m_t can be moved along the bridge and its position is given by s. The trolley carries a winch with a radius r_w and a moment of inertia J_w. The winch is connected to the load via a massless, longitudinally stiff cable with the length l. It is assumed that the radius r_w is negligible to the length l. The load is modeled as mass point with its mass m. Cable and mass can rotate free in space, which can be described e.g. by the rotation angles φ_2 and φ_3 or the Cartesian coordinates x, y, z. Similar examples of such a crane can also be found e.g. in [13, 26, 52]. A DAE-approach, which is based on a rotationless formulation, is presented in [13]. The equations of motion, which are related to redundant coordinates, are presented in [26]. Furthermore, a slightly different DAE-approach is presented in this work. A flatness-based control of a similar rotating tower crane is presented in [52]. However, independent coordinates are used, which leads to a more complicated derivation of the control law than that which is presented in section 7.3.4.

The goal of this example is to move the load along a trajectory in order to avoid an obstacle as illustrated in Fig. 7.18. The system can either be modeled by

7.3. 3D ROTARY CRANE

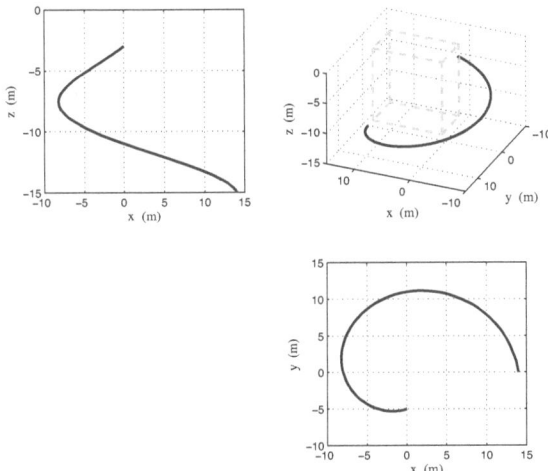

Figure 7.18: Desired trajectory of the load of the 3D rotary crane

five independent coordinates $\mathbf{q} = [\varphi_1, s, l, \varphi_2, \varphi_3]^T$ or by six dependent coordinates $\mathbf{q} = [\varphi_1, s, l, x, y, z]^T$. The three input variables are the torque acting at the tower, the force acting at the trolley and the torque acting at the winch $\mathbf{u} = [M_b, F, M_w]^T$. Hence, the rotary crane is an underactuated multibody system. The following numerical values are used: $J_b = 30000 kgm^2$, $m_t = 50kg$, $m = 500kg$, $J_w = 0.1 kgm^2$, $r_w = 0.1m$.

7.3.2 Equations of Motion

Formulation with generalized (independent) coordinates:

The equations of motion with the generalized coordinates $\mathbf{q} = [\varphi_1, s, l, \varphi_2, \varphi_3]^T$ are derived from Lagrange's equations of the second kind (2.55). Therefore, the coordinate systems as depicted in Fig. 7.19 are introduced. Then the rotation matrices are derived.

$$\mathbf{A}_{10} = \begin{bmatrix} \cos\varphi_1 & -\sin\varphi_1 & 0 \\ \sin\varphi_1 & \cos\varphi_1 & 0 \\ 0 & 0 & 1 \end{bmatrix} \quad (7.53)$$

$$\mathbf{A}_{21} = \begin{bmatrix} \cos\varphi_2 & 0 & -\sin\varphi_2 \\ 0 & 1 & 0 \\ \sin\varphi_2 & 0 & \cos\varphi_2 \end{bmatrix} \quad (7.54)$$

$$\mathbf{A}_{32} = \begin{bmatrix} 1 & 0 & 0 \\ 0 & \cos\varphi_3 & -\sin\varphi_3 \\ 0 & \sin\varphi_3 & \cos\varphi_3 \end{bmatrix} \quad (7.55)$$

7.3. 3D ROTARY CRANE

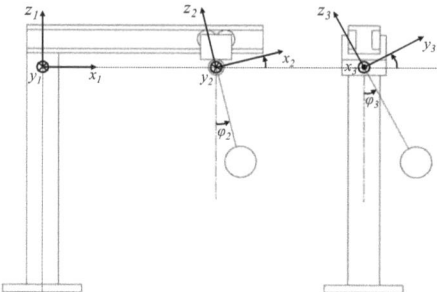

Figure 7.19: Coordinate systems of the rotary crane

Hence, the positions of the trolley $\mathbf{r}_{t,0}$ and the load $\mathbf{r}_{l,0}$ with respect to the inertial frame can be calculated.

$$\mathbf{r}_{t,0} = \mathbf{A}_{10}\mathbf{r}_{t,1} = \begin{bmatrix} s\cos\varphi_1 \\ s\sin\varphi_1 \\ 0 \end{bmatrix} \quad (7.56)$$

$$\mathbf{r}_{l,0} = \mathbf{A}_{10}\mathbf{r}_{t,1} + \mathbf{A}_{10}\mathbf{A}_{21}\mathbf{A}_{32}\mathbf{r}_{l,3} = \begin{bmatrix} s\cos\varphi_1 - l\left(\sin\varphi_1\sin\varphi_3 - \cos\varphi_1\sin\varphi_2\cos\varphi_3\right) \\ s\sin\varphi_1 - l\left(-\cos\varphi_1\sin\varphi_3 - \sin\varphi_1\sin\varphi_2\cos\varphi_3\right) \\ -l\cos\varphi_2\cos\varphi_3 \end{bmatrix} \quad (7.57)$$

Hence, the cable length with respect to the inertial frame can be calculated as difference between trolley and load position.

$$\mathbf{r}_{\text{diff},0} = \mathbf{r}_{t,0} - \mathbf{r}_{l,0} = \begin{bmatrix} l\left(\sin\varphi_1\sin\varphi_3 - \cos\varphi_1\sin\varphi_2\cos\varphi_3\right) \\ l\left(-\cos\varphi_1\sin\varphi_3 - \sin\varphi_1\sin\varphi_2\cos\varphi_3\right) \\ l\cos\varphi_2\cos\varphi_3 \end{bmatrix} \quad (7.58)$$

Consequently, the translational velocities can be calculated. The skew-symmetric matrices of the rotational velocities can be calculated by $\tilde{\mathbf{\Omega}} = \mathbf{A}^T\dot{\mathbf{A}}$, Eq. (2.20). As a result the velocity vectors read as

$$\boldsymbol{\omega}_{10} = \begin{bmatrix} 0 \\ 0 \\ \dot{\varphi}_1 \end{bmatrix} \quad (7.59)$$

$$\boldsymbol{\omega}_{20} = \boldsymbol{\omega}_{10} + \mathbf{A}_{10}\boldsymbol{\Omega}_{21} = \begin{bmatrix} \dot{\varphi}_2\sin\varphi_1 \\ -\dot{\varphi}_2\cos\varphi_1 \\ \dot{\varphi}_1 \end{bmatrix} \quad (7.60)$$

$$\boldsymbol{\omega}_{30} = \boldsymbol{\omega}_{10} + \mathbf{A}_{10}\boldsymbol{\Omega}_{31} = \begin{bmatrix} \dot{\varphi}_3\cos\varphi_1\cos\varphi_2 + \dot{\varphi}_2\sin\varphi_1 \\ \dot{\varphi}_3\sin\varphi_1\cos\varphi_2 - \dot{\varphi}_2\cos\varphi_1 \\ \dot{\varphi}_1 + \dot{\varphi}_3\sin\varphi_2 \end{bmatrix} \quad (7.61)$$

7.3. 3D ROTARY CRANE

Consequently, kinetic and potential energy can be calculated. The kinetic energy has to be formulated for four parts with mass: (1) tower and bridge, (2) trolley, (3) winch and (4) load. Due to the coordinate systems in Fig. 7.19, the potential energy has to be calculated only for the load.

All the symbolic computations regarding to kinematics and the derivation of the equations of motion are done in *Maple*. They are not presented here because of the lengthly expressions. Some important intermediate results should be depicted here. The kinetic energy yields

$$\begin{aligned}T = {} & T_{b,rot} + T_{t,trans} + T_{w,rot} + T_{l,trans} = \\
& \tfrac{1}{2}\dot\varphi_1 J_b + \tfrac{1}{2}m_t(\dot s + \dot\varphi_1^2 s^2) - \tfrac{1}{2}m(-2s\cos\varphi_2\cos\varphi_3 l\dot\varphi_2 + 2\dot s l\dot\varphi_3 \sin\varphi_2\sin\varphi_3 - \\
& 2\dot\varphi_1^2 sl\sin\varphi_2\cos\varphi_3 - \dot\varphi_1^2 s^2 - \dot s^2 - \cos^2\varphi_3 l^2 \dot\varphi_2^2 + \dot\varphi_1^2 l^2 \cos^2\varphi_3 \cos^2\varphi_2 - \dot l^2 - \\
& 2\dot s \dot l\sin\varphi_2\cos\varphi_3 + 2\dot s\dot\varphi_1 l\sin\varphi_3 - 2\dot l\sin\varphi_3\dot\varphi_1 s - \dot\varphi_1^2 l^2 - l^2\dot\varphi_3^2 - \\
& 2\dot\varphi_1 sl\dot\varphi_3\cos\varphi_3 - 2\dot\varphi_1 l^2 \dot\varphi_3 \sin\varphi_2 + 2\dot\varphi_1 l^2 \sin\varphi_3 \cos\varphi_2 \cos\varphi_3 \dot\varphi_2) + \\
& \tfrac{1}{2r_w^2}(J_w(\cos^2\varphi_3 l^2 \dot\varphi_2^2 - \dot\varphi_1^2 l^2 \cos^2\varphi_3\cos^2\varphi_2 + \dot l^2 + \dot\varphi_1^2 l^2 + l^2\dot\varphi_3^2 + \\
& 2\dot\varphi_1 l^2 \dot\varphi_3 \sin\varphi_2 - 2\dot\varphi_1 l^2 \sin\varphi_3 \cos\varphi_2 \cos\varphi_3 \dot\varphi_2))\end{aligned} \quad (7.62)$$

The potential energy simply reads as

$$V = -mgl\cos\varphi_2\cos\varphi_3 \quad (7.63)$$

The necessary partial derivatives for Lagrange's equations are computed in *Maple*. The resulting ODEs are formulated in the form

$$\mathbf{M}(\mathbf{q})\ddot{\mathbf{q}} + \mathbf{\Phi}(\mathbf{q},\dot{\mathbf{q}})\dot{\mathbf{q}} + \mathbf{b}(\mathbf{q}) = \mathbf{Q} \quad (7.64)$$

where the matrix $\mathbf{\Phi}(\mathbf{q},\dot{\mathbf{q}})$ denotes the gyroscopic matrix, the vector $\mathbf{b}(\mathbf{q})$ the conservative forces, which result from a potential and the vector $\mathbf{Q}$ denotes the generalized applied forces. The mass matrix $\mathbf{M}(\mathbf{q})$ and the gyroscopic matrix $\mathbf{\Phi}(\mathbf{q},\dot{\mathbf{q}})$ are not detailed here because they consist of lengthly expressions. The vector of conservative forces yields

$$\mathbf{b}(\mathbf{q}) = [0,\ -mg\cos\varphi_2\cos\varphi_3,\ 0,\ mgl\sin\varphi_2\cos\varphi_3,\ mgl\cos\varphi_2\sin\varphi_3]^T \quad (7.65)$$

The vector of generalized applied forces reads as

$$\mathbf{Q} = \left[F,\ \frac{M_w}{r_w},\ M_b,\ 0,\ 0\right]^T \quad (7.66)$$

It can be seen that the derivation of the equations of motion with a generalized coordinates formulation is quite laborious.

7.3. 3D ROTARY CRANE

Formulation with redundant (dependent) coordinates:

The equations of motion expressed in dependent coordinates $\mathbf{q} = [\varphi_1, s, l, x, y, z]^T$ are formulated in the form (4.2b) and read as:

$$\begin{bmatrix} J_b + m_t s^2 & 0 & 0 & 0 & 0 & 0 \\ 0 & m_t & 0 & 0 & 0 & 0 \\ 0 & 0 & \frac{J_w}{r_w^2} & 0 & 0 & 0 \\ 0 & 0 & 0 & m & 0 & 0 \\ 0 & 0 & 0 & 0 & m & 0 \\ 0 & 0 & 0 & 0 & 0 & m \end{bmatrix} \cdot \begin{bmatrix} \ddot{\varphi}_1 \\ \ddot{s} \\ \ddot{l} \\ \ddot{x} \\ \ddot{y} \\ \ddot{z} \end{bmatrix} - \begin{bmatrix} -2 m_t s \dot{s} \dot{\varphi}_1 \\ 0 \\ 0 \\ 0 \\ 0 \\ -mg \end{bmatrix} +$$

$$+ \frac{\lambda}{L} \begin{bmatrix} (x \sin \varphi_1 - y \cos \varphi_1)s \\ s - x \cos \varphi_1 - y \sin \varphi_1 \\ -L \\ x - s \cos \varphi_1 \\ y - s \sin \varphi_1 \\ z \end{bmatrix} + \begin{bmatrix} -1 & 0 & 0 \\ 0 & -1 & 0 \\ 0 & 0 & -\frac{1}{r_w} \\ 0 & 0 & 0 \\ 0 & 0 & 0 \\ 0 & 0 & 0 \end{bmatrix} \cdot \begin{bmatrix} M_b \\ F \\ M_w \end{bmatrix} = \begin{bmatrix} 0 \\ 0 \\ 0 \\ 0 \\ 0 \\ 0 \end{bmatrix} \quad (7.67)$$

$$\Leftrightarrow \mathbf{M}\dot{\mathbf{v}} - \mathbf{f} + \mathbf{G}^T \boldsymbol{\lambda} + \mathbf{B}^T \mathbf{u} = \mathbf{0}$$

The corresponding geometric constraint (4.2d) reads as

$$\mathbf{g}(\mathbf{q}) = L - l = \sqrt{(x - s \cos \varphi_1)^2 + (y - s \sin \varphi_1)^2 + z^2} - l = 0 \quad (7.68)$$

The constraint Jacobian is calculated by $\mathbf{G} = D\mathbf{g}(\mathbf{q})$ and yields

$$\mathbf{G} = \left[\frac{s(x \sin \varphi_1 - y \cos \varphi_1)}{L}, \frac{s - x \cos \varphi_1 - y \sin \varphi_1}{L}, -1, \frac{x - s \cos \varphi_1}{L}, \frac{y - s \sin \varphi_1}{L}, \frac{z}{L} \right] \quad (7.69)$$

The constraint matrix is introduced in combination with the Lagrange multiplier λ in Eq. (7.67). The Lagrange multiplier corresponds to the cable tension force. The input transformation matrix is directly obtained from (7.67).

$$\mathbf{B} = \begin{bmatrix} -1 & 0 & 0 & 0 & 0 & 0 \\ 0 & -1 & 0 & 0 & 0 & 0 \\ 0 & 0 & -\frac{1}{r_w} & 0 & 0 & 0 \end{bmatrix} \quad (7.70)$$

Comparison of the Different Formulations:

It can be seen that the formulation with generalized coordinates has a much more complex form. Especially the mass matrix and the gyroscopic matrix are fully occupied and highly nonlinear. The formulation with dependent coordinates is much easier. The mass matrix is positive definite and diagonal, but not constant due to the variable s. The only drawback of this formulation is the computation of the geometric constraint (7.68) at velocity and acceleration level.

7.3. 3D ROTARY CRANE

7.3.3 DAE Approach with Control Constraints

The control constraints are formulated in order that the load moves from the initial position $[14, 0, -15]^T$ m along a specific trajectory to the final position $[0, -5, -3]^T$ m to avoid an obstacle as illustrated in Fig. 7.18. The desired trajectory is taken from [26]. The control constraints (4.10d) read as

$$\mathbf{c}(\mathbf{q}, t) = \mathbf{\Phi}(\mathbf{q}) - \boldsymbol{\gamma}(t) = \begin{bmatrix} x_d \\ y_d \\ z_d \end{bmatrix} - \{\boldsymbol{\gamma}_0 + (\boldsymbol{\gamma}_f - \boldsymbol{\gamma}_0)s(t)\} = \mathbf{0} \qquad (7.71)$$

The trajectory $\boldsymbol{\gamma}(t)$ is modeled in cylindrical coordinates $\boldsymbol{\gamma}_c = [r, \varphi, z]^T$ and defined as $\boldsymbol{\gamma}(t) = \boldsymbol{\gamma}_0 + (\boldsymbol{\gamma}_f - \boldsymbol{\gamma}_0)s(t)$, where $s(t)$ is a specific polynomial function. This function and the desired outputs have to be continuously differentiable, if they are used for a flatness-based parameterization. The initial position is given by

$$\boldsymbol{\gamma}_{c,0} = [r_0, \varphi_0, z_0]^T = [14m, 0°, -15m]^T$$

and the final position by

$$\boldsymbol{\gamma}_{c,f} = [r_f, \varphi_f, z_f]^T = [5m, 270°, -3m]^T$$

The desired outputs in cylindrical coordinates read as

$$r_d = r_0 + (r_f - r_0)s(t), \quad \varphi_d = \varphi_0 + (\varphi_f - \varphi_0)s(t), \quad z_d = z_0 + (z_f - z_0)s(t)$$

These variables can be differentiated with respect to time, depending on the polynomial function $s(t)$. However, the position of the load should be formulated in Cartesian coordinates $x_d(t)$, $y_d(t)$ and $z_d(t)$. These variables and their derivatives can simply be calculated from the cylindrical coordinates.

$$x_d(t) = r_d \cos \varphi_d, \quad y_d(t) = r_d \sin \varphi_d, \quad z_d(t) = z_d \qquad (7.72)$$

The derivatives of these functions are not explicitly pointed out here. The polynomial function $s(t)$ is split into an acceleration phase $s_1(t)$, a phase of constant velocity $s_2(t)$ and a deceleration phase $s_3(t)$.

$$s_1(t) = \frac{1}{\tau - \tau_0} \left(-\frac{5t^8}{2\tau_0^7} + \frac{10t^7}{\tau_0^6} - \frac{14t^6}{\tau_0^5} + \frac{7t^5}{\tau_0^4} \right)$$

$$s_2(t) = \frac{1}{\tau - \tau_0} \left(t - \frac{\tau_0}{2} \right)$$

$$s_3(t) = 1 + \frac{1}{\tau - \tau_0} \left(\frac{5(\tau - t)^8}{2\tau_0^7} - \frac{10(\tau - t)^7}{\tau_0^6} + \frac{14(\tau - t)^6}{\tau_0^5} - \frac{7(\tau - t)^5}{\tau_0^4} \right)$$

(7.73a)

(7.73b)

(7.73c)

7.3. 3D ROTARY CRANE

The whole maneuver should be completed within $\tau = 40s$ and the acceleration and deceleration times are set to $\tau_0 = 10s$. Details of the DAE-approach with control constraints regarding to this specific rotary crane with the desired trajectory (7.71), (7.73) can be found in [26].

In the next step the projection matrices $\mathbf{C}$ and $\mathbf{D}$ for the index reduction procedure have to be found.

$$\mathbf{C} = D\mathbf{\Phi}(\mathbf{q}) = \begin{bmatrix} 0 & 0 & 0 & 1 & 0 & 0 \\ 0 & 0 & 0 & 0 & 1 & 0 \\ 0 & 0 & 0 & 0 & 0 & 1 \end{bmatrix} \qquad (7.74)$$

The matrix $\mathbf{D}$ which fulfills the complementarity condition (4.8) can be formulated by

$$\mathbf{D} = \begin{bmatrix} 1 & 0 & 0 \\ 0 & 1 & 0 \\ 0 & 0 & 1 \\ 0 & 0 & 0 \\ 0 & 0 & 0 \\ 0 & 0 & 0 \end{bmatrix} \qquad (7.75)$$

The projection matrices (7.74) and (7.75) are simple sparse matrices as in the previous example of the planar overhead crane. Furthermore, a tangent control constraint realization occurs.

$$p = \text{rank}(\mathbf{CM}^{-1}\mathbf{B}^T) = 0 \qquad (7.76)$$

Hence, the third row in Table 4.1 is fulfilled.

The index reduction procedure (4.10) is applied to the system by using the projection matrices $\mathbf{C}$ and $\mathbf{D}$. The resulting system is numerically integrated by the implicit Euler method, where a step size of $\Delta t = 0.1\,s$ is used.

7.3.4 Flatness-Based Trajectory Tracking

A flatness-based control design with a similar crane, which is formulated with independent coordinates, is derived in [52]. However, the DAE-formulation (7.67), (7.68), (7.71) is better suited regarding the inverse dynamics problem. A well suited parameterization is published e.g. in [26]. The algebraic system (4.10c) results in

$$\ddot{x}_d + \frac{\lambda}{L}\frac{x_d - s\cos\varphi_1}{m} = 0 \qquad (7.77a)$$

$$\ddot{y}_d + \frac{\lambda}{L}\frac{y_d - s\sin\varphi_1}{m} = 0 \qquad (7.77b)$$

$$\ddot{z}_d + g + \frac{\lambda}{L}\frac{z_d}{m} = 0 \qquad (7.77c)$$

It should be recalled that $L = \sqrt{(x_d - s\cos\varphi_1)^2 + (y_d - s\sin\varphi_1)^2 + z_d^2}$. The system (7.77) can be solved for $\lambda_d(t)$, $s_d(t)$ and $\varphi_{1,d}(t)$.

$$\lambda_d = m\sqrt{\ddot{x}_d^2 + \ddot{y}_d^2 + (\ddot{z}_d + g)^2} \tag{7.78a}$$

$$\varphi_{1,d} = \arctan\left(\frac{\frac{y_d - z_d \ddot{y}_d}{\ddot{z}_d + g}}{\frac{x_d - z_d \ddot{x}_d}{\ddot{z}_d + g}}\right) \tag{7.78b}$$

$$s_d = \sqrt{\left(\frac{x_d - z_d\ddot{x}_d}{\ddot{z}_d + g}\right)^2 + \left(\frac{y_d - z_d\ddot{y}_d}{\ddot{z}_d + g}\right)^2} \tag{7.78c}$$

Subsequently, the cable length $l_d(t)$ can be be calculated from the geometric constraint (4.10e), i.e. from Eq. (7.68).

$$l_d = z_d \frac{\sqrt{\ddot{x}_d^2 + \ddot{y}_d^2 + (\ddot{z}_d + g)^2}}{\ddot{z}_d + g} \tag{7.79}$$

Finally, the control inputs can be calculated from the differential equations (4.10b).

$$M_{b,d} = (J_b + m_t s_d^2)\ddot{\varphi}_{1,d} + 2m_t s_d \dot{s}_d \dot{\varphi}_{1,d} + \lambda_d \frac{s_d}{l_d}(x_d\sin\varphi_{1,d} - y_d\cos\varphi_{1,d}) \tag{7.80a}$$

$$F_d = m_t \ddot{s}_d + \frac{\lambda_d}{l_d}(s_d - x_d\cos\varphi_{1,d} - y_d\sin\varphi_{1,d}) \tag{7.80b}$$

$$M_{w,d} = r_w\left(\frac{J_w \ddot{l}_d}{r_w^2} - \lambda_d\right) \tag{7.80c}$$

This feedforward control can be extended by a feedback controller, e.g. a linear PI-controller. As a result disturbances are compensated. In the parameterizations of the control inputs (7.80) it can be seen that the desired outputs must be differentiated with respect to time up to fourth order.

$$M_{b,d} = M_{b,d}(x_d, \dot{x}_d, \ldots, x_d^{(4)}, y_d, \dot{y}_d, \ldots, y_d^{(4)}, z_d, \dot{z}_d, \ldots, z_d^{(4)}) \tag{7.81a}$$

$$F_d = F_d(x_d, \dot{x}_d, \ldots, x_d^{(4)}, y_d, \dot{y}_d, \ldots, y_d^{(4)}, z_d, \dot{z}_d, \ldots, z_d^{(4)}) \tag{7.81b}$$

$$M_{w,d} = M_{w,d}(x_d, \dot{x}_d, \ldots, x_d^{(4)}, y_d, \dot{y}_d, \ldots, y_d^{(4)}, z_d, \dot{z}_d, \ldots, z_d^{(4)}) \tag{7.81c}$$

Hence, all state variables and input variables can be parameterized by the flat outputs x_d, y_d and z_d and their time derivatives up to a certain order. As a consequence, it can be stated that the system is differentially flat. By using definition (6.58) it can be shown that the DAEs with control constraints (7.67), (7.68), (7.71) are index 5 DAEs.

$$i = \beta + 1 = 5 \tag{7.82}$$

The polynomial function $s(t)$ and its derivatives up to fourth order are shown in Fig. 7.20.

7.3. 3D ROTARY CRANE

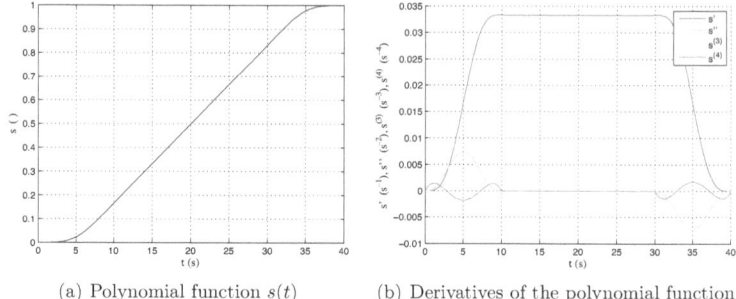

(a) Polynomial function $s(t)$

(b) Derivatives of the polynomial function

Figure 7.20: Polynomial for the desired trajectory with its time derivatives

7.3.5 Optimal Control

The Kelley-Bryson method for constrained systems is applied to the equations of motion (7.67). The state variables $q_1 = \varphi_1$, $q_2 = s$, $q_3 = l$, $q_4 = x$, $q_5 = y$, $q_6 = z$, $q_7 = \dot{\varphi}_1$, $q_8 = \dot{s}$, $q_9 = \dot{l}$, $q_{10} = \dot{x}$, $q_{11} = \dot{y}$, $q_{12} = \dot{z}$ are introduced. The constraint equation (7.68) is reformulated by

$$g(\mathbf{q}) = \frac{1}{2}\left[(q_4 - q_2 \cos q_1)^2 + (q_5 - q_2 \sin q_1)^2 + q_6^2 - q_3^2\right] = 0 \qquad (7.83)$$

In the following section the terms $\sin q_1$ and $\cos q_1$ are abbreviated by sq_1 and cq_1, respectively. Symbolic computations are performed with *Mathematica*.
The constraint Jacobian reads as

$$\mathbf{G}^T = \begin{bmatrix} q_2(q_4 sq_1 - q_5 cq_1) \\ q_2 - q_4 cq_1 - q_5 sq_1 \\ -q_3 \\ q_4 - q_2 cq_1 \\ q_5 - q_2 sq_1 \\ q_6 \end{bmatrix} \qquad (7.84)$$

7.3. 3D ROTARY CRANE

As a consequence, the GGL-formulation (5.105) yield

$$(J_b + m_t q_2^2)\dot{q}_1 = (J_b + m_t q_2^2)v_1 - [q_2(q_4 s q_1 - q_5 c q_1)]\nu \tag{7.85a}$$

$$m_t \dot{q}_2 = m_t v_2 - [q_2 - q_4 c q_1 - q_5 s q_1]\nu \tag{7.85b}$$

$$\frac{J_w}{r_w^2}\dot{q}_3 = \frac{J_w}{r_w^2}v_3 + q_3\nu \tag{7.85c}$$

$$m\dot{q}_4 = mv_4 - (q_4 - q_2 c q_1)\nu \tag{7.85d}$$

$$m\dot{q}_5 = mv_5 - (q_5 - q_2 s q_1)\nu \tag{7.85e}$$

$$m\dot{q}_6 = mv_6 - q_6\nu \tag{7.85f}$$

$$(J_b + m_t q_2^2)\dot{v}_1 = M_b - 2m_t q_2 v_2 v_1 - [q_2(q_4 s q_1 - q_5 c q_1)]\lambda \tag{7.85g}$$

$$m_t \dot{v}_2 = F - [q_2 - q_4 c q_1 - q_5 s q_1]\lambda \tag{7.85h}$$

$$\frac{J_w}{r_w^2}\dot{v}_3 = \frac{M_w}{r_w} + q_3\lambda \tag{7.85i}$$

$$m\dot{v}_4 = -(q_4 - q_2 c q_1)\lambda \tag{7.85j}$$

$$m\dot{v}_5 = -(q_5 - q_2 s q_1)\lambda \tag{7.85k}$$

$$m\dot{v}_6 = -mg - q_6\lambda \tag{7.85l}$$

$$0 = \frac{1}{2}\left[(q_4 - q_2 c q_1)^2 + (q_5 - q_2 s q_1)^2 + q_6^2 - q_3^2\right] \tag{7.85m}$$

$$0 = [q_2(q_4 s q_1 - q_5 c q_1)]v_1 + [q_2 - q_4 c q_1 - q_5 s q_1]v_2 - \cdots \\ q_3 v_3 + (q_4 - q_2 c q_1)v_4 + (q_5 - q_2 s q_1)v_5 + q_6 v_6 \tag{7.85n}$$

The DAE-system of index 2 (7.85) can be integrated forwards in time. This results in the coordinates $\mathbf{q}(t)$, the velocities $\mathbf{v}(t)$ and the Lagrange multipliers $\lambda(t)$ and $\nu(t)$.

The integral part of the cost functional (5.75) is formulated by

$$\mathcal{L}(\mathbf{x},t) = \begin{array}{l} \frac{1}{2}(q_4 - x_d)^2 + \frac{1}{2}(q_5 - y_d)^2 + \frac{1}{2}(q_6 - z_d)^2 + \cdots \\ \frac{\chi}{2}(v_4 - \dot{x}_d)^2 + \frac{\chi}{2}(v_5 - \dot{y}_d)^2 + \frac{\chi}{2}(v_6 - \dot{z}_d)^2 \end{array} \tag{7.86}$$

The gradients of $\mathcal{L}$ simply read as

$$\mathcal{L}_\mathbf{q} = \begin{bmatrix} 0 \\ 0 \\ 0 \\ q_4 - x_d \\ q_5 - y_d \\ q_6 - z_d \end{bmatrix}, \quad \mathcal{L}_\mathbf{v} = \begin{bmatrix} 0 \\ 0 \\ 0 \\ \chi(v_4 - \dot{x}_d) \\ \chi(v_5 - \dot{y}_d) \\ \chi(v_6 - \dot{z}_d) \end{bmatrix} \tag{7.87}$$

Furthermore, the matrices $\mathbf{R} = D_{\mathbf{q}}(\mathbf{G}(\mathbf{q})\mathbf{v})$ and $\mathcal{V}(\mathbf{q},\mathbf{v},\dot{\mathbf{v}}) = \mathbf{f} - \mathbf{G}^T\boldsymbol{\lambda} - \mathbf{M}\dot{\mathbf{v}}$ are needed to derive the adjoint equations (5.115).

$$\mathbf{R}^T = \begin{bmatrix} (q_2 q_4 v_1 - q_5 v_2 - q_2 v_5)cq_1 + (q_4 v_2 + q_2(q_5 v_1 + v_4))sq_1 \\ v_2 - (q_5 v_1 + v_4)cq_1 + (q_4 v_1 - v_5)sq_1 \\ -v_3 \\ v_4 - v_2 cq_1 + q_2 v_1 sq_1 \\ v_5 - q_2 v_1 cq_1 - v_2 sq_1 \\ v_6 \end{bmatrix} \quad (7.88)$$

$$\mathcal{V} = \begin{bmatrix} M_b - 2m_t q_2 v_2 v_1 - [q_2(q_4 sq_1 - q_5 cq_1)]\lambda - (J_b + m_t q_2^2)\dot{v}_1 \\ F - [q_2 - q_4 cq_1 - q_5 sq_1]\lambda - m_t \dot{v}_2 \\ \dfrac{M_w}{r_w} + q_3 \lambda - \dfrac{J_w}{r_w^2}\dot{v}_3 \\ -(q_4 - q_2 cq_1)\lambda - m\dot{v}_4 \\ -(q_5 - q_2 sq_1)\lambda - m\dot{v}_5 \\ -mg - q_6 \lambda - m\dot{v}_6 \end{bmatrix} \quad (7.89)$$

The Jacobian matrices of $\mathcal{V}$ read as

$$\mathcal{V}_{\mathbf{q}} = \begin{bmatrix} -q_2\lambda(q_4 cq_1 + q_5 sq_1) & -2m_t(q_2\dot{v}_1 + v_1 v_2) + q_5\lambda cq_1 - q_4\lambda sq_1 & 0 & -q_2\lambda sq_1 & q_2\lambda cq_1 & 0 \\ \lambda(q_5 cq_1 - q_4 sq_1) & -\lambda & 0 & \lambda cq_1 & \lambda sq_1 & 0 \\ 0 & 0 & \lambda & 0 & 0 & 0 \\ -q_2\lambda sq_1 & \lambda cq_1 & 0 & -\lambda & 0 & 0 \\ q_2\lambda cq_1 & \lambda sq_1 & 0 & 0 & -\lambda & 0 \\ 0 & 0 & 0 & 0 & 0 & -\lambda \end{bmatrix} \quad (7.90)$$

$$\mathcal{V}_{\mathbf{v}} = \begin{bmatrix} -2m_t q_2 v_2 & -2m_t q_2 v_1 & 0 & 0 & 0 & 0 \\ 0 & 0 & 0 & 0 & 0 & 0 \\ 0 & 0 & 0 & 0 & 0 & 0 \\ 0 & 0 & 0 & 0 & 0 & 0 \\ 0 & 0 & 0 & 0 & 0 & 0 \\ 0 & 0 & 0 & 0 & 0 & 0 \end{bmatrix} \quad (7.91)$$

7.3. 3D ROTARY CRANE

Now the costate equations (5.115) can be formulated.

$$\dot{p}_1 = \frac{1}{m\,m_t(J_b + m_t q_2^2)} [(-m_t q_2(J_b + m_t q_2^2)w_5\lambda + \cdots$$
$$m\,(m_t q_2 q_4 w_1 \lambda - J_b q_5 w_2 \lambda - m_t q_2^2 q_5 w_2 \lambda + J_b m_t q_2 q_5 \mu + \cdots$$
$$m_t^2 q_2^3 q_5 \mu - m_t (J_b + m_t q_2^2)(q_2 q_4 v_1 - q_5 v_2 - q_2 v_5)\xi)\cos(q_1) - \cdots$$
$$(-m_t q_2 (J_b + m_t q_2^2) w_4 \lambda + m\,(-m_t q_2 q_5 w_1 \lambda - J_b q_4 w_2 \lambda - \cdots$$
$$m_t q_2^2 q_4 w_2 \lambda + J_b m_t q_2 q_4 \mu + m_t^2 q_2^3 q_4 \mu + \cdots$$
$$m_t (J_b + m_t q_2^2)(q_4 v_2 + q_2 (q_5 v_1 + v_4))\xi))\sin(q_1)]$$

$$\dot{p}_2 = \frac{1}{m\,m_t(J_b + m_t q_2^2)} [-m\,(-J_b w_2 \lambda + m_t\,(-q_2^2 w_2 \lambda + J_b q_2 \mu + \cdots$$
$$J_b v_2 \xi + m_t(-2q_2 \dot{v}_1 w_1 - 2v_1 v_2 w_1 + q_2^3 \mu + q_2^2 v_2 \xi))) + \cdots$$
$$m_t\,(-(J_b + m_t q_2^2) w_4 \lambda + m\,(-q_5 w_1 \lambda + (J_b + m_t q_2^2) q_4 \mu + \cdots$$
$$(J_b + m_t q_2^2)(q_5 v_1 + v_4)\xi))\cos(q_1) + \cdots$$
$$m_t\,(-(J_b + m_t q_2^2) w_5 \lambda + m\,(q_4 w_1 \lambda - (J_b + m_t q_2^2) q_4 v_1 \xi + \cdots$$
$$(J_b + m_t q_2^2)(q_5 \mu + v_5 \xi))) \sin(q_1)]$$

$$\dot{p}_3 = -\frac{r_w^2 w_3 \lambda}{J_w} + q_3 \mu + v_3 \xi$$

$$\dot{p}_4 = x_d + \frac{w_4 \lambda}{m} - q_4(1 + \mu) - v_4 \xi + \left(-\frac{w_2 \lambda}{m_t} + q_2 \mu + v_2 \xi\right)\cos(q_1) - \cdots$$
$$\frac{q_2(-w_1 \lambda + (J_b + m_t q_2^2) v_1 \xi)\sin(q_1)}{J_b + m_t q_2^2}$$

$$\dot{p}_5 = y_d + \frac{w_5 \lambda}{m} - q_5(1 + \mu) - v_5 \xi + \left(-\frac{w_2 \lambda}{m_t} + q_2 \mu + v_2 \xi\right)\sin(q_1) + \cdots$$
$$\frac{q_2(-w_1 \lambda + (J_b + m_t q_2^2) v_1 \xi)\cos(q_1)}{J_b + m_t q_2^2}$$

$$\dot{p}_6 = z_d + \frac{w_6 \lambda}{m} - q_6(1 + \mu) - v_6 \xi$$

$$\dot{w}_1 = -p_1 + \frac{2m_t q_2 v_2 w_1}{J_b + m_t q_2^2} + q_2 q_5 \xi \cos(q_1) - q_2 q_4 \xi \sin(q_1)$$

$$\dot{w}_2 = -p_2 + \frac{2m_t q_2 v_1 w_1}{J_b + m_t q_2^2} - q_2 \xi + q_4 \xi \cos(q_1) + q_5 \xi \sin(q_1)$$

$$\dot{w}_3 = -p_3 + q_3 \xi$$

$$\dot{w}_4 = -p_4 - q_4 \xi - v_4 \chi + \dot{x}_d \chi + q_2 \xi \cos(q_1)$$

$$\dot{w}_5 = -p_5 - q_5 \xi - v_5 \chi + \dot{y}_d \chi + q_2 \xi \sin(q_1)$$

$$\dot{w}_6 = -p_6 - q_6 \xi - v_6 \chi + \dot{z}_d \chi$$

$$0 = \frac{1}{J_w\,m\,m_t(J_b + m_t q_2^2)}[(J_b + m_t q_2^2)(-m\,m_t q_3 r_w^2 w_3 + J_w(m\,q_2 w_2 + \cdots$$
$$m_t q_4 w_4 + m_t q_5 w_5 + m_t q_6 w_6)) - J_w\,(m\,m_t q_2 q_5 w_1 + m(J_b + m_t q_2^2)\cdots$$
$$q_4 w_2 + m_t q_2 (J_b + m_t q_2^2) w_4)\cos(q_1) - J_w(-m\,m_t q_2 q_4 w_1 + \cdots$$
$$m(J_b + m_t q_2^2) q_5 w_2 + m_t q_2 (J_b + m_t q_2^2) w_5)\sin(q_1)]$$

$$0 = \frac{1}{J_w\,m\,m_t(J_b + m_t q_2^2)}[(J_b + m_t q_2^2)(-m\,m_t q_3 r_w^2 p_3 + J_w(m\,q_2 p_2 + \cdots$$
$$m_t q_4 p_4 + m_t q_5 p_5 + m_t q_6 p_6)) - J_w\,(m\,m_t q_2 q_5 p_1 + m(J_b + m_t q_2^2)\cdots$$
$$q_4 p_2 + m_t q_2 (J_b + m_t q_2^2) p_4)\cos(q_1) - J_w(-m\,m_t q_2 q_4 p_1 + \cdots$$
$$m(J_b + m_t q_2^2) q_5 p_2 + m_t q_2 (J_b + m_t q_2^2) p_5)\sin(q_1)]$$

(7.92a)

7.3. 3D ROTARY CRANE

This adjoint system of index 2 (7.92) is solved backwards in time and results in the costates $\mathbf{p}(t)$ and $\mathbf{w}(t)$ as well as in the adjoint Lagrange multipliers $\mu(t)$ and $\xi(t)$. Due to

$$\mathbf{f}_\mathbf{u}^T = \begin{bmatrix} 1 & 0 & 0 & 0 & 0 & 0 \\ 0 & 1 & 0 & 0 & 0 & 0 \\ 0 & 0 & \frac{1}{r_w} & 0 & 0 & 0 \end{bmatrix}$$

the control update (5.116) results in

$$\delta M_b = -\kappa w_1 \tag{7.93a}$$
$$\delta F = -\kappa w_2 \tag{7.93b}$$
$$\delta M_w = -\kappa \frac{w_3}{r_w} \tag{7.93c}$$

7.3.6 Results

Fig. 7.21(a) presents the inversely computed control inputs M_b, F and M_w. The inputs are calculated by the DAE approach with control constraints and the flatness based parameterization (7.80), respectively. It can be seen that the bridge torque M_b and the trolley force F start and end at zero. This results due to the desired trajectory, where initial and final velocity, acceleration and higher derivatives are equal to zero, Fig. 7.20(b). Initial and final winch torques are equal to $M_w = -490.5\,Nm$. This is the static torque $M_{w,s} = -mgr_w$ which is required to hold the load. Fig.

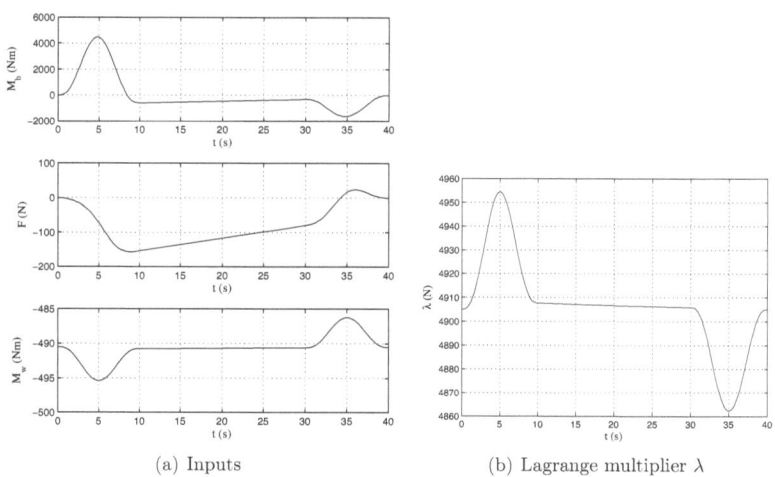

(a) Inputs (b) Lagrange multiplier λ

Figure 7.21: Inputs and Lagrange multiplier of the rotary crane

7.21(b) shows the computed Lagrange multiplier λ. Due to the formulation of the geometric constraints (7.68), it is identical to the cable tension force. It can be seen

7.3. 3D ROTARY CRANE

that the initial and final values of λ are identical to the static force, which results from the load $\lambda_s = mg = 4905\,N$. Again, the computed inputs are used in a forward dynamics simulation. However, the formulation with independent coordinates is used in order that an ODE-solver can be used in *Matlab*. In Fig. 7.22(a) the redundant coordinates φ_1, s and l are shown. Fig. 7.22(b) illustrates the redundant coordinates x, y and z, which are the outputs of the multibody system. The outputs are identical to the desired trajectories. The results, which are obtained from the

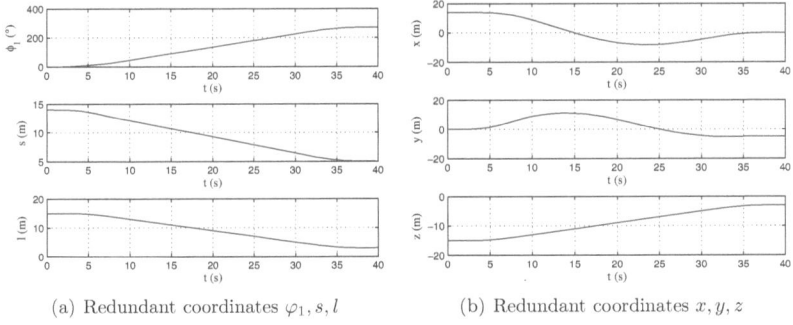

(a) Redundant coordinates φ_1, s, l (b) Redundant coordinates x, y, z

Figure 7.22: Redundant coordinates and outputs of the rotary crane

forward numerical integration, are animated in *Matlab/Simulink*. Fig. 7.23 shows three snapshots of the motion of the 3D rotary crane. It can be seen that the load is moved exactly along the desired trajectory from Fig. 7.18.

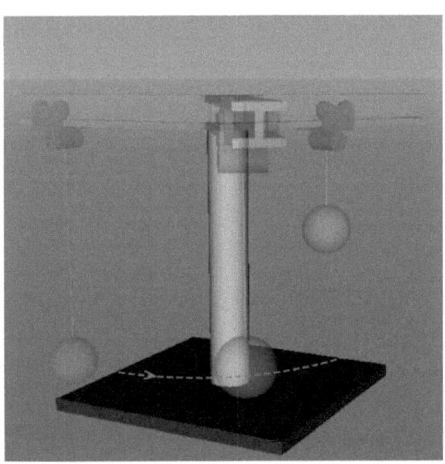

Figure 7.23: Snapshots of the motion of the rotary crane

7.3.7 Discussion

In this example, of a three-dimensional multibody system, big differences occur in the formulation of the equations of motion. The formulation with independent coordinates is time-consuming and requires many symbolic computations. These computations are done in *Maple*. On the other side, the formulation with dependent coordinates can be done rather in a straightforward way. The only disadvantage is the formulation of the geometric constraint (7.68) which results in more complex derivatives at velocity and acceleration level. However, the DAE approach with control constraints can be applied straightforward. The flatness-based parameterization (7.80) with redundant coordinates is much easier than with generalized coordinates, cf. [52]. In addition, the resulting control laws are less complex and hence numerically more efficient. The equations of the Kelley-Bryson method are also much easier, if redundant coordinates are used. However, the DAEs of index 2 have to be integrated. By applying a further index reduction, the system can be transfered to an index 1 system, which ca be integrated by the solver *ode15s*.

On the other side, the forward dynamics simulation is easier with the generalized coordinates formulation. The standard ODE-solver *ode45* is used to integrate the equations of motion (7.64). Merely, larger matrices, which are fully occupied and not constant have to be handled. The animation of the rotary crane in *Matlab/Simulink* is seen as big advantage in order to illustrate the motion and to have a better insight in the multibody system.

> *Virtual prototypes are a way of thinking out loud. You want the right people to think out loud with you!*
>
> Paul MacCready, inventor of the first practical flying machine powered by a human being

Chapter 8
Examples from Industrial Applications

This section is focused on three industrial examples regarding to nonlinear inverse problems. In the first example the vibrations of a steel converter are considered and the excitation forces are computed in an inverse calculation by using the method of virtual iteration. The second and third example illustrate a virtual test rig where agricultural machines are considered. The drive signals of the test rig are computed by the virtual iteration method. The steel converter is modeled as finite element model with the commercial software *Abaqus*. The agricultural machines are modeled as flexible multibody systems in the software *Adams*. All systems have in common that the equations of motion are not given in a symbolic form. Hence, the DAE-approach with control constraints, the optimal control algorithm and the flatness-based trajectory tracking control cannot be applied. Furthermore, the measured target signals present a stochastic behavior and they are not sufficiently smooth functions that are continuously differentiable. However, the method of virtual iteration is shown as excellent method to compute the inputs of these large hybrid multibody systems.

8.1 AOD Converter

The vibrations of an AOD (argon oxygen decarburization) converter are considered. Considerable oscillations are observed during the blowing process. This causes wear in the tilting drive and furthermore vibrations are introduced in the foundation and the surrounding infrastructure like offices or control rooms. The aim of the present study is to compute the forces that cause the converter to vibrate, i.e. the excitation forces caused by the AOD process. The AOD converter under consideration is shown in Fig. 8.1. The core components are the converter vessel and the blowing lance. The vessel is linked to the trunnion ring, which is supported by two ball bearing joints. The tilting drive allows to tilt and empty the vessel. Two tension rods and a torsion bar link the tilting drive and the foundation. Consequently, rotations of the drive are disabled. However, translations due to weight or temperature loads are allowed.
The converter is equipped with accelerometers and strain gauges to measure the

8.1. AOD CONVERTER

Figure 8.1: AOD converter (Source: Siemens VAI)

vibrations during the AOD process. This allows the study of the influence of process and heat parameters on one hand and to verify and calibrate the numerical model on the other hand.

8.1.1 Model Description

A finite element model of the AOD converter is created in the software *Abaqus*. The main parts of the model are shown in Fig. 8.2. Vessel, lining and liquid content are modeled as mass point with appropriate mass and inertia properties. The vessel is linked to the trunnion ring via MPC (multi physics constraints) beams, illustrated by the dashed blue lines. For modeling the trunnion ring (green), Timoshenko beam elements are used. Brick elements are used for the main bearings (yellow) and tetrahedral elements for the consoles of the bearings (blue). Gear wheels and gear box are mass points with a so-called display body (gray). The display body does not have physical properties but is is rather used to illustrate the motion. The tension rods are truss elements (white) and the torsion bar is a Timoshenko beam element (red). Bearings and consoles of the torsion bar are modeled with tetrahedral elements (orange, dark red). The gear box is connected to the tension rods via MPC connector elements. The gear wheel and the shaft are connected via kinematic couplings. Gear wheel and gear box are coupled via torsional springs. The values for masses, torsional spring stiffness, etc. are identified in [145]. The main components are coupled to the foundation, as shown in Fig. 8.3. The foundation is modeled with brick elements, where a coarser mesh is used. The foundation consists of a reinforced concrete, which is not modeled in detail. Rather an average value for the Young's modulus is used to restrain the complexity of the model. A tie contact is used to connect all the different parts. Linear material models are applied to the different parts and hence the whole model is a linear finite element model. The basis of the model has been previously developed at *Siemens VAI metals technologies*

8.1. AOD CONVERTER

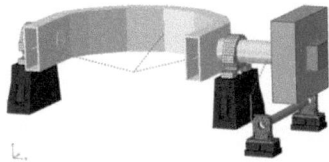

Figure 8.2: Main parts of AOD converter

GmbH and in the work of [145]. The final model consists of approximately 280000 DOFs.

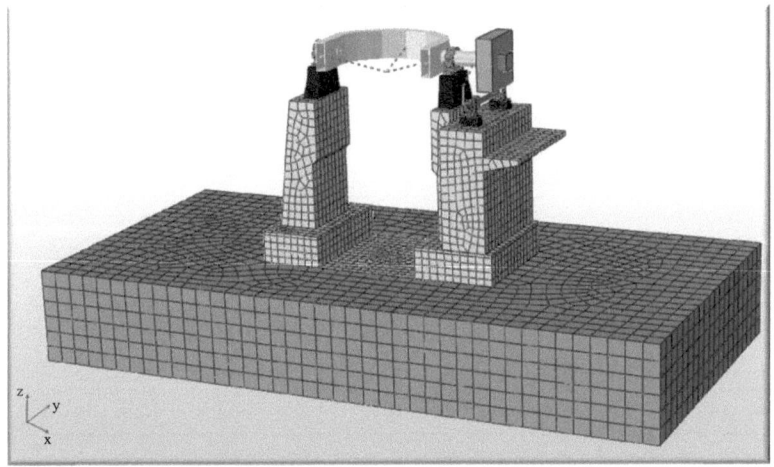

Figure 8.3: Finite element model of the AOD converter

8.1.2 Measuring Setup

The sensors are placed in such a way that the mode shapes and the motions of the converter can be observed. Two strains e_{Fr}, e_{Fl} and seven accelerations a_{yL}, a_{xL}, a_{zFl}, a_{axFl}, a_{zFr}, a_{xFr}, a_{yFm} are considered. The subscripts x, y, z denote the direction according to the coordinate system shown in Fig. 8.3 and Fig. 8.4. The two strain gauges are mounted at the tension rods. Five accelerometers are mounted at the gear box and two accelerometers are placed at the outer ball bearing to measure longitudinal and transversal oscillations. The measured frequency range is $0...12.5$ Hz, which is the band of major interest. In the numerical model the "virtual sensors" are placed exactly in the same way. The positions can be seen in Fig. 8.4.

8.1. AOD CONVERTER

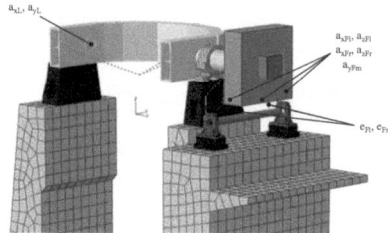

Figure 8.4: Positions of accelerometers and strain gauges

8.1.3 Modal Analysis

The first numerical computation is a modal analysis to compare the measured and simulated eigenfrequencies and mode shapes. Undamped and damped eigenfrequencies are calculated. Rayleigh damping is used for the complex eigenvalue analysis. Fig. 8.5 shows the pole zero map for the first 20 eigenvalues. The highest plotted eigenvalue corresponds to a frequency of 60 Hz (the imaginary axis illustrates the circular frequency in $1/s$). It can be seen that all real parts are negative and hence the equilibrium point is asymptotically stable. The first 10 undamped eigenfrequen-

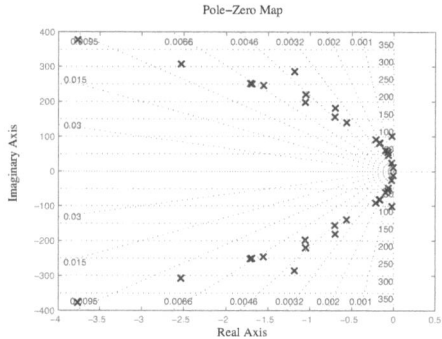

Figure 8.5: Pole zero map of the AOD converter

cies and mode shapes can be seen in Fig. 8.6. The first mode at 2 Hz is a critical torsional mode because it is a rotation around the tilting axis. The second mode at 3.9 Hz is a bending mode in the longitudinal direction and the third mode at 7.5 Hz is a bending mode in the transversal direction. The fourth mode at 8.8 Hz is a bending mode in the vertical direction. Fifth and sixth mode are higher order transversal bending modes. The other modes are out of the frequency range of interest.

The first two eigenfrequencies correspond with measured peaks in the frequency spectra from different process steps. Higher eigenfrequencies could not be found in the measured spectra. Reasons can be found in damping properties or in the characteristics of the foundation. As already mentioned the reinforced concrete in the foundation is not modeled in detail.

In [145] two simplified analytical rigid body models are derived. The first model consists of three DOFs and the second model of 21 DOFs. The eigenfrequencies of these analytical models correspond well with the measured and predicted eigenfrequencies.

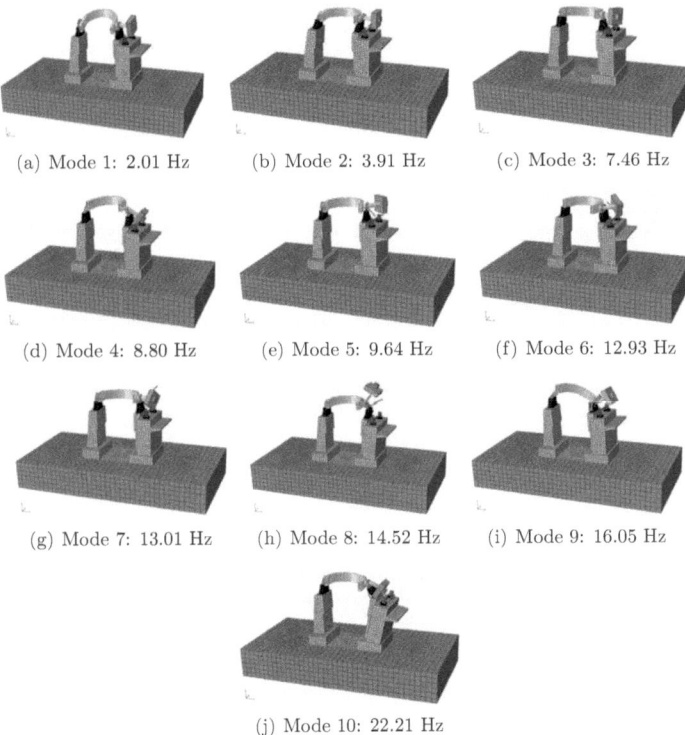

(a) Mode 1: 2.01 Hz (b) Mode 2: 3.91 Hz (c) Mode 3: 7.46 Hz

(d) Mode 4: 8.80 Hz (e) Mode 5: 9.64 Hz (f) Mode 6: 12.93 Hz

(g) Mode 7: 13.01 Hz (h) Mode 8: 14.52 Hz (i) Mode 9: 16.05 Hz

(j) Mode 10: 22.21 Hz

Figure 8.6: Eigenfrequencies and mode shapes of the AOD converter

8.1.4 Transfer Functions

The transfer functions of the linear FEM are calculated in an *Abaqus "steady-state dynamics"* step. Inputs are the forces and torques in each direction F_x, F_y, F_z, M_x, M_y, M_z in the vessel's center of mass. The outputs are identical to the measured signals. This results in a (6×9) transfer matrix $\mathbf{G}(i\omega)$. The frequency range $0...12.5$ Hz is sampled with 1000 discretization points. In the steady-state dynamics step a modal damping of 1% is used. Fig. 8.7 shows the magnitude plots and Fig. 8.8 the phase plots of the transfer matrix.

8.1. AOD CONVERTER

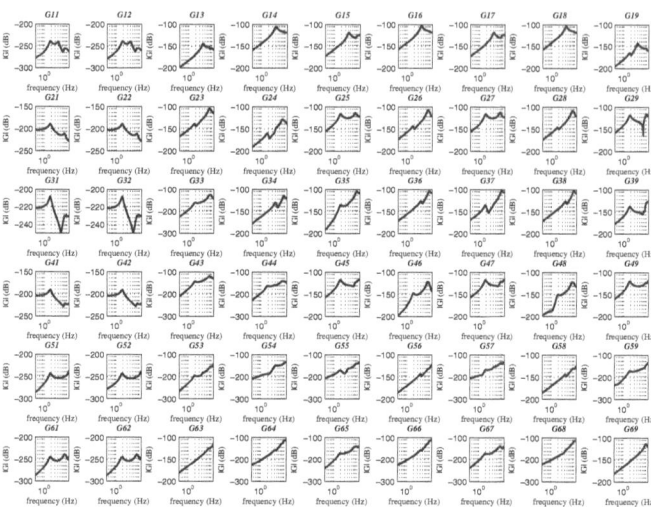

Figure 8.7: Magnitude plots of the transfer matrix, AOD converter

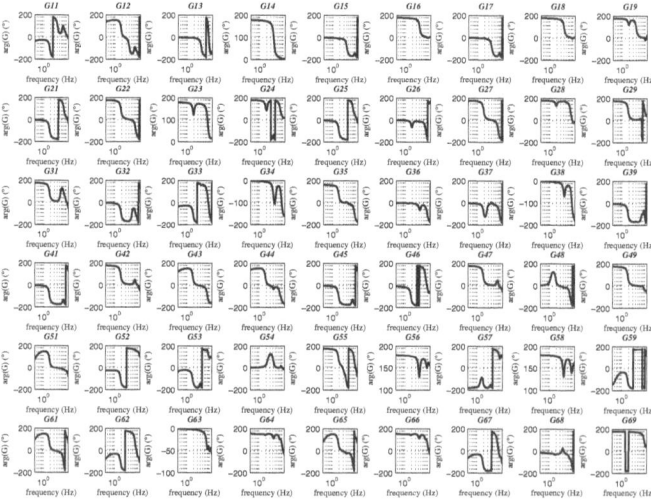

Figure 8.8: Phase plots of the transfer matrix, AOD converter

8.1.5 Inverse Computation of the Excitations

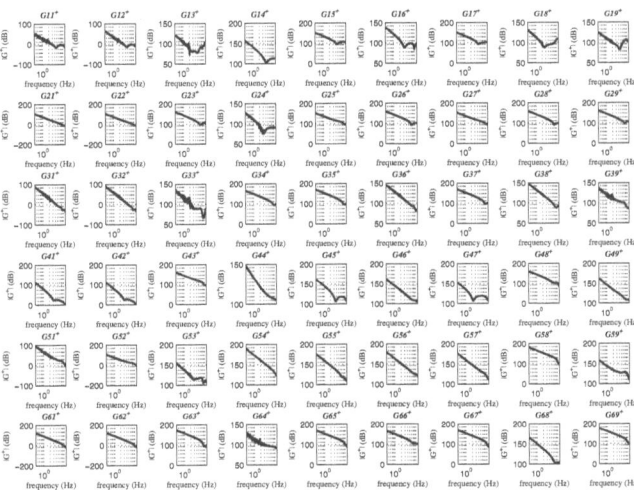

Figure 8.9: Magnitude plots of the pseudoinverse of the transfer matrix, AOD converter

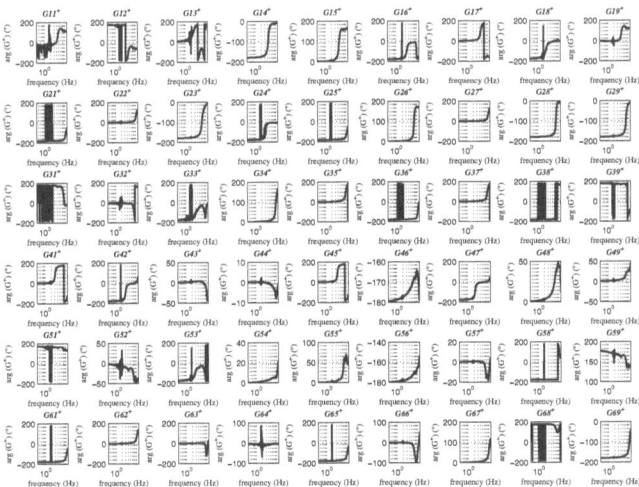

Figure 8.10: Phase plots of the pseudoinverse of the transfer matrix, AOD converter

8.1. AOD CONVERTER

The Moore-Penrose pseudoinverse of the transfer matrix is shown in Fig. 8.9 and 8.10, respectively.

Target signals are measured during different process steps. An illustrative segment is shown in Fig. 8.11. All measured signals are presented as relative values which are related to the individual maxima.

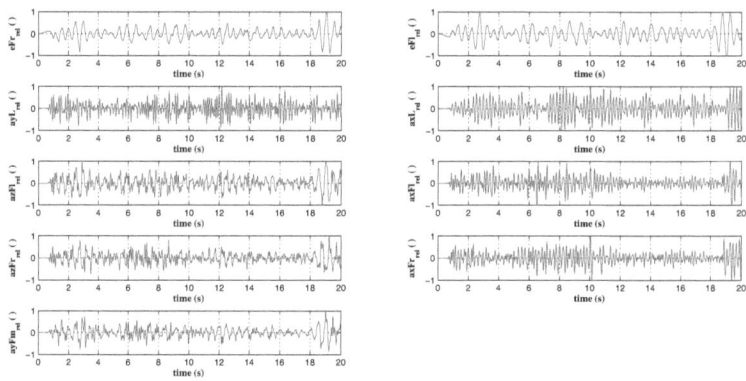

Figure 8.11: Measured target signals of the AOD converter

The measured targets from Fig. 8.11 are transformed into the frequency range via a FFT. The corresponding frequency spectra can be seen in Fig. 8.12. In a detailed

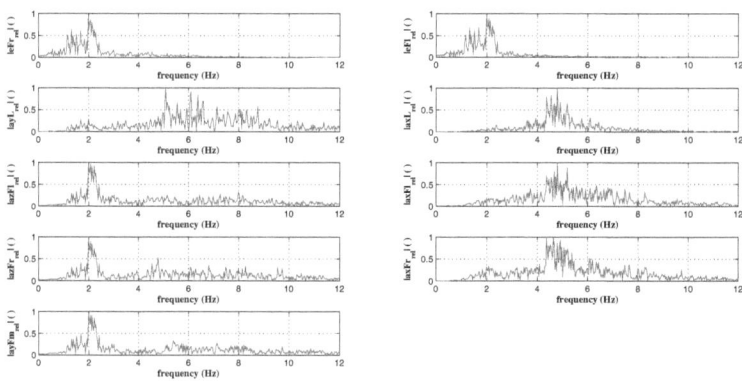

Figure 8.12: Frequency spectra of target signals, AOD converter

consideration of the targets a drift phenomenon can be seen in the strain signals. In the frequency spectra this effect can be seen in the low-frequency range, Fig. 8.13. However, these static effects introduce an error in the inverse computation. Hence, the signals are filtered and the spectra below a cut-off frequency of 0.4 Hz are deleted.

8.1. AOD CONVERTER

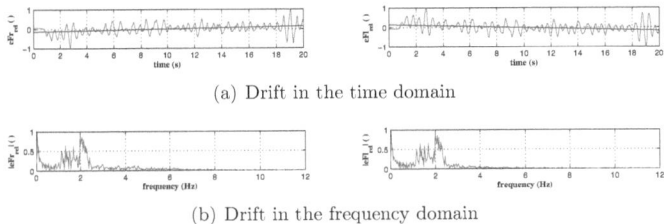

(a) Drift in the time domain

(b) Drift in the frequency domain

Figure 8.13: Measured unfiltered strains with drift effect

The excitations are computed by using the filtered targets and the pseudoinverse of the transfer matrix. The first drives are already the final solution due to the linearity of the model. Subsequent iterations are not required in this example. The computed excitation forces and torques are shown in Fig. 8.14. Again, the values are normalized with respect to their maximum forces and torques, respectively. The

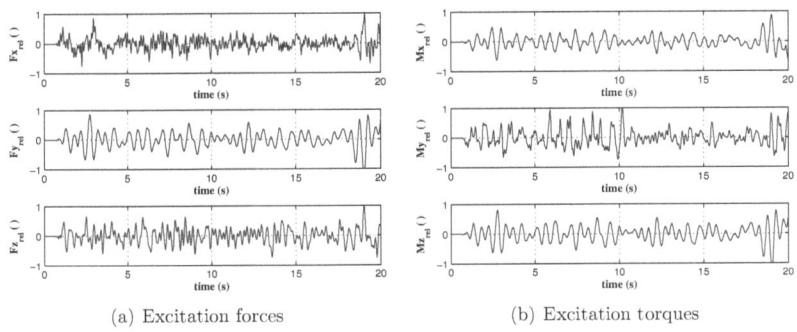

(a) Excitation forces (b) Excitation torques

Figure 8.14: Computed excitations in the AOD converter

frequency spectra of the excitations are of more interest, Fig. 8.15. The inverse computation is performed for different AOD process steps and hence it can be seen which frequencies are excited in the specific steps. The results of the different steps show a specific trend, which can also be seen in Fig. 8.15. In the frequency spectra of the transversal force F_y an increase between $1.0\ldots1.8$ Hz and especially at 2 Hz can be seen. Similar effects occur in the torques M_x and M_z. This indicates a pendular movement of the vessel. However, the excited frequency is critical because the first eigenfrequency is also 2 Hz.

Without signal filtering of the strain signals the computed excitations would be superimposed to a low-frequency oscillation.

8.1.6 Verification

In order to verify the computed excitations a forward dynamics analysis is performed. A "Modal Dynamics step" can be used in Abaqus because the model is fully linear. If

8.1. AOD CONVERTER

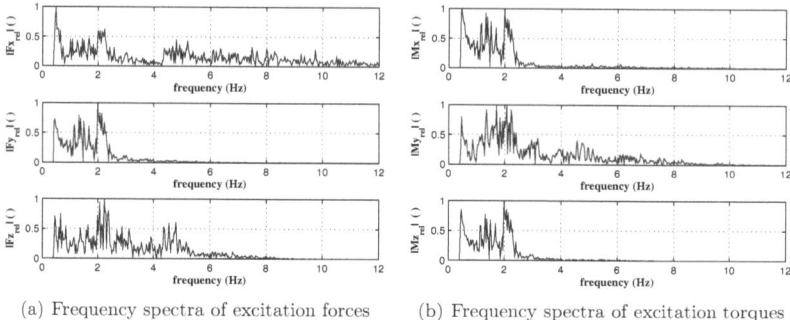

(a) Frequency spectra of excitation forces   (b) Frequency spectra of excitation torques

Figure 8.15: Frequency spectra of the excitations in the AOD converter

the model would be nonlinear, an implicit time integration could be used. However, this would require much more computational effort.

The outputs of the forward dynamics simulation are compared with the measured targets. The results are shown in Fig. 8.16 and 8.17, respectively. It can be seen that simulation outputs and targets coincide. Small deviations can be seen in the spectra of the acceleration $|ayL_{rel}|$. However, the absolute values of this output channel are very low and hence it can be stated that these deviations are negligible numerical effects. Therefore, the results of the inverse computation are verified.

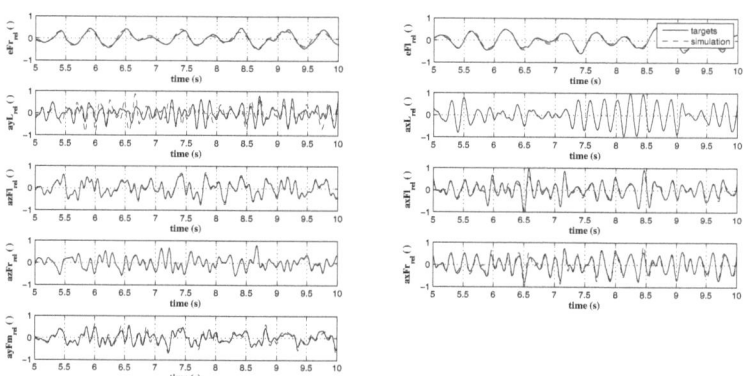

Figure 8.16: Comparison between measured targets and simulation outputs, AOD converter

8.1.7 Sensitivity Analysis

Modal damping is a simplification of the real physical behavior and therefore a sensitivity analysis regarding to the modal damping parameter is performed. The focus

8.1. AOD CONVERTER

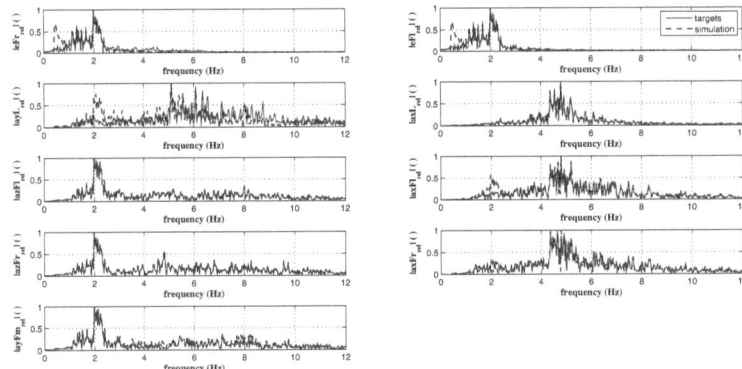

Figure 8.17: Comparison of frequency spectra between measured targets and simulation outputs, AOD converter

is put on the first eigenfrequency, which is critical regarding the computed excitations, Fig. 8.15(a), 8.15(b). The damping parameter is varied between $\xi = 0.5\%$ and $\xi = 10\%$ and the RMS-error between system outputs and targets is evaluated. The transfer function $G_{11}(i\omega)$ is examplarily shown in Fig. 8.18 for the different damping factors. An ideal value of $\xi = 5\%$ is found. In addition an analytical esti-

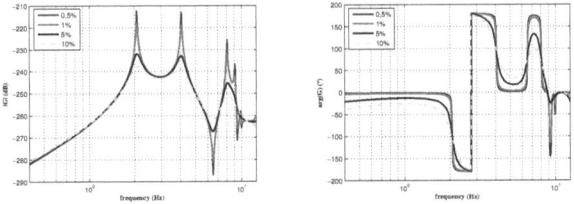

(a) Magnitude plots of G_{11} for specific modal damping parameters

(b) Phase plots of G_{11} for specific modal damping parameters

Figure 8.18: Sensitivity of modal damping

mation of the modal damping is carried out for the first eigenfrequency, Fig. 8.19. This estimation consideres the width of the amplitde, Eq. (8.1) and results in an ideal value of $\xi = 4\%$.

$$\xi_i = \frac{\nu_{i2} - \nu_{i1}}{\nu_{i2} + \nu_{i1}} \tag{8.1}$$

8.1. AOD CONVERTER

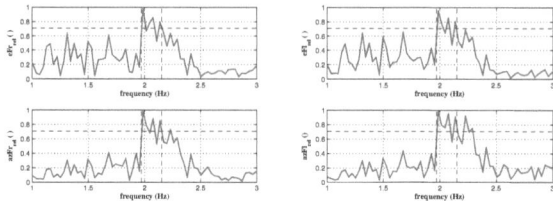

Figure 8.19: Estimation of modal damping for 1st Eigenfrequency

8.1.8 Calculation of Foundation Forces

The calculated excitations are the basis for further computations. In a forward dynamics analysis (*Modal dynamics*) the foundation forces are calculated at specific points. Forces and torques are evaluated at the left and right bearing of the trunnion ring. Furthermore, forces and torques at the bearings of the torsion bar are considered. The results are shown in Fig. 8.20, 8.21, 8.22 and 8.23. It can be seen that the forces that are introduced in the foundation are a superposition of static forces due to the mass of the vessel with the liquid steel bath and the dynamic forces due to the AOD process.

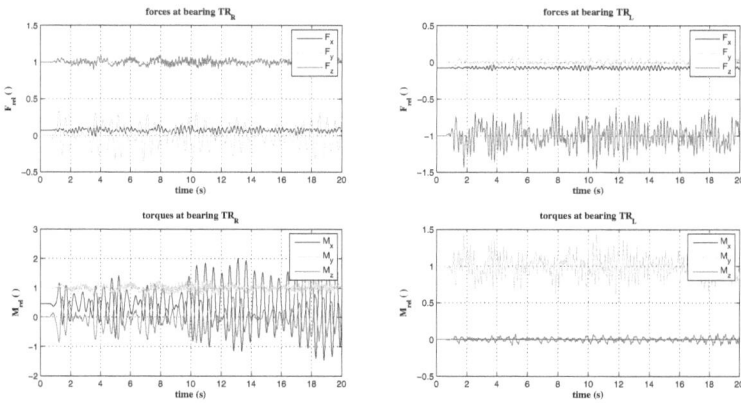

Figure 8.20: Forces and torques at the trunnion ring

8.1.9 Movement of the Converter

On the basis of the computed excitations, the converter movements are considered. Especially the movement of the vessel is of great interest, Fig. 8.24. The peak at 2 Hz can be seen clearly in the transversal motion $|v_{rel}|$, Fig. 8.24(c) and the rotation around the tilting axis $|\Phi_{rel}|$, Fig. 8.24(d).

8.1. AOD CONVERTER

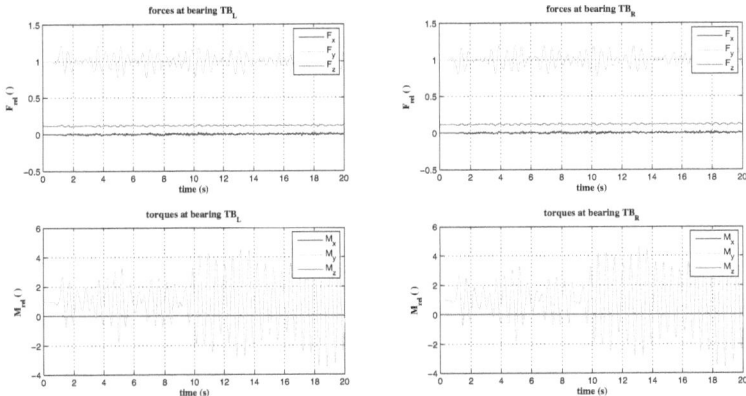

Figure 8.21: Forces and torques at the torsion bar

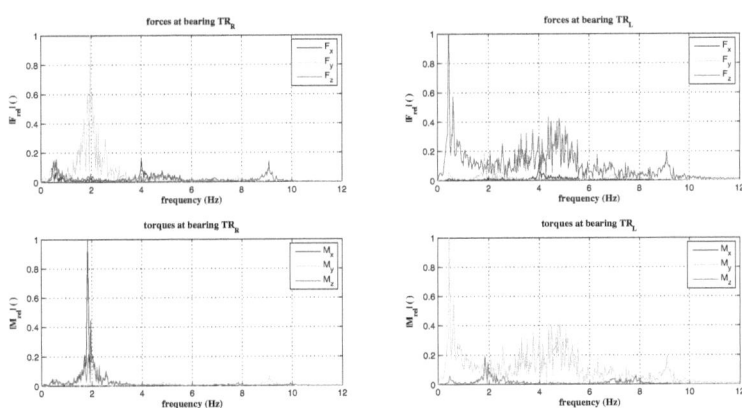

Figure 8.22: Frequency spectra of forces and torques at the trunnion ring

8.1. AOD CONVERTER

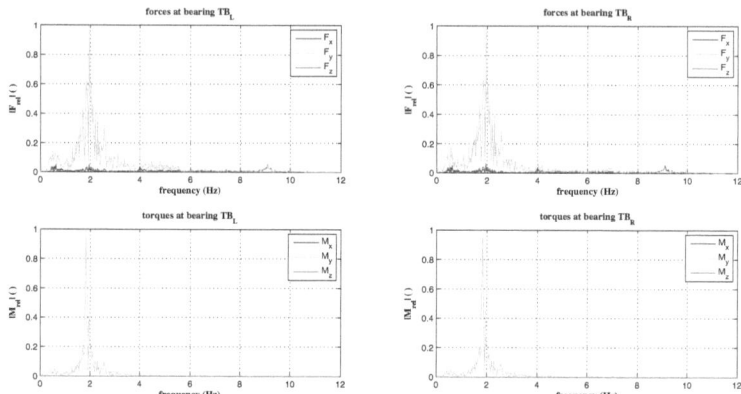

Figure 8.23: Frequency spectra of forces and torques at the torsion bar

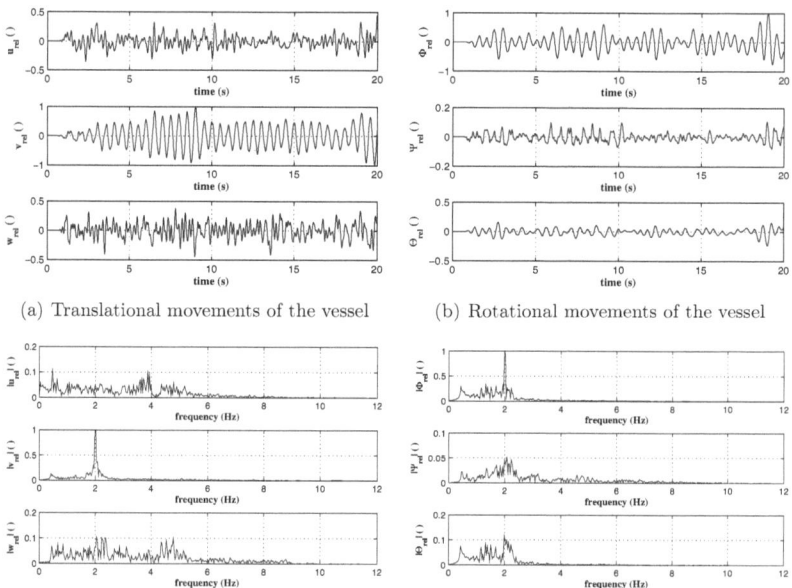

(a) Translational movements of the vessel

(b) Rotational movements of the vessel

(c) Frequency spectra of translational movements of the vessel

(d) Frequency spectra of rotational movements of the vessel

Figure 8.24: Movement of the vessel

8.1. AOD CONVERTER

8.1.10 Discussion

The method of virtual iteration is shown as suitable method to inversely compute the excitations of a large finite element model. Due to the linearity of the FEM the inverse computation can be done in one step without subsequent iterations. It is shown that the converter movement can be captured with nine measuring points. The over-determined system can be inverted by using the Moore-Penrose pseudoinverse. The inverse calculation is carried out for different process steps and the results are statistically analyzed. However, this is out of the scope of this dissertation and therefore these results are not shown here.
It is shown that a critical frequency of 2 Hz is excited, which is identical to the first eigenfrequency of the mechanical system. As a consequence, a pendular movement of the converter is excited around its tilting axis. In specific measurements a beat can be found which results from the eigenfrequency at 2Hz and an excitation in the area of 2 Hz.

It should be mentioned that the model of the vessel, which is a simple mass point, is a rough simplification. In further investigations the motion of the liquid steel bath should be included in the computation as well. For that reason the liquid steel is studied with CFD (computational fluid dynamics) methods at the institute of fluid mechanics and heat transfer at the Johannes Kepler university. The first studies consider a rotating tank with identical physical properties than the vessel with liquid steel. It is shown that the first eigenfrequency is 2 Hz, which correlates with the results of the computed excitations, Fig. 8.15. If the eigenfrequency of the steel bath and the mechanical system are close together, it results in a beat. Hence, the characteristics of the measured target signals can be physically interpreted. Parts of the results are published in [127].
In future work the FEM and the CFD-models will be coupled and the resulting system will be considered.

8.2 Trailed Cultivator: Synkro 6003T

The considered system is a trailed cultivator, named "Synkro 6003T" of the company Alois Pöttinger Maschinenfabrik GmbH, Fig. 8.25. A test drive on a representative test track is performed with the real physical machine and specific target signals are measured. Subsequently, the cultivator is put on a 3-poster test rig in the laboratory. The servo-hydraulic cylinders excite the system at the tractor linkage drawbar and the two wheels in the vertical direction. The goal is to compute the drive signals in order to reproduce the measured targets. It should be mentioned that the pack ring roller and the front wheels, which can be seen in Fig. 8.25, are not mounted on the considered machine.

Figure 8.25: Trailed cultivator: Synkro 6003T (Source: Pöttinger)

8.2.1 Model Description

The whole system is modeled as a MBS in the software *Adams*, Fig. 8.26. The main components are included as flexible bodies due to their bending and torsional stiffness which result from the frame construction of such agricultural machines. In a comparative study it is shown that a rigid body model does not provide satisfying results [124]. Hence, finite element models of the the main parts (i) drawbar, (ii) central framework, (iii) left and right folding parts, (iv) connection beam and (v) rear axle are created. All parts are made of steel, which is modeled as linear elastic material. Mesh and boundary conditions are prepared in *Ansys* and *I-Deas* and the modal reduction is performed in *Nastran*. The whole FEM consists of approximately 550000 nodes, 270000 tetrahedral elements and therefore 1.65 million DOFs. By using a Craig-Bampton reduction and a CMS, cf. section 2.9.5, the number of DOFs in the resulting MBS can be reduced to 213. Interface nodes for the MBS are created at each joint and the constraint modes corresponding to the DOFs of the joint are calculated. 10 fixed-boundary normal modes are chosen to be sufficient

8.2. TRAILED CULTIVATOR: SYNKRO 6003T

for each individual part. In a sensitivity analysis of several dynamic simulations an ideal value of 0.5% is found for modal damping of the first 10 modes.

The headland position, where the folding parts are fully lowered, is considered at

Figure 8.26: MBS model of the virtual test rig with the trailed cultivator Synkro 6003T

the test rig. In the MBS hydraulic cylinders and flexible parts are connected with revolute joints. Cylinder and piston rod of the hydraulic cylinders are also connected by revolute joints. As a consequence, the relative rotation is possible but the translational movement is locked. This is modeled because all hydraulic cylinders in the real machine are locked by multi-port-valves during operation. Furthermore, it is assumed that the oil is absolutely incompressible.

A very important modeling detail is the front linkage that connects the test rig with the machine, Fig. 8.26 (gray and yellow). This connection is used in the real test rig and hence it has to be included in the model as well. The parts are rigid bodies and identical joints as in the real test rig are used.

The tires are modeled by bushing elements (in *Adams* called VFORCE) with (non)linear spring and damper characteristics in all three Cartesian coordinates.

8.2.2 Tire Modeling

A tire test rig, which is developed by Pöttinger, is used to determine the tire characteristics, Fig. 8.27. The most important characteristics regarding to the test rig are (nonlinear) radial stiffness and damping. The stiffness is measured by a defined force which results in a deflection of the tire. Typically, a progressive spring characteristics occurs. The linear damping factor is determined in a vibration test. In a first step a linearized stiffness of $c_z = 550\,N/mm$ and a damping factor of $k_z = 5Ns/mm$ are used in the vertical direction. A sensitivity analysis in dynamic

8.2. TRAILED CULTIVATOR: SYNKRO 6003T

simulations results in a linear stiffness of $c_y = 275 N/mm$ and a damping coefficient of $k_y = 2.5\, Ns/mm$ in the transversal direction.

Furthermore, the stiffness in longitudinal direction is important in the virtual test rig with the trailed cultivator. In the real test rig the machine is not braked but wheel chocks are used, which makes the system more complicated. This is modeled by a nonlinear spring characteristic. It allows the machine to move almost freely in the longitudinal direction until the tires roll up onto the chock. If this case occurs, the stiffness is increased in a suitable way. The approximation of the tires by using

(a) Tire test rig at laboratory   (b) MBS model of tire test rig

Figure 8.27: Tire test rig (Source: Pöttinger)

bushing elements is chosen to be sufficiently accurate for the simulation of the virtual test rig. However, in a full-system simulation, where the agricultural machine is driven over a cart track, more sophisticated tire models are required. A so-called "adaptive footprint" model is developed as user-written subroutine [126]. However, the details of tire modeling are not in the scope of this dissertation and therefore they are not be described here in detail.

Nevertheless, some important references are given. Modeling details of agricultural tires can be found e.g. in [2, 16, 55, 56, 88, 89, 98, 99, 100, 147, 155, 156]. A detailed review of tire models in the automotive industry is given in [118].

It should be briefly mentioned that a driver model is also needed in such a full-system simulation. Full system simulations in the automotive industry are published e.g. in [81, 97, 101, 120, 129], suitable approaches for driver models in [49, 51, 154, 159, 160].

8.2.3 Measuring Setup

The positions of the used sensors can be seen in Fig. 8.28. Ten accelerometers are used to capture the target signals during the test drive (a_{1x}, a_{2z}, a_{3z}, a_{4x}, a_{5z}, a_{6Ry}, a_{7Rz}, a_{8Az}, a_{9Iz}, a_{10z}). The subscripts x, y, z denote the orientation in the Cartesian coordinate system in Fig. 8.28. Measurement points are at the drawbar,

8.2. TRAILED CULTIVATOR: SYNKRO 6003T

the central framework, the folding parts and at the rear axle. The most important sensors measure the acceleration in the vertical direction. The vertical direction is specifically excited by three servo-hydraulic cylinders. The measuring points in the model are identical to the real system. In the measured targets the relevant frequency range is between 0.9...4.9 Hz.

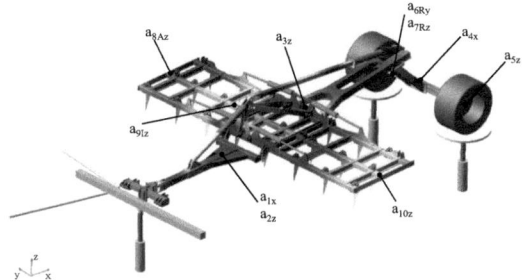

Figure 8.28: Sensor positions of the Synkro 6003T

8.2.4 Modal Analysis, Verification

A numerical modal analysis is carried out for validation purposes of the virtual test rig. The static equilibrium of the virtual test rig is computed and the model is linearized at this point. Fig. 8.29 shows the pole zero map of all 213 calculated eigenvalues. The highest plotted eigenvalue corresponds to an eigenfrequency of 167 Hz. It can be seen that all real parts of the eigenvalues are in the negative half plane and hence the equilibrium point is asymptotically stable.

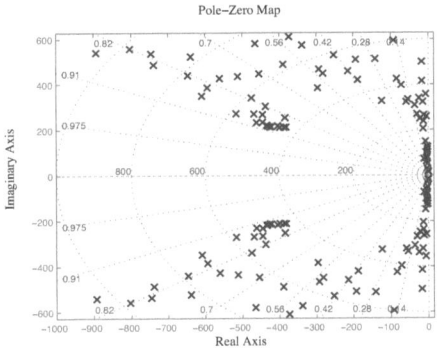

Figure 8.29: Pole zero map of the Synkro 6003T

The eigenfrequencies from the real system are also determined by a modal analysis from a stroke of a hammer. 90 individual frequency spectra are considered and the

8.2. TRAILED CULTIVATOR: SYNKRO 6003T

resulting boxplots are shown in Fig. 8.30. Furthermore, the measured and calculated eigenfrequencies are compared in Fig. 8.30. Simulated eigenfrequencies are presented by the green dots. The frequency range till 25 Hz is evaluated. It can be seen that the measured and simulated eigenfrequencies correspond well. Therefore, it can be stated that the linearized model of the virtual test rig is verified.

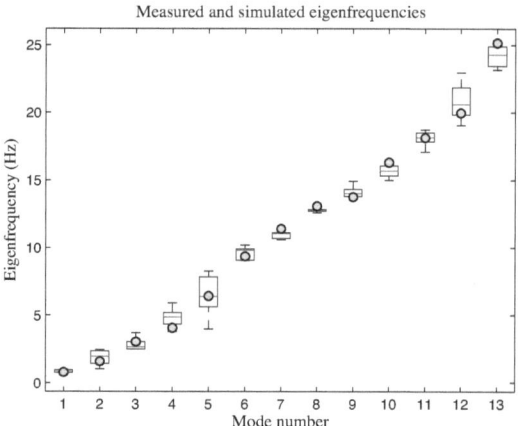

Figure 8.30: Measured and simulated Eigenfrequencies of Synkro 6003T

The mode shapes which correspond to the eigenvalues are shown in Fig. 8.31. The first mode at 0.79 Hz shows a movement in the longitudinal direction. Its eigenfrequency is mainly influenced by the longitudinal stiffness of the tires. The second mode at 1.59 Hz illustrates a pendular movement around the longitudinal axis and is specified by the vertical (radial) stiffness of the tires. A transversal motion occurs in the third mode at 3.05 Hz. This results from the transversal stiffness in the tires. In the fourth mode at 4.07 Hz bending of the drawbar, the connection beam and the rear axle can be seen. The higher modes are already out of the relevant frequency range. Generally a significant movement of the front linkage is observed in all modes. It is important to note that the lower eigenfrequencies within the range of the excitations are mainly determined by the stiffness values of the tires.

8.2.5 Transfer Functions

In the iteration process the two targets which measure the acceleration in x-direction a_{1x}, a_{4x} are not considered. The resulting outputs are $y_1 = a_{2z}$, $y_2 = a_{3z}$, $y_3 = a_{5Lz}$, $y_4 = a_{6Ry}$, $y_5 = a_{7Rz}$, $y_6 = a_{8Az}$, $y_7 = a_{9Iz}$, $y_8 = a_{10z}$. This results from very low amplitudes in the vibrations, which can be explained by the vertical excitation of the system. The three drive signals are the cylinder displacements at the front u_1 and the excitations at the left wheel u_2 and the right wheel u_3, respectively. As a

8.2. TRAILED CULTIVATOR: SYNKRO 6003T

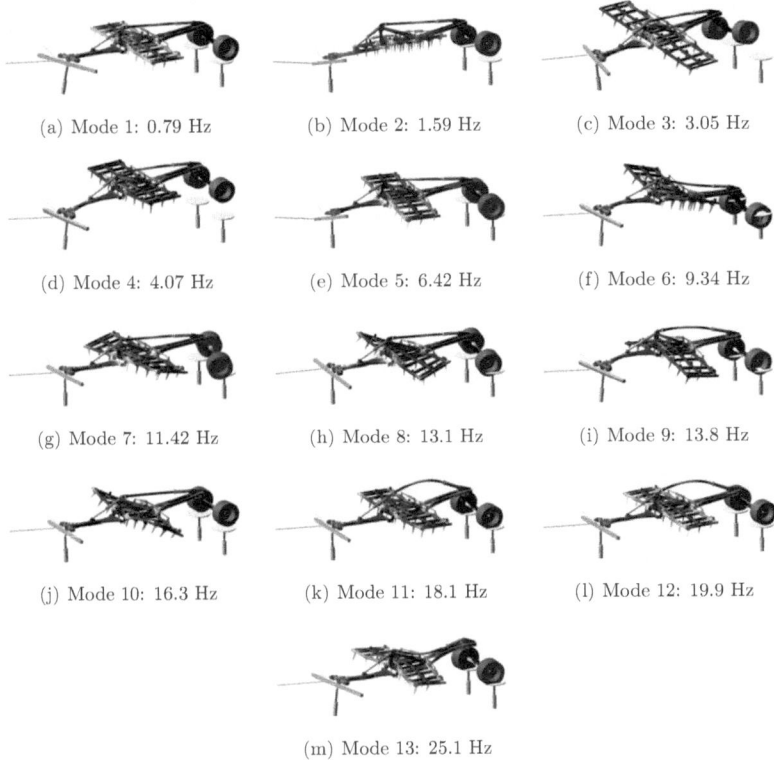

Figure 8.31: Eigenfrequencies and mode shapes of the Synkro6003T

consequence, a (3×8) transfer matrix is obtained. The state matrices **A**, **B**, **C** and **D** are calculated in *Adams* and the transfer matrix is assembled in *Matlab*. The individual transfer functions are sampled with 500 points in the frequency range between $0.5\ldots10$ Hz.

The magnitude plots of the transfer matrix can be seen in Fig. 8.32 and the phase plots in Fig. 8.33.

8.2. TRAILED CULTIVATOR: SYNKRO 6003T

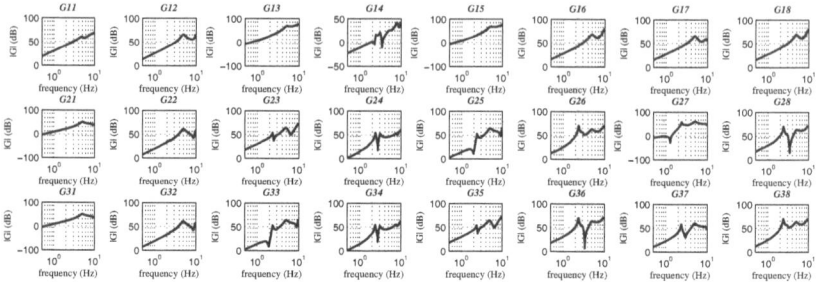

Figure 8.32: Magnitude plots of the transfer matrix, Synkro 6003T

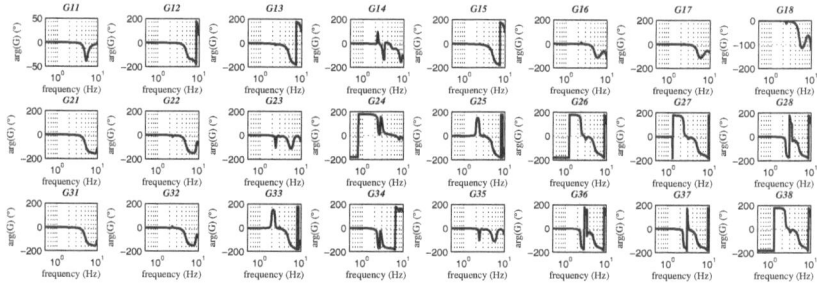

Figure 8.33: Phase plots of the transfer matrix, Synkro 6003T

8.2. TRAILED CULTIVATOR: SYNKRO 6003T

The Moore-Penrose pseudoinverse of the transfer matrix is illustrated in Fig. 8.34 and 8.35.

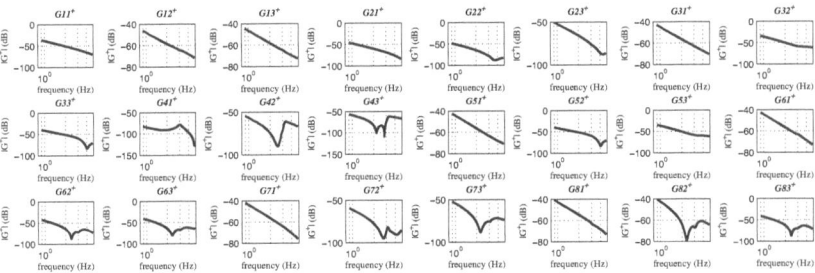

Figure 8.34: Magnitude plots of the pseudoinverse transfer matrix, Synkro 6003T

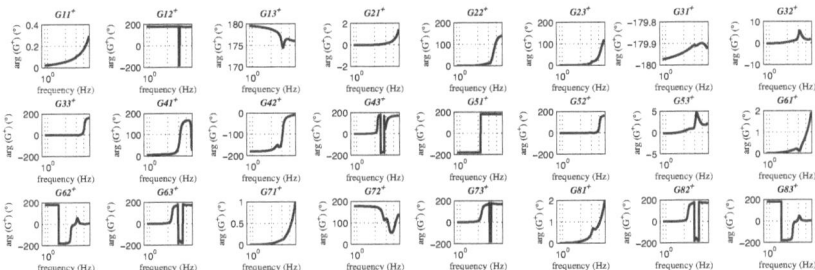

Figure 8.35: Phase plots of the pseudoinverse transfer matrix, Synkro 6003T

8.2. TRAILED CULTIVATOR: SYNKRO 6003T

8.2.6 Virtual Iteration

Target signals are captured in a sequence of 60s, as shown in Fig. 8.36. The frequency spectra are calculated via a FFT and presented in Fig. 8.37. It can be seen that the excitation from the test track is a stochastic signal. Peaks can be seen in the area of 3 Hz and 4 Hz which correspond to the third and fourth eigenfrequency. The excited frequency range is between 0.9...4.9 Hz. Hence, the transfer matrix is re-sampled.

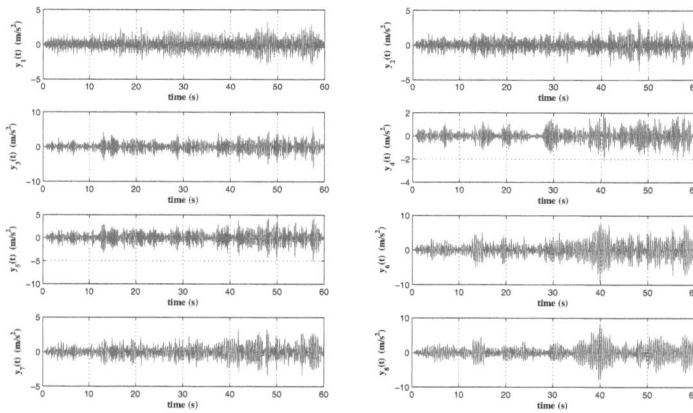

Figure 8.36: Measured targets of the Synkro 6003T

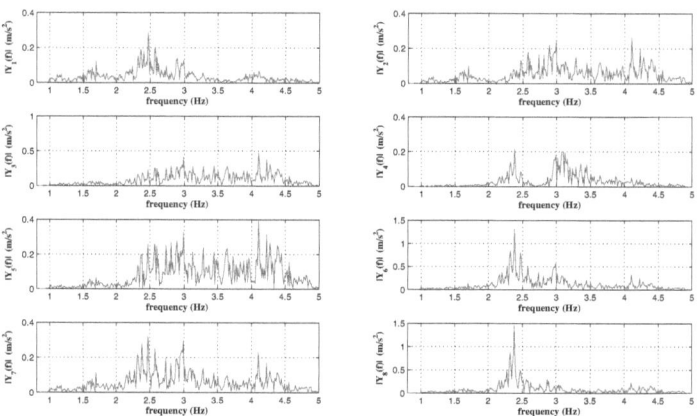

Figure 8.37: Frequency spectra of the target signals of the Synkro 6003T

8.2. TRAILED CULTIVATOR: SYNKRO 6003T

The virtual iteration algorithm is applied to the system. Due to nonlinearities in the tires and the front linkage several iterations are required. The forward dynamics simulation is performed with the *Adams*-solver GSTIFF I3. The scalar factor α^{n+1} in Eq. (3.37) is chosen to be equal to one. Fig. 8.38 shows the outputs of the simulation (blue), the targets (black) and the error (red) in the first iteration. An illustrative segment between 20s and 25s is plotted in order to zoom into the results. Fig. 8.39 shows the frequency spectra in the first iteration. It can be seen that the error due to the linearization of the nonlinear model is significant. As a consequence, subsequent iterations are carried out.

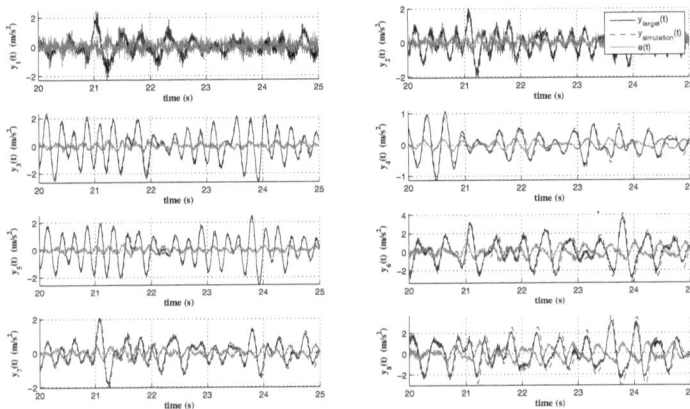

Figure 8.38: Targets, simulation results and error in the 1st iteration, Synkro 6003T

The error in the fifth iteration is presented in Fig. 8.40 and the corresponding frequency spectra in Fig. 8.41. It can be seen that the error is reduced compared to the first iteration, Fig. 8.38. However, the first two outputs $y_1 = a_{2z}$ and $y_2 = a_{3z}$ do not convergence as fast as the other outputs. The iterative procedure is stopped after the fifth iteration. A segment of the final drive signals is shown in Fig. 8.42. The iteration is also performed on the real test rig. Therefore, the drives from the virtual test rig and the real test rig can be compared. It can be seen that they correlate well. The frequency spectra of the virtually computed drives are presented in Fig. 8.43. The error between targets and system outputs is measured in each iteration. The RMS-error and the error of maximum deviation are summarized to an illustrative error indicator, which is calculated by Eq. (3.41). The RMS-error, the MAX-error and the resulting error indicator are presented for each output channel in every iteration, Fig. 8.44.

8.2.7 Discussion

It is shown that the implementation of flexible bodies is crucial for modeling an agricultural machine as the trailed cultivator Synkro 6003T. The Craig-Bampton

8.2. TRAILED CULTIVATOR: SYNKRO 6003T

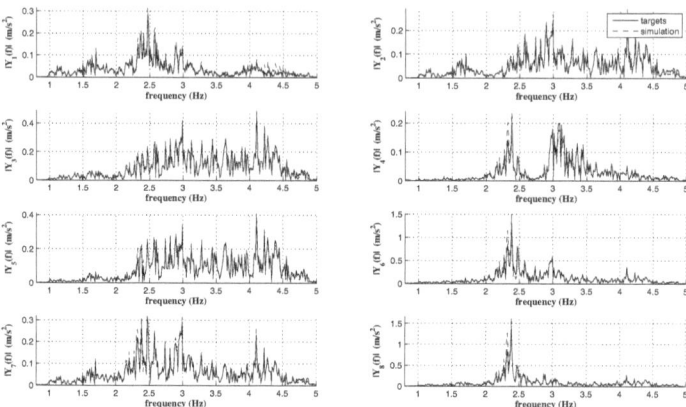

Figure 8.39: Frequency spectra of targets and system outputs in the 1st iteration, Synkro 6003T

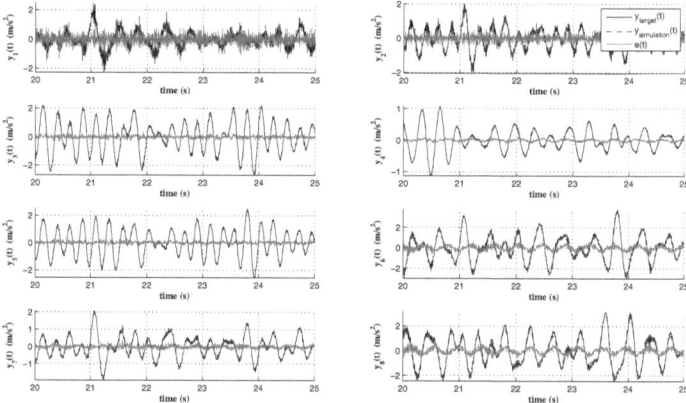

Figure 8.40: Targets, simulation results and error in the 5th iteration, Synkro 6003T

method is suitable for the implementation. It is also shown that the number of DOFs can be reduced from 1.65 million in the original FEM to 213 in the resulting MBS. Nonlinearities as the spring characteristics in the tires can be included in the MBS. The approximation of bushing elements to describe the tire behavior is sufficiently accurate for a virtual test rig as in Fig. 8.26 but not detailed enough for a full system simulation on a test track.

It is shown that the virtual iteration procedure converges and that the error between targets and system outputs is decreased in each iteration. 8 of 10 output channels are included in the iteration. However, the two channels that are not considered,

8.2. TRAILED CULTIVATOR: SYNKRO 6003T

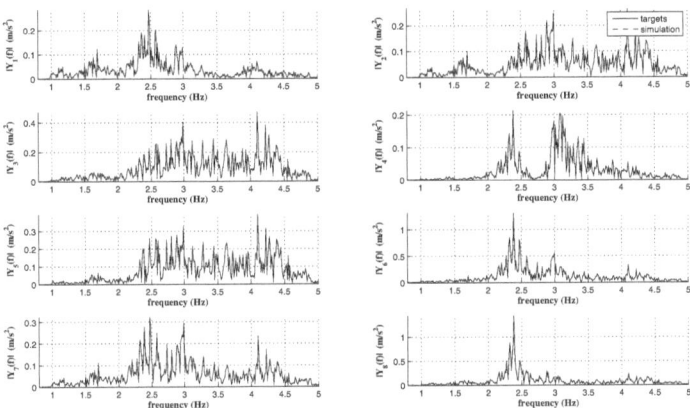

Figure 8.41: Frequency spectra of targets and system outputs in the 5th iteration, Synkro 6003T

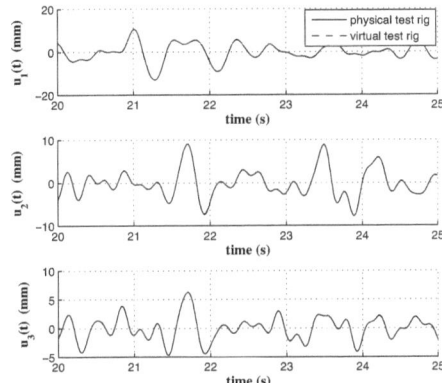

Figure 8.42: Computed drives signals and drives from the physical test rig, Synkro 6003T

converge as well. This is not explicitly shown here. The error indicator is between 22% and 58% in the first iteration. In the fifth iteration the error is reduced to values between 4% and 15%. The convergence behavior could not be improved by including the output channels a_{1x} and a_{4x} in the iteration. The error could further be reduced by additional iterations. However, the convergence speed decreases with each iteration, as it can be seen in Fig. 8.44.

8.2. TRAILED CULTIVATOR: SYNKRO 6003T

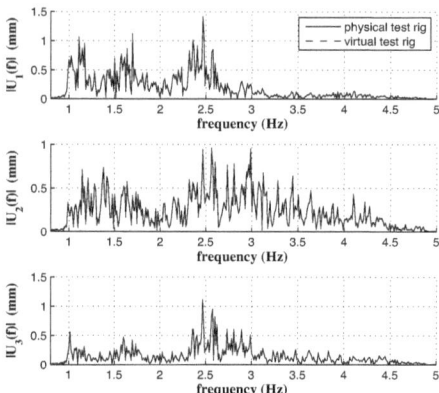

Figure 8.43: Frequency spectra of the drive signals, Synkro 6003T

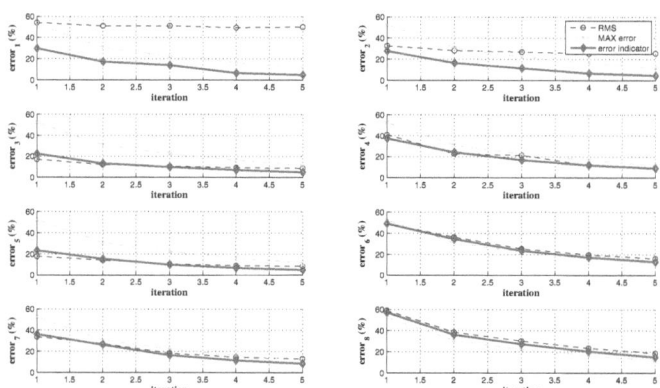

Figure 8.44: Convergence curves of the Synkro 6003T

8.3 Plough: Servo 6.50

In this section a large plough with eight shares, named "Servo 6.50" is the object under consideration. This agricultural machine is developed and produced by the company Alois Pöttinger Maschinenfabrik Gmbh, Fig. 8.45. The objective in this example is similar to the previous example of the Synkro 6003T, chapter 8.2. A test drive on a specific track should be reproduced on a real physical test rig as well as on a virtual test rig.

Figure 8.45: Plough: Servo 6.50 (Source: Pöttinger)

8.3.1 Model Description

The virtual test rig is modeled equivalent to the real physical 2-poster test rig. The MBS is modeled in *Adams* and the FEM is created in *Ansys* and *I-Deas*, Fig. 8.46. Due to the length and the heavy shares of the machine it is important to include flexible bodies. The main components are split into four flexible sub-models, (i) the headstock to the lower link of the tractor, (ii) the headstock with the revolute joint, (iii) the main body including all beams and shares and (iv) the wheel carrier. The transport position as shown in Fig. 8.45 and 8.46 is considered on the test rig. The hydraulic cylinders in the front, which turn the plough, are locked. The rear cylinders that adjust the width are locked as well. This is modeled directly in the FEM by using beam elements with appropriate stiffness properties. The whole FEM consists of 1.36 million nodes and over 700000 tetrahedral elements. This results in more than 4 million DOFs. Again, the flexible bodies are implemented via a Craig-Bampton reduction. The final MBS consists of 90 DOFs. Due to the large finite element models, the number of interface nodes is reduced compared to the model of the Synkro 6003T. For each FEM 10 fixed-boundary normal modes are computed. Modal damping is set to 0.5%.

The leverage in the front of the test rig is modeled by rigid parts. The excitation

8.3. PLOUGH: SERVO 6.50

Figure 8.46: MBS model of the virtual test rig with the plough Servo 6.50

in the front u_1 is modeled by a point motion and the rear cylinder u_2 is modeled as motion in a translational joint.
The tires are modeled by using bushing elements with $c_z = 525\,N/mm$, $k_z = 5\,Ns/mm$ in the vertical direction and $c_x = 80\,N/mm$, $k_x = 2.5\,Ns/mm$ in the transversal direction. Stiffness and damping in the longitudinal direction are set to zero because the plough is not braked at the test rig and no wheel chocks are used. The hydraulic cylinder that connects the headstock with its revolute joint to the main part is not locked. Hence, a nonlinear force element is included [16]. In a first step a linearized value of 23.5 kN is used. Furthermore, a contact problem arises between main part and wheel carrier. On the real test rig it is observed that the contact opens at high peaks in the excitations. However, a full contact model is not included in the MBS. Rather a stiff spring with $c = 10^6\,N/mm$ is included between the two bodies.

8.3.2 Measuring Setup

The plough is equipped with five accelerometers a_{1x}, a_{2z}, a_{3z}, a_{4z}, a_{5z} and three strain gauges ε_6^b, ε_7^b, ε_8^t. The subscripts x, y, z denote the orientation in the Cartesian coordinates as shown in Fig. 8.47. The superscripts b and t denote bending and torsional strains. It can be seen that most sensors measure the vibration in the vertical direction, which results from the excitations in z-direction. The relevant frequency band of the targets is in the range of $0.7\ldots 3.0$ Hz.

8.3. PLOUGH: SERVO 6.50

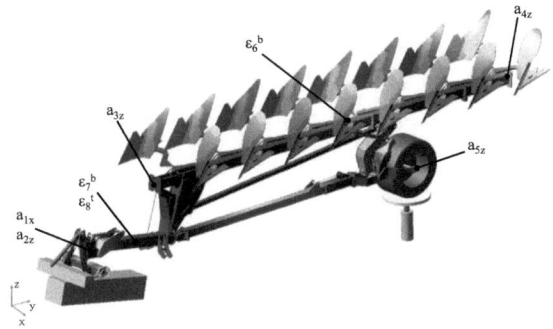

Figure 8.47: Sensor positions of Servo 6.50

8.3.3 Modal Analysis, Verification

The multibody system is linearized at its equilibrium point and a modal analysis is carried out. The pole zero map of all 90 eigenvalues is presented in Fig. 8.48. The highest eigenvalue corresponds to an eigenfrequency of 8.35 kHz. All real parts of the eigenvalues are negative and therefore the equilibrium point is asymptotically stable. The simulated eigenfrequencies are compared with the measured eigenfrequencies,

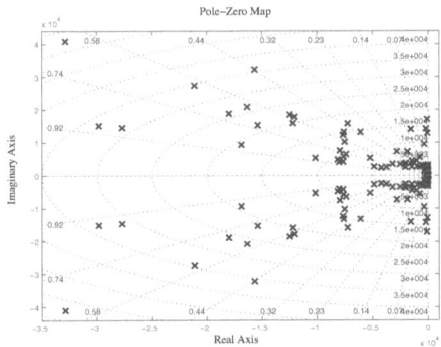

Figure 8.48: Pole zero map of Servo 6.50

which are determined from 16 independent measurements. The resulting boxplots and the simulated eigenfrequencies are presented in Fig. 8.49, where the simulation results are shown as green dots. A frequency range from 0 to 25 Hz is evaluated. It can be seen that calculated and measured results of the first two eigenfrequencies are identical. The third computed eigenfrequency is slightly higher than the measured value. This effect tends to increase till 14 Hz. However, these frequencies are out of the range regarding to the spectra of the excitations. The reason for the higher computed eigenvalues can be seen in the mass distribution. Several small attachment

8.3. PLOUGH: SERVO 6.50

parts like screws, bolts, bubble storages or hydraulic components of the real machine, Fig. 8.45 are not considered in the model. As a consequence the mass of the model is to low at specific points, which results in higher eigenfrequencies. Despite these small deviations it can be stated that the model is verified. The first mode at 0.96

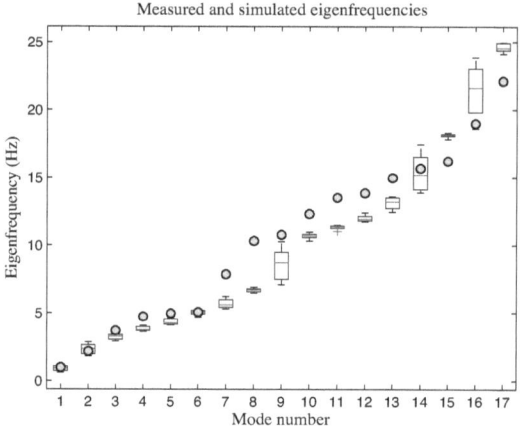

Figure 8.49: Measured and simulated Eigenfrequencies of Servo 6.50

Hz shows a transversal movement which is influenced by the transversal stiffness of the tire. A vertical motion combined with bending of the lower rectangular shaped tube occurs in the second mode at 2.15 Hz. The third mode at 3.70 Hz is a bending mode about the vertical axis and the fourth mode at 4.73 Hz is a torsional mode around the longitudinal axis. In the higher modes movements of the wheel carrier, the main shaped tube and the shares can be seen. It is worth mentioning that the lower eigenfrequencies, which are in the range of the excitations, are mainly influenced by the tire characteristics.

8.3. PLOUGH: SERVO 6.50

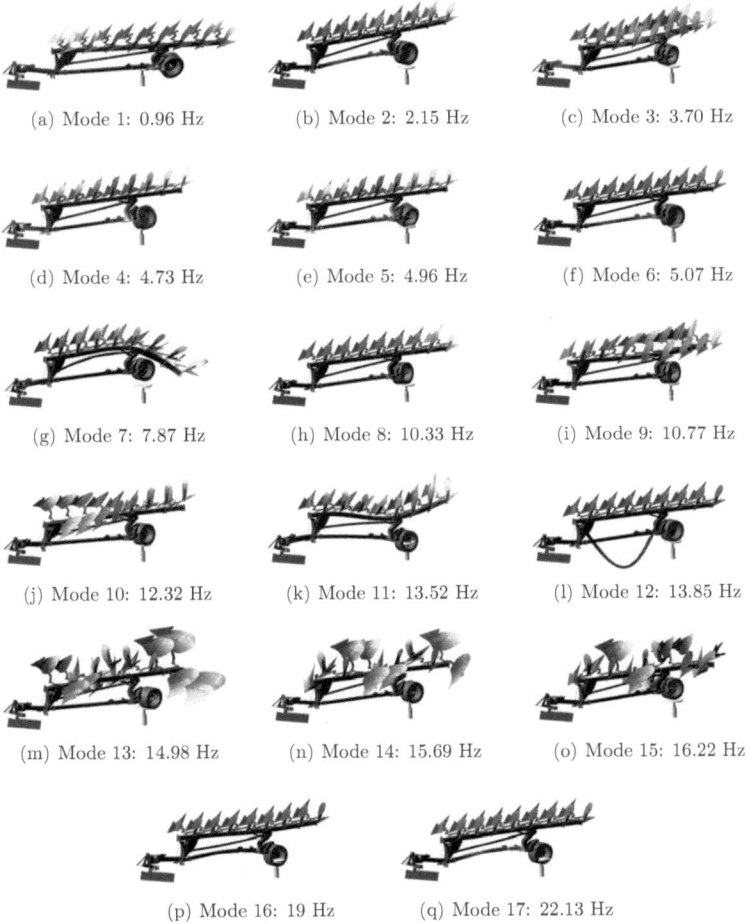

(a) Mode 1: 0.96 Hz (b) Mode 2: 2.15 Hz (c) Mode 3: 3.70 Hz
(d) Mode 4: 4.73 Hz (e) Mode 5: 4.96 Hz (f) Mode 6: 5.07 Hz
(g) Mode 7: 7.87 Hz (h) Mode 8: 10.33 Hz (i) Mode 9: 10.77 Hz
(j) Mode 10: 12.32 Hz (k) Mode 11: 13.52 Hz (l) Mode 12: 13.85 Hz
(m) Mode 13: 14.98 Hz (n) Mode 14: 15.69 Hz (o) Mode 15: 16.22 Hz
(p) Mode 16: 19 Hz (q) Mode 17: 22.13 Hz

Figure 8.50: Eigenfrequencies and mode shapes of Servo 6.50

8.3.4 Transfer Functions

Seven outputs signals, which are $y_1 = a_{2z}$, $y_2 = a_{1x}$, $y_3 = a_{3z}$, $y_4 = a_{5z}$, $y_5 = a_{4z}$, $y_6 = \varepsilon_8^t$ and $y_7 = \varepsilon_6^b$, i.e. five accelerations and two strains are considered in the virtual iteration. The output channel ε_7^b is not considered due to errors in the measurements. This results in a (2×7) transfer matrix, which is computed on the basis of the linearized model and the state matrices **A**, **B**, **C** and **D**. The transfer functions are sampled with 500 points in the frequency band between 0.5 and 10 Hz. The magnitude plots are shown in Fig. 8.51 and the phase plots in Fig. 8.52. The Moore-Penrose pseudoinverse is presented in Fig. 8.53 and 8.54, respectively.

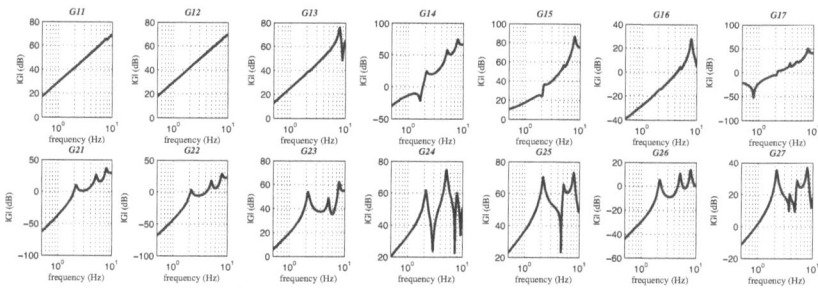

Figure 8.51: Magnitude plots of the transfer matrix, Servo 6.50

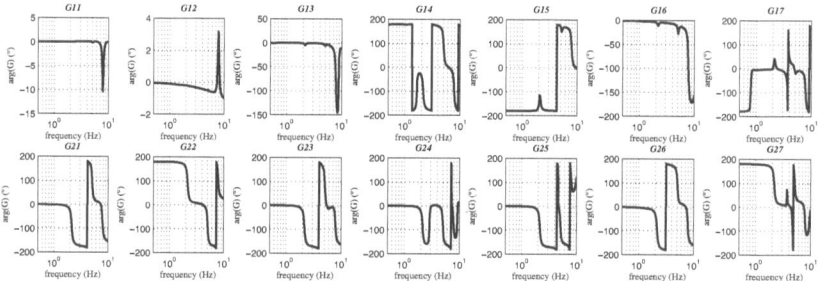

Figure 8.52: Phase plots of the transfer matrix, Servo 6.50

8.3.5 Virtual Iteration

The measured target signals are shown in Fig. 8.55 and the corresponding frequency spectra in Fig. 8.56. The excited frequencies are in the band $0.7 \ldots 3.0$ Hz. The linearized model does not coincide with the nonlinear model due to nonlinearities in the front hydraulic cylinder. As a consequence, iterative loops are required to reduce the error between targets and system outputs and to find the final solution of the drive signals. The forward dynamics simulation is performed with the HHT-solver

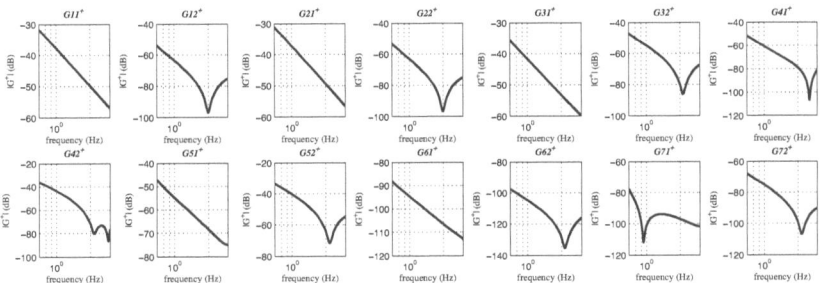

Figure 8.53: Magnitude plots of pseudoinverse transfer matrix, Servo 6.50

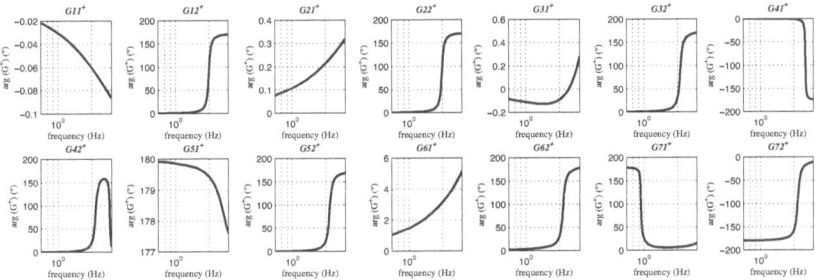

Figure 8.54: Phase plots of pseudoinverse transfer matrix, Servo 6.50

in *Adams*. The scalar weighting factor α^{n+1} in the iterative algorithm is chosen to one in each iteration.

The outputs, the targets and the error in the first iteration are shown in Fig. 8.57. An illustrative segment between 20s and 25s is chosen in order to zoom into the curves. The corresponding frequency spectra can be seen in Fig. 8.58. The results after three iterations are shown in Fig. 8.59 and 8.60, respectively.

It can be seen that the error is nearly zero and hence the algorithm is stopped after the third iteration.

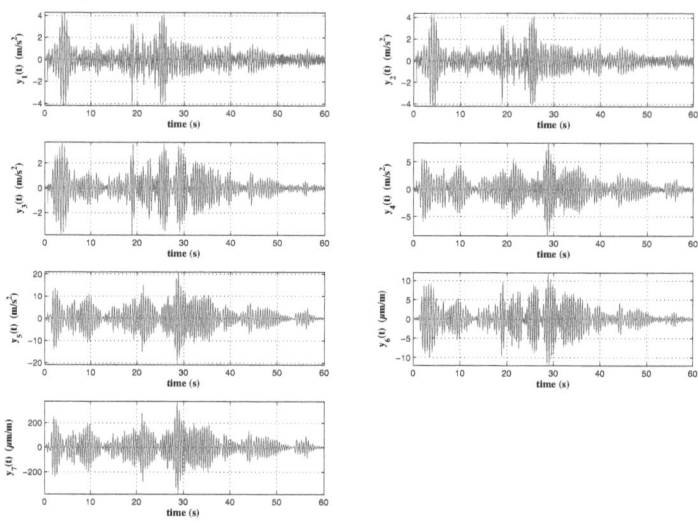

Figure 8.55: Measured target signals of Servo 6.50

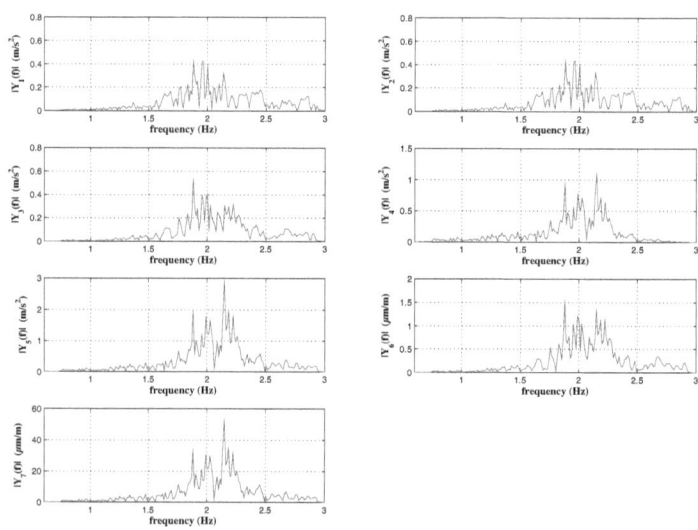

Figure 8.56: Frequency spectra of the measured target signals, Servo 6.50

8.3. PLOUGH: SERVO 6.50

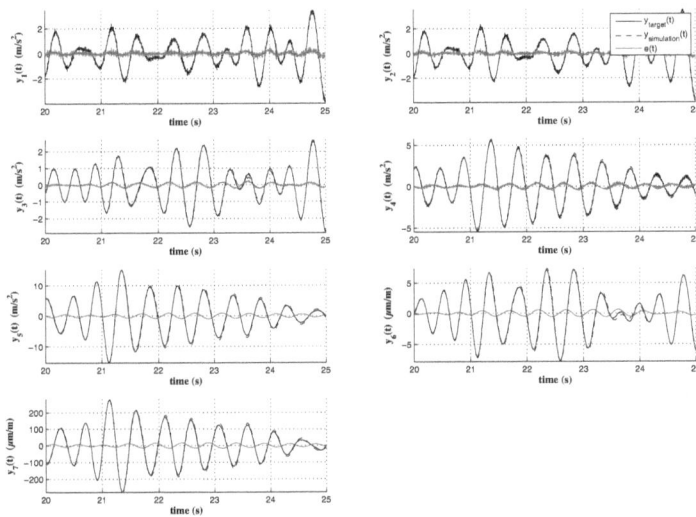

Figure 8.57: Targets, simulation results and error in the 1st iteration, Servo 6.50

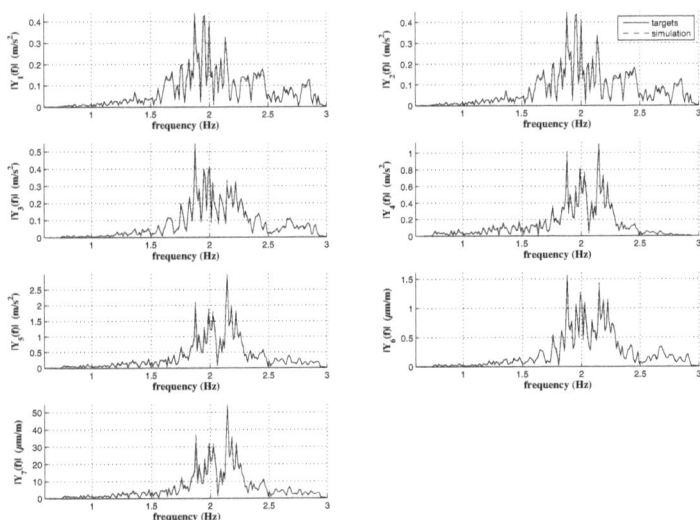

Figure 8.58: Frequency spectra of targets and system outputs in the 1st iteration, Servo 6.50

8.3. PLOUGH: SERVO 6.50

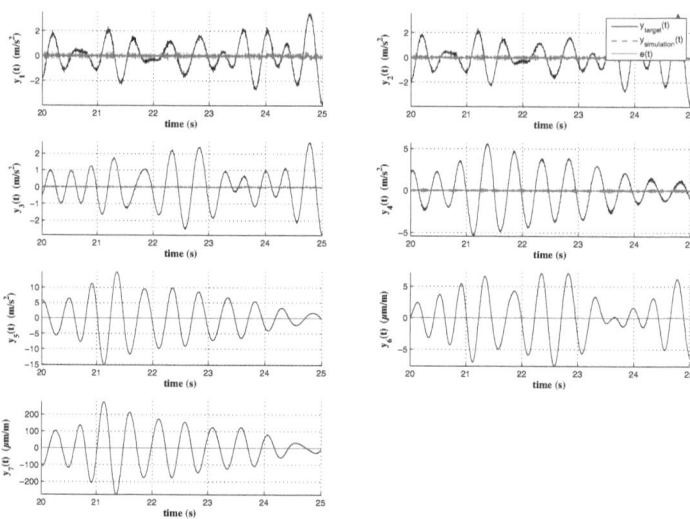

Figure 8.59: Targets, simulation results and error in the 3rd iteration, Servo 6.50

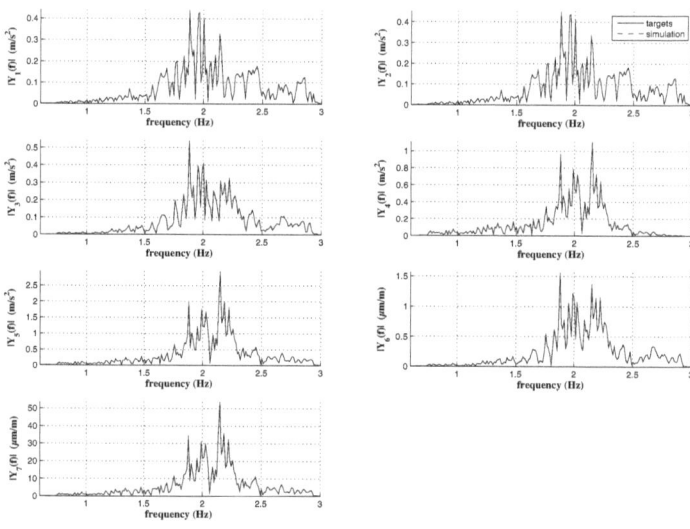

Figure 8.60: Frequency spectra of targets and system outputs in the 3rd iteration, Servo 6.50

8.3. PLOUGH: SERVO 6.50

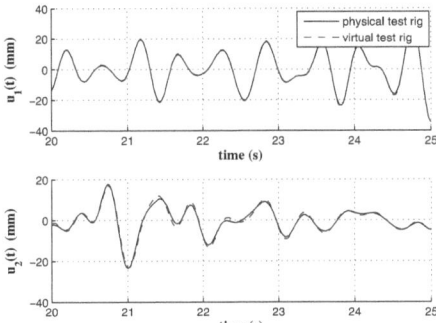

Figure 8.61: Computed drives signals and drives from the physical test rig, Servo 6.50

The resulting drive signals are presented in Fig. 8.62. The computed results are compared with the solution from the physical test rig. It can be seen that they are congruent. The frequency spectra of the drives are shown in Fig. 8.63.
Fig. 8.63 illustrates the convergence behavior of the virtual iteration of the Servo 6.50. It can be seen that an error of 7% occurs in the first iteration. In the third iteration the error indicator is reduced to values between 1% and 1.5%.

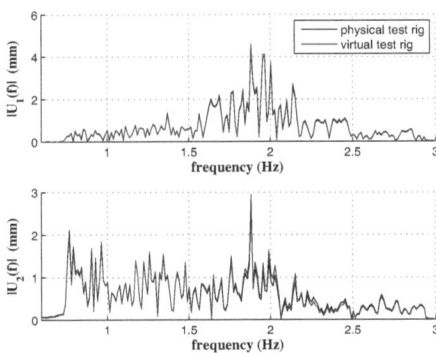

Figure 8.62: Frequency spectra of the drive signals, Servo 6.50

8.3.6 Discussion

The modeling techniques of the Servo 6.50 are very similar to that of the Synkro 6003T. An essential part are the flexible bodies in order to describe the bending and torsional stiffness of the machine. By using the Craig-Bampton reduction with a

8.3. PLOUGH: SERVO 6.50

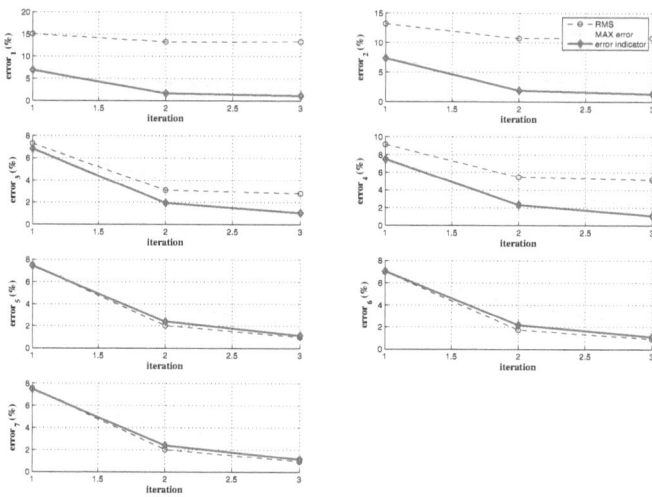

Figure 8.63: Convergence curves of the Servo 6.50

CMS the number of DOFs is dramatically reduced from 4 million to 90. Nonlinear characteristics of the front hydraulic cylinder are included in the multibody system. The rest of the model is fully linear. As a consequence, the virtual iteration converges rapidly. The relative error between targets and system outputs is reduced to less than 1.5% after three iterations. It should be mentioned that the contact problem between main part and wheel carrier is not fully modeled. A linear stiff spring element is used instead of a nonlinear contact element. The contact opens itself, if high peaks occur in the excitation. These peak loads affect the targets and can be seen at 18s and 28s, Fig. 8.55. Such effects cannot be reproduced by the linear spring. Furthermore, it should be mentioned that such peaks are smoothed in the FFT and as a result they do not appear so distinctivly in the drives. This effect can possibly introduce an error in a subsequent fatigue computation.

All intelligent thoughts have already been thought; what is necessary is only to try to think them again.

Johann Wolfgang von Goethe

Chapter 9
Conclusion

The content of this dissertation is a contribution to inverse problems of underactuated multibody systems. It was shown that the control of underactuated systems is much more challenging than that of fully actuated systems. It has to be distinguished between MBS where the equations of motion are given in a symbolic form and MBS that are modeled in commercial software.
If the equations of motion are available, the mathematical methods related to the inverse problem can go into the details of these equations. Specific mathematical methods take advantage of the particular formulation of the differential equations, cf. chapters 4, 5 and 6. It was shown that the formulation with redundant coordinates is beneficial with respect to the trajectory tracking problem. This was already stated for the DAE approach with control constraints by [25].
The methods considered in this thesis are not restricted to MBS where the equations of motion are symbolically obtainable. The method of virtual iteration was applied to large multibody systems including flexible bodies and a large finite element model. Such systems from industrial applications are modeled in commercial MBS- or FEM-software. As a consequence, the model is very detailed and the physical behavior, with its nonlinearities, can be considered in an appropriate way. However, the user of such a tool usually does not have access to the dynamic equations. As a consequence the system has to be treated as black box with respect to the inverse problem.

Four different approaches were studied, namely (i) the virtual iteration, (ii) the DAE approach with control constraints, (iii) the optimal control approaches and (iv) the flatness-based trajectory tracking. It was shown that all methods are suitable for a specific class of the problem. The methods under consideration are compared in the following paragraph.

Virtual iteration, chapter 3:
The method of virtual iteration is based on a linearization of the nonlinear system at an equilibrium point. The transfer matrix between inputs and outputs is computed from the linearized system. The inverse or the pseudoinverse of the transfer matrix is computed and the computation of the inputs is performed in the frequency domain. It was shown that the Moore-Penrose pseudoinverse is suitable to handle overdetermined systems. The results are used in a forward simulation in the time

domain and the error between targets and system outputs is calculated. An iterative loop is carried out until the error is reduced to a minimum tolerance value.

The method was applied to large flexible multibody systems and a finite element model with hundred thousands to millions of degrees of freedom. It was shown that the method of virtual iteration can handle such large systems. However, the nonlinear characteristics of the systems should not be too distinctive in order that the algorithm converges.

Virtual iteration is characterized by an inverse computation in the frequency domain and a forward computation in the time domain. Due to the linearization and the Fourier transform it can occur that some single peaks are undetected. However, such sharp peaks in the excitations can be crucial in a fatigue calculation. Hence, the root mean square error has to be evaluated as well as the maximum deviations.

DAE approach with control constraints, chapter 4:
The equations of motion can either be formulated by independent (generalized) or by dependent (redundant) coordinates. The formulation with independent coordinates results in ODEs and the formulation with dependent coordinates in DAEs. Both types of dynamic equations can be extended by so called control or servo constraints. This results in DAEs with an index, which is generally higher than three. The redundant coordinates formulation with control constraints yields index 5 problems. Such systems are beneficial regarding the inverse problem, even if they cannot be solved directly. A suitable index reduction procedure was applied to the system and the resulting index 3 problem was solved numerically. For that reason an implicit Euler algorithm was implemented.

The DAE approach with control constraints is an excellent method for inverse problems, if the equations of motion are given in a symbolic form. Arbitrary signals can be used as desired targets. In contrast to the flatness-based trajectory tracking, the targets must be differentiated only twice in the index reduction procedure. In the flatness-based approach the flat outputs have to be identified and these outputs must be continuously differentiable up to a higher order. Furthermore, the number of inputs must be identical to the number of outputs. Overdetermined systems cannot be treated as in the virtual iteration algorithm.

Optimal control, chapter 5:
It was shown how to formulate the inverse dynamics problem as an optimal control problem. A cost functional, which includes the error between targets and system outputs, has to be minimized. The resulting system can be solved in different ways. Indirect methods formulate the necessary optimality conditions. This can be done efficiently by introducing a Hamiltonian based on state and costate variables. In addition, boundary and transversality conditions must be fulfilled. This results in a two-point boundary value problem that can be challenging to solve. Collocation methods or (multiple) shooting methods are possible solution strategies. Another method is the gradient method, called Kelley-Bryson method. The variation of the control variable was derived in order to minimize the cost functional. The method is characterized by an integration of the states forwards in time and an integration of the costates backwards in time. It was shown that the method can be applied

to ODE-systems as well as to DAE-systems. For DAE-systems the index 2 Gear-Gupta-Leimkuhler formulation is preferred.

Direct methods discretize the system and reformulate the optimal control problem to a finite dimensional static optimization problem (NLP problem). In a partly discretization the control variable is split into many sub-intervals in time. In a full discretization the differential equations are discretized as well. The considered approach treats the MBS as black box where inputs, outputs and state variables can be exchanged via an interface. The optimization algorithm runs externally. The idea behind that method was to apply it to MBS that are modeled in commercial software. Unfortunately, the interface of such commercial tools is very limited and as a consequence the algorithm is very inefficient. However, the method was applied to systems that are modeled in *Matlab/Simulink*. It was shown that a Tikhonov regularization term in the cost functional can improve the solution of the control inputs.

In contrast to the DAE-approach with control constraints a very short step size was required in order to minimize the iteration in each sub-interval. It was shown that the DAE approach with control constraints is numerically more efficient. The Kelley-Bryson method is also seen as more advantageous compared to the considered direct optimal control algorithm. In contrast to the DAE approach with control constraints and the flatness-based trajectory tracking, the optimal control formulation is not limited to fully determined systems where the number of inputs is identical to the number of outputs.

Exact linearization and flatness-based trajectory tracking, chapter 6:

Exact linearization was applied to affine input systems. Derivatives of the system outputs with respect to time are calculated until the inputs explicitly appears. For that reason symbolic computations of the Lie-derivatives are required. By using a local diffeomorphism the system can be transferred to the so-called Byrnes-Isidori normal form. This formulation splits the system in a controllable, observable system and a non-observable system. It was shown that the closed-loop circuit can only be stabilized with a stable zero dynamics, i.e. the system is a minimum phase system. The high effort in the symbolic computations of the Lie-derivatives are avoided in a flatness-based parameterization. It was shown that the state variables and the input variables can be parameterized by the outputs and their time derivatives up to a certain order. Such systems are called differentially flat and the outputs are known as flat outputs. The challenge is to identify such flat outputs. In the considered examples of a planar overhead crane and a 3D rotary crane it was shown that the position of the load is always a flat output.

The big advantage of the flatness-based parameterization is that an analytical (nonlinear) control law is derived for the inputs. The DAE approach with control constraints and the optimal control approach yield a pure numerical solution of the input variables. The desired outputs in a flatness-based trajectory control must be sufficiently smooth, i.e. continuously differentiable. Arbitrary trajectories or measured signals have to be filtered in order to obtain sufficiently smooth functions. As a consequence, this method cannot be applied to all classes of inverse problems. The flatness-based approach is also limited to systems where the number of inputs is identical to the number of outputs.

The methods of the DAE approach with control constraints, the optimal control approaches and the flatness-based trajectory tracking were applied to three academic examples. The first example considered a nonlinear oscillator, which represented a fully actuated system. In the second example an underactuated planar overhead crane was the object under consideration. This example showed the advantages of a redundant coordinates formulation and illustrated the characteristics of the applied methods. The third example was a more complicated multibody system of a 3D rotary crane.
The method of virtual iteration was successfully applied to a finite element model of an AOD-steel converter and multibody systems of a trailed cultivator and a plough.

In summary it can be stated that the dissertation includes inverse methods that are suitable for academic examples where the equations of motion can be symbolically derived as well as methods that can be applied to large systems in industrial applications.

Future Perspectives:
The optimal control approaches show perspectives for further developments. It was shown that the gradient method can be applied to multibody systems that are formulated with redundant coordinates. The resulting state and costate equations are formed by index 2 DAEs, which have to be integrated numerically. Hence, it is of great interest to find appropriate solvers that can handle these index 2 systems.

The considered direct optimal control approach, which treats the MBS as black box, was applied to systems that are modeled in *Matlab/Simulink*. The goal is to apply the algorithm to systems that are modeled in commercial MBS or FEM tools. Up to now this has been limited by the interface of such software. The co-simulation between *Matlab* and *Adams* is seen as inefficient by now. If the interface will be more open in future, it would be useful to couple the optimal control algorithm with a MBS-software. This would allow the user to handle more complicated systems with nonlinearities, flexible bodies and all possible modeling techniques. It would also be interesting to compare the presented optimal control algorithms with the software presented in section 5.5.4.

The DAE approach with control constraints, the indirect optimal control and the flatness-based trajectory tracking are based on a symbolic form of the equations of motion. A couple of multibody codes, which formulate the equations of motion not pure numerically but rather symbolically, are available in the academic community. It would be of great interest to evaluate such code regarding larger multibody systems. As a consequence, the equations of motion are symbolically derived in an automatic algorithm, which is a big advantage. It would be desirable if these symbolic equations of motion can be used in several numerical software for further computations.

Appendix A

The Gradient of a Functional

A given linear functional (or a linearized nonlinear functional) $F(u)$ has to be minimized by a function $u(t)$.

$$F(u) = \int_a^b f(t)u(t)\,dt \tag{A.1}$$

It is assumed that the function $f(t)$ is given. The function $u(t)$ is constrained by a norm

$$\|u\|^2 = \int_a^b u^2(t)\,dt = c \tag{A.2}$$

The function $u(t)$ can be calculated by applying the calculus of variations. It states that the variation of the modified functional

$$\hat{F} = \int_a^b f(t)u(t)\,dt + \lambda \left(\int_a^b u^2(t)\,dt - c \right) \tag{A.3}$$

becomes stationary [93]:

$$\delta\hat{F} = \int_a^b f(t)\delta u\,dt + \lambda \int_a^b 2u\delta u\,dt + \delta\lambda \left(\int_a^b u^2(t)\,dt - c \right) = 0 \tag{A.4}$$

Hence,

$$\delta\hat{F} = \int_a^b \left(f(t) + 2\lambda u(t) \right) \delta u\,dt + \delta\lambda \left(\int_a^b u^2(t)\,dt - c \right) = 0 \tag{A.5}$$

As a consequence, the conditions for stationarity are:

$$f(t) + 2\lambda u(t) = 0 \tag{A.6a}$$

$$\int_a^b u^2(t)\,dt - c = 0 \tag{A.6b}$$

Now the function $u(t)$ can be calculated from the stationarity condition (A.6a)

$$u(t) = -\frac{1}{2\lambda}f(t) = -\kappa f(t) \qquad (A.7)$$

This shows that $u(t)$ is just a scaling of the function $f(t)$.

Bibliography

[1] ABDEL-RAHMAN, E. M., NAYFEH, A. H., AND MASOUD, Z. N. Dynamics and Control of Cranes: A Review. *Journal of Vibration and Control 9* (2003), 863–908.

[2] AHMED, O. B., AND GOUPILLON, J. F. Predicting the Ride Vibration of an Agricultural Tractor. *Journal of Terramechanics 34 (1)* (1997), 1–11.

[3] ARNOLD, V. I. *Gewöhnliche Differentialgleichungen*, 2 ed. Springer, 2001.

[4] ASCHER, U. M., AND PETZOLD, L. R. *Computer Methods for Ordinary Differential Equations and Differential-Algebraic Equations.* Siam Society for Industrial and Applied Mathematics, 1998.

[5] BASTOS, G. J., AND BRÜLS, O. Trajectory Optimization of Flexible Robots Using an Optimal Control Approach. In *Proceedings of the 1st Joint International Conference on Multibody System Dynamics, Lappeenranta, Finland* (2010).

[6] BATHE, K.-J. *Finite Element Procedures.* Prentice Hall, 1996.

[7] BAUMEISTER, J. *Stable Solution of Inverse Problems.* Friedr. Vieweg & Sohn Braunschweig/Wiesbaden, 1987.

[8] BENKER, H. *Mathematische Optimierung mit Computeralgebrasystemen.* Springer, 2003.

[9] BESTLE, D. *Analyse und Optimierung von Mehrkörpersystemen.* Springer-Verlag, 1994.

[10] BETSCH, P. The discrete null space method for the energy consistent integration of constrained mechanical systems Part I: Holonomic constraints. *Computer methods in applied mechanics and engineering 194* (2005), 5159–5190.

[11] BETSCH, P. Energy-consistent numerical integration of mechanical systems with mixed holonomic and nonholonomic constraints. *Computer methods in applied mechanics and engineering 195* (2006), 7020–7035.

[12] BETSCH, P., QUASEM, M., AND UHLAR, S. Numerical integration of discrete mechanical systems with mixed holonomic and control constraints. *Journal of Mechanical Science and Technology 23* (2009), 1012–1018.

[13] BETSCH, P., UHLAR, S., AND QUASEM, M. Numerical Integration of Mechanical Systems with Mixed Holonomic and Control Constraints. In *Proceedings of the ECCOMAS Thematic Conference on Multibody Dynamics, Warsaw, Poland* (2009).

[14] BETTS, J. T. *Practical Methods for Optimal Control and Estimation Using Nonlinear Programming*. Siam Society for Industrial and Applied Mathematics. Advances in Design and Control, 2010.

[15] BHATTI, M. A. *Practical Optimization Methods: With Mathematica Applications*. Springer-Verlag New York, 2000.

[16] BÖHLER, H. *Traktormodell zur Simulation der dynamischen Belastungen bei Transportfahrten*. PhD thesis, Technische Universität München, 2001.

[17] BIERBAUM, F., DIETZE, S., EICH-SOELLNER, E., GERLACH, W., AND PÖNISCH, G. *Numerische Mathematik*. Fachbuchverlag Leipzig im Carl Hanser Verlag, 2001.

[18] BIRK, O., AND BÄCKER, M. Schnittlastenermittlung durch MKS-Simulation des ungefesselten Fahrzeugs unter Verwendung von Referenz-Radnabenkräften. In *29. Tagung des DVM-Arbeitskreises Betriebsfestigkeit. Fahrwerke und Betriebsfestigkeit, 09./10. Oktober 2002, Osnabrück* (2002).

[19] BITSCH, G., DRESSLER, K., MARQUARDT, A., NIKELAY, I., AND GÖLZER, M. Computing digital road profiles for agricultural vehicle simulations. *VDI-Berichte 2001* (2007), 533–538.

[20] BLAJER, W. Some Methods for Constraint Violation Stabilization/Elimination in Numerical Simulation of Constrained Multibody Systems: a Comparative Study. In *Proceedings of the 1st Joint International Conference on Multibody System Dynamics, Lappeenranta, Finland* (2010).

[21] BLAJER, W., AND KOLODZIEJCZYK, K. A Geometric Approach to Solving Problems of Control Constraints: Theory and a DAE framework. *Multibody System Dynamics 11* (2004), 343–364.

[22] BLAJER, W., AND KOLODZIEJCZYK, K. Control of underactuated mechanical systems with servo-constraints. *Nonlinear Dynamics 50* (2007), 781–791.

[23] BLAJER, W., AND KOLODZIEJCZYK, K. Motion planning and control of gantry cranes in cluttered work environment. *IET Control Theory and Applications 1 (5)* (2007), 1370–1379.

[24] BLAJER, W., AND KOLODZIEJCZYK, K. Modeling of Underactuated Mechanical Systems in Partly Specified Motion. *Journal of Theoretical and Applied Mechanics 46 (2)* (2008), 383–394.

[25] BLAJER, W., AND KOLODZIEJCZYK, K. Dependent Versus Independent Variable Formulation for the Dynamics and Control of Cranes. *Solid State Phenomena 147-149* (2009), 221–230.

BIBLIOGRAPHY

[26] BLAJER, W., AND KOLODZIEJCZYK, K. Improved DAE Formulation for Inverse Dynamics Simulation of Cranes. In *Proceedings of the ECCOMAS Thematic Conference on Multibody Dynamics, Warsaw, Poland* (2009).

[27] BLUNDELL, M., AND HARTY, D. *The Multibody Systems Approach to Vehicle Dynamics*. Elsevier Butterworth-Heinemann, 2004.

[28] BOTTASSO, C. L., AND CROCE, A. On the Solution of Inverse Dynamics and Trajectory Optimization Problems for Multibody Systems. *Multibody System Dynamics 11* (2004), 1–22.

[29] BOTTASSO, C. L., AND CROCE, A. Optimal Control of Multibody Systems Using an Energy Preserving Direct Transcription Method. *Multibody System Dynamics 12* (2004), 17–45.

[30] BOTTASSO, C. L., LURAGHI, F., AND MAISANO, G. A Unified Approach to Trajectory Optimization and Parameter Estimation in Vehicle Dynamics. In *International Symposium on Coupled Methods in Numerical Dynamics FESB, Split, Croatia* (2009).

[31] BOTTASSO, C. L., MAISANO, G., AND SCORCELLETTI, F. Trajectory Optimization Procedures for Rotorcraft Vehicles, their Software Implementation and Applicability to Models of Varying Complexity. In *American Helicopter Society* 64th *Annual Forum, Montreal, Canada* (2008).

[32] BOTTASSO, C. L., MAISANO, G., AND SCORCELLETTI, F. Maneuvering Multibody Dynamics - New Developments for Models with Fast Solution Scales and Pilot-in-the-Loop Effects. In *Proceedings of the ECCOMAS Thematic Conference on Multibody Dynamics, Warsaw, Poland* (2009).

[33] BOTTASSO, C. L., MAISANO, G., AND SCORCELLETTI, F. Maneuvering Multibody Dynamics - New Developments for Models with Fast Solution Scales and Pilot-in-the-Loop Effects. *Multibody Dynamics: Computational Methods and Applications, Computational Methods in Applied Sciences 23* (2010), 29–48.

[34] BOTTASSO, C. L., RIVIELLO, L., RUZZENE, M., AND SCORCELLETTI, F. A Direct Multiple Shooting Approach to the Solution of Optimal Control Problems with Time-Delays. *Journal of Optimization Theory and Applications (under review)* (2008), 1–32.

[35] BRENAN, K. E., CAMPBELL, S. L., AND PETZOLD, L. R. *Numerical Solution of Initial-Value Problems in Differential-Algebraic Equations*. Elsevier Science Publishing Co., Inc., 1989.

[36] BRONSTEIN, I. N., SEMENDJAJEW, K. A., MUSIOL, G., AND MÜHLIG, H. *Taschenbuch der Mathematik*, 4. auflage ed. Verlag Harri Deutsch, 1999.

[37] BRYSON, A. E., AND HO, Y.-C. *Applied Optimal Control. Optimization, Estimation, and Control*. Hemisphere Publishing Corporation, 1975.

[38] BÜSKENS, C. Direkte Optimierungsmethoden zur numerischen Berechnung optimaler Steuerungen. Diplomarbeit, Westfälische Wilhelms-Universität Münster, Institut für numerische und instrumentelle Mathematik, 1993.

[39] BURGER, M., DRESSLER, K., MARQUARDT, A., AND SPECKERT, M. Calculating Invariant Loads for System Simulation in Vehicle Engineering. In *Proceedings of the ECCOMAS Thematic Conference on Multibody Dynamics, Warsaw, Poland* (2009).

[40] BURGER, M., SPECKERT, M., AND DRESSLER, K. Optimal Control Methods for the Calculation of Invariant Excitation Signals for Multibody Systems. In *Proceedings of the 1st Joint International Conference on Multibody System Dynamics, Lappeenranta, Finland* (2010).

[41] CHACHUAT, B. C. Nonlinear and Dynamic Optimization: From Theory to Practice. http://lawww.epfl.ch/page4234.html, 2007.

[42] CRAIG, R. R. J., AND BAMPTON, M. C. C. Coupling of Substructures for Dynamics Analyses. *AIAA Journal 6 (7)* (1968), 1313–1319.

[43] CRAIG, R. R. J., AND KURDILA, A. J. *Fundamentals of Structural Dynamics*, Second Edition ed. John Wiley & Sons, 2006.

[44] CROCKER, M. J., Ed. *Handbook of Noise and Vibration Control*. John Wiley & Sons, 2007.

[45] DANNBAUER, H., GATTRINGER, O., AND STEINBATZ, M. Integrating Virtual Test Methods and Physical Testing to Assure Accuracy and to Reduce Effort and Time. In *SAE World Congress & Exhibition* (2005).

[46] DASSAULT SYSTEMS SIMULIA. *Transient modal dynamic analysis, Section 6.3.7 of the Abaqus Analysis User's Manual*, 2010.

[47] DE CUYPER, J. *Linear Feedback Control for Durability Test Rigs in the Automotive Industry*. PhD thesis, Katholieke Universiteit Leuven, Departement Werktuigkunde Afdeling Productietechnieken, Machinebouw en Automatisering, 2006.

[48] DRESIG, H., AND HOLZWEISSIG, F. *Maschinendynamik*, 8. Auflage ed. Springer, 2007.

[49] EDELMANN, J., AND PLÖCHL, M. A combined driver model for lateral and longitudinal vehicle dynamics. In *Simpack User-Meeting, Baden-Baden* (2006).

[50] EICH-SOELLNER, E., AND FÜHRER, C. *Numerical Methods in Multibody Dynamics*. B.G. Teubner Stuttgart, 1998.

[51] ENNE, R. Fahrermodell für die Regelung der Längsdynamik eines Kraftfahrzeuges. Master's thesis, Technische Universität Wien, Institut für Mechanik und Mechatronik, 2006.

BIBLIOGRAPHY

[52] ETZELSTORFER, M. Modellierung und flachheitsbasierte Regelung eines Labor-Turmdrehkranes. Diplomarbeit, Institut für Automatisierungs- & Regelungstechnik, Technische Universität Wien, 2010.

[53] EWINS, D. J. *Modal Testing : theory, practice and application*, 2 ed. John Wiley & Sons, 2000.

[54] FABIEN, B. C. Direct optimization of dynamic systems described by differential-algebraic equations. *Optimal Control Applications and Methods 29* (2008), 445–466.

[55] FERHADBEGOVIC, B. *Entwicklung und Applikation eines instationären Reifenmodells zur Fahrdynamiksimulation von Ackerschleppern*. PhD thesis, Universität Hohenheim, Institut für Agrartechnik, 2008.

[56] FERHADBEGOVIC, B., BRINKMANN, C., AND KUTZBACH, H. D. Dynamic Longitudinal Model for Agricultural Tyres. In *Proceedings of the 15th International Conference of the ISTVS* (2005).

[57] FLIESS, M., LÉVINE, J., AND ROUCHON, P. A Simplified Approach of Crane Control via a generalized State-Space Model. In *Proceedings of the 30th Conference on Decision and Control, Brighton, England* (1991).

[58] FLIESS, M., LÉVINE, J., AND ROUCHON, P. Generalized state variable representation for a simplified crane description. *International Journal of Control 58 (2)* (1993), 277–283.

[59] FÖLLINER, O., SARTORIUS, H., AND KREBS, V. *Optimale Regelung und Steuerung*. R. Oldenbourg Verlag München Wien, 1994.

[60] FÖLLINGER, O. *Nichtlineare Regelungen 1*. R. Oldenbourg Verlag München Wien, 1998.

[61] FREGOLENT, A. Inverse problems in structural dynamics. SICON training course coordinated by Fabrizio Vestroni and Guido De Roeck, Dipartimento di Meccanica e Aeronautica, Universita di Roma "La Sapienza", Italy, 2008.

[62] GATTRINGER, O., PUCHNER, K., AND RIENER, H. Virtueller Prüfstand - Versuch und Simulation rücken zusammen. In *chassies.tech TÜV Süd, München* (2007).

[63] GATTRINGER, O., RIENER, H., AND DANNBAUER, H. Integration von Prüfstandsmethoden in die Simulation. In *34. Tagung des DVM-Arbeitskreises Betriebsfestigkeit. Lastannahmen und Betriebsfestigkeit, 10./11. Oktober2007, Wolfsburg* (2007).

[64] GERDTS, M. *Numerische Methoden optimaler Steuerprozesse mit differential-algebraischen Gleichungssystemen höheren Indexes und ihre Anwendungen in der Kraftfahrzeugsimulation und Mechanik*. PhD thesis, Mathematisches Institut der Universität Bayreuth, 2001.

[65] GERDTS, M. Direct Shooting Method for the Numerical Solution of Higher-Index DAE Optimal Control Problems. *Journal of Optimization Theory and Applications 117 (2)* (2003), 267–294.

[66] GERDTS, M. Optimal control and real-time optimization of mechanical multibody systems. *ZAMM - Journal of Applied Mathematics and Mechanics 83 (10)* (2003), 705–719.

[67] GERDTS, M. Gradient evaluation in DAE optimal control problems by sensitivity equations and adjoint equations. *PAMM - Proceedings in Applied Mathematics and Mechanics 5* (2005), 43–46.

[68] GERDTS, M. Representation of the Lagrange Multipliers for Optimal Control Problems Subject to Differential-Algebraic Equations of Index Two. *Journal of Optimization Theory and Applications 130 (2)* (2006), 231–251.

[69] GERDTS, M., AND KUNKEL, M. A globally convergent semi-smooth Newton method for control-state constrained DAE optimal control problems. *Computational Optimization and Applications DOI: 10.1007/s10589-009-9275-0* (2009), 1–33.

[70] GERSTMAYR, J. *A Solution Strategy for Elasto-Plastic Multibody Systems and Related Problems*. PhD thesis, Johannes Kepler Universität Linz, Technisch-Naturwissenschaftliche Fakultät, Institut für Mechanik und Maschinenlehre, 2001.

[71] GHOSH, T. Component Mode Synthesis of Structures with Geometric Stiffening in MSC/Nastran. In *Proceedings of the MSC Aerospace Users' Conference* (1999).

[72] GÖLLES, M. Vibrationsanalyse. Diplomarbeit, Institut für elektrische Messtechnik und Messsignalverarbeitung, Technische Universität Graz, 2003.

[73] GOICOLEA, J. M., AND ORDEN, J. C. G. Dynamic analysis of rigid and deformable multibody systems with penalty methods and energy-momentum schemes. *Computer methods in applied mechanics and engineering 188* (2000), 789–804.

[74] GOLDSTEIN, H. *Klassische Mechanik*. Aula Verlag Wiesbaden, 1991.

[75] GONZALEZ, O. Exact energy and momentum conserving algorithms for general models in nonlinear elasticity. *Computer methods in applied mechanics and engineering 190* (2000), 1763–1783.

[76] GONZALEZ, O., AND SIMO, J. C. On the stability of symplectic and energy-momentum algorithms for non-linear Hamiltonian systems with symmetry. *Computer methods in applied mechanics and engineering 134* (1996), 197–222.

[77] GRAICHEN, K. Optimierung. Vorlesungsskriptum. Vienna University of Technology, Automation and Control Institute, WS 2009/2010.

[78] HAIRER, E., LUBICH, C., AND ROCHE, M. *The Numerical Solution of Differential-Algebraic Systems by Runge-Kutta Methods.* Springer, 1989.

[79] HAIRER, E., LUBICH, C., AND WANNER, G. *Geometric Numerical Integration. Structure-Preserving Algorithms for Ordinary Differential Equations.* Springer, 2002.

[80] HEINE, S., HACKMAIR, C., SCHNEIDER, P., WÖLFL, A., JUNG, G., AND BÄCKER, M. Optimierung von Prüfsystemen mit Hilfe von virtuellen Methoden am Beispiel des Fahrdynamischen Fahrwerksprüfstands. In *33. Tagung des DVM-Arbeitskreises Betriebsfestigkeit. Betriebsfestigkeit in der virtuellen Produktentwicklung* (2006).

[81] HEINZL, P. *Regelung und Simulation dynamischer Systeme: zwei Beispiele aus der Mehrkörpersystem- und Fahrzeugdynamik.* PhD thesis, Technische Universität Wien, Institut für Mechanik, Abteilung für angewandte Mechanik, 2001.

[82] HEYDEN, T., AND WOERNLE, C. Dynamcis and flatness-based control of a kinematically undetermined cable suspension manipulator. *Multibody System Dynamics 16* (2006), 155–177.

[83] ISIDORI, A. *Nonlinear Control Systems.* Springer, 1995.

[84] JUNGE, O., MARSDEN, J. E., AND OBER-BLÖBAUM, S. Discrete Mechanics and Optimal Control. In *Proceedings of the 16th IFAC World Congress, Prague, Czech Republic* (2005).

[85] KANE, C., MARSDEN, J. E., AND ORTIZ, M. Symplectic-Energy-Momentum Preserving Variational Integrators. *Journal of Mathematical Physics 40* (1999), 3353–3371.

[86] KELLEY, H. J. *Method of Gradients. Optimization Techniques with Applications to Aerospace Systems.* Academic Press, New York, 1962, Dover Publications, 1998.

[87] KIRK, D. E. *Optimal Control Theory. An Introduction.* Prentice Hall Electrical Engineering Series, 1970.

[88] KISING, A., AND GÖHLICH, H. Ackerschlepper-Reifendynamik. *Grundlagen der Landtechnik 38 (4)* (1988), 137–143.

[89] KISING, A., AND GÖHLICH, H. Dynamic Characteristics of Large Tyres. *Journal of Agricultural Engineering Research 43* (1989), 11–21.

[90] KISS, B., LÉVINE, J., AND MÜLLHAUPT, P. Modeling, Flatness and Simulation of a Class of Cranes. *Periodica Polytechnica, Ser. Electrical Engineering 43 (3)* (1999), 215–225.

[91] KUGI, A. Regelungssysteme. Vorlesungsskriptum. Vienna University of Technology, Automation and Control Institute, SS 2010.

BIBLIOGRAPHY

[92] KUNKEL, P., AND MEHRMANN, V. *Differential - Algebraic Equations. Analysis and Numerical Solution*, vol. VIII. European Mathematical Society, 2006.

[93] LANCZOS, C. *The Variational Principles of Mechanics*. Dover Publications, New York, 1986.

[94] LEVENBERG, K. A Method for the Solution of Certain Problems in Least Squares. *Quarterly of Applied Mathematics 2* (1944), 164–168.

[95] LÉVINE, J. *Analysis and Control of Nonlinear Systems. A Flatness-based Approach*. Springer Dordrecht Heidelberg London New York, 2009.

[96] LEYENDECKER, S., OBER-BLÖBAUM, S., MARSDEN, J. E., AND ORTIZ, M. Discrete mechanics and optimal control for constrained systems. *Optimal Control Applications and Methods Published online in Wiley InterScience (www.interscience.wiley.com), DOI: 10.1002/oca.912* (2009), 1–24.

[97] LIEVEN, W., WARNECKE, U., SCHICK, A., AND LE-THE, H. Festigkeits- und Betriebsfestigkeitsanalyse von PKW-Karosserien auf Basis virtuell ermittelter Betriebslasten. In *33. Tagung des DVM-Arbeitskreises Betriebsfestigkeit. Betriebsfestigkeit in der virtuellen Produktentwicklung* (2006).

[98] LINES, J. A., AND MURPHY, K. The Radial Damping of Agricultural Tractor Tyres. *Journal of Terramechanics 28 (2/3)* (1991), 229–241.

[99] LINES, J. A., AND MURPHY, K. The Stiffness of Agricultural Tractor Tyres. *Journal of Terramechanics 28 (1)* (1991), 49–64.

[100] LINES, J. A., AND YOUNG, N. A. A Machine for Measuring the Suspension Characteristics of Agricultural Tyres. *Journal of Terramechanics 26 (3/4)* (1989), 201–210.

[101] LION, A., AND EICHLER, M. Gesamtfahrzeugsimulation auf Prüfstrecken zur Bestimmung von Lastkollektiven. *VDI Berichte 1559* (2000), 369–398.

[102] LOUIS, A. *Inverse und schlecht gestellte Probleme*. B.G. Teubner Stuttgart, 1989.

[103] LUTZ, H., AND WENDT, W. *Taschenbuch der Regelungstechnik*. Verlag Harri Deutsch, 2000.

[104] MAAS, R., AND LEYENDECKER, S. Structure preserving optimal control simulation of index finger dynamics. In *Proceedings of the 1st Joint International Conference on Multibody System Dynamics, Lappeenranta, Finland* (2010).

[105] MAIA, N. M. M., AND SILVA, J. M. M., Eds. *Theoretical and Experimental Modal Analysis*. John Wiley & Sons, 1998.

[106] MARITI, L., PENNESTRI, E., VELENTINI, P. P., AND BELFIORE, N. P. Review and Comparison of Solution Strategies for Multibody Dynamics Equations. In *Proceedings of the 1st Joint International Conference on Multibody System Dynamics, Lappeenranta, Finland* (2010).

BIBLIOGRAPHY

[107] MAUCH, H., AHMADI, A., ZHANG, G., AND KERSTEN, T. Numerische Simulation von Prüfsystemen. In *34. Tagung des DVM-Arbeitskreises Betriebsfestigkeit. Lastannahmen und Betriebsfestigkeit, 10./11. Oktober2007, Wolfsburg* (2007).

[108] MORÉ, J. The Levenberg-Marquardt algorithm: Implementation and theory. In *Numerical Analysis*, G. Watson, Ed., vol. 630. Springer Berlin / Heidelberg, 1978, pp. 105–116.

[109] MSC.SOFTWARE CORPORATION. *Adams/Flex Theory Manual*, 2003.

[110] MSC.SOFTWARE CORPORATION. *Adams/Vibration Theory Manual*, 2007.

[111] NEGRUT, D., AND ORTIZ, J. L. On an Approach for the Linearization of the Differential Algebraic Equations of Multibody Dynamics. In *Proceedings of IDETC/MESA, ASME/IEEE International Conference on Mechatronic and Embedded Systems and Applications* (2005).

[112] NEUPERT, J., HILDEBRANDT, A., SAWODNY, O., AND SCHNEIDER, K. Trajectory Tracking For Boom Cranes Using A Flatness Based Approach. In *SICE-ICASE International Joint Conference, Bexco, Busan, Korea* (2006).

[113] NEUWIRTH, E., HUNTER, K., DITTMANN, K.-J., AND SINGH, P. Operativer Einsatz des Virtuellen Prüfstandes zur Darstellung von Betriebsfestigkeitsprüfungen an Gesamtfahrzeugen mithilfe Mehrkörpersimulation (MKS). *VDI-Berichte 1846* (2004), 381–407.

[114] NIKRAVESH, P. E. *Planar Multibody Dynamics: Formulation, Programming and Applications.* CRC Press/Taylor & Francis, 2008.

[115] OBER-BLÖBAUM, S. *Discrete Mechanics and Optimal Control.* PhD thesis, Universität Paderborn, Fakultät für Elektrotechnik, Informatik und Mathematik, 2008.

[116] OPPERMANN, H., BÄCKER, M., AND LANGTHALER, T. Drivesignalgenerierung am virtuellen Prüfstand - ein weiterer Schritt zur CAE-basierten Lebensdaueranalyse. *VDI Berichte 1701* (2002), 701–719.

[117] OPPERMANN, H., HACKMAIR, C., OLBRICH, M., BÄCKER, M., AND LANGTHALER, T. Virtueller Betriebslastennachfahrversuch am BMW-Gesamtfahrzeugprüfstand. In *VDI Fahrzeug- und Verkehrstechnik, Erprobung und Simulation in der Fahrzeugentwicklung - Mess und Versuchstechnik* (2005).

[118] PACEJKA, H. B. *Tyre and Vehicle Dynamics.* Elsevier, 2006.

[119] QUASEM, M., UHLAR, S., AND BETSCH, P. Inverse dynamics of underactuated multibody systems. In *Proceedings of the 80th Annual Meeting of the International Association of Applied Mathematics and Mechanics (GAMM), Gdansk, Poland* (2009).

[120] RABL, M. F. Fahren auf der Schlechtwegstrecke. Master's thesis, Technische Universität Wien, Fakultät für Maschinenbau, 2001.

[121] REICHL, S., AND STEINER, W. Trajectory Tracking of Underactuated Multibody Systems. In 8th *International Conference on Multibody Systems, Nonlinear Dynamics, and Control (MSNDC), Proceedings of the ASME 2011 International Design Engineering Technical Conferences & Computers and Information in Engineering Conference, Washington, DC, USA* (2011).

[122] REICHL, S., STEINER, W., AND STEINBATZ, M. Optimal Control Methods for the Computation of Excitation Signals in Multibody Systems. In *Proceedings of the 1st Joint International Conference on Multibody System Dynamics, Lappeenranta, Finland* (2010).

[123] REICHL, S., STEINER, W., STEINBATZ, M., AND HOFER, M. Evaluation of Different Methods to Compute Excitation Signals Based on Measured Targets in Agricultural Machines. In *Proceedings of the 80th Annual Meeting of the International Association of Applied Mathematics and Mechanics (GAMM), Gdansk, Poland* (2009).

[124] REICHL, S., STEINER, W., STEINBATZ, M., AND HOFER, M. Practical Approaches for Inverse Calculations of Drive Signals in a Virtual Test Rig with Regard to Agricultural Machines. In *Proceedings of the ECCOMAS Thematic Conference on Multibody Dynamics, Warsaw, Poland* (2009).

[125] REICHL, S., STEINER, W., STEINBATZ, M., AND HOFER, M. Praktische Ansätze zur inversen Berechnung von Drivesignalen für Landmaschinen auf dem virtuellen Prüfstand. In *Tagungsband Internationales Forum Mechatronik, Linz, Austria* (2009).

[126] REICHL, S., STEINER, W., STEINBATZ, M., AND HOFER, M. *Energieeffiziente Mobilität*. Shaker Verlag, 2010, ch. Integration von Leichtbau-Berechnungsmethoden in die Produktentwicklung von Landmaschinen, pp. 141–149.

[127] REICHL, S., WIMMER, G., REITER, T., EGGER, M., AND STEINER, W. Inverse Computations of Causative Vibrations in an AOD Converter. In *Abaqus User Meeting, Vienna, Austria* (2010).

[128] RIENER, H., MAYR, A., AND STEINBATZ, M. Verbesserungspotentiale bei Zeit und Kosten durch Integration des "virtuellen Prüfstands" in den Simulations- und Prüfprozess. In *29. Tagung DVM-Arbeitskreis Betriebsfestigkeit, 09./10. Oktober 2002, Osnabrück* (2002).

[129] RILL, G. *Simulation von Kraftfahrzeugen*. Vieweg-Verlag, 2007.

[130] RILL, G., AND SCHAEFFER, T. *Grundlagen und Methodik der Mehrkörpersimulation mit Anwedungsbeispielen*. Vieweg+Teubner Verlag, 2010.

BIBLIOGRAPHY

[131] ROTHFUSS, R. Anwendung der flachheitsbasierten Analyse und Regelung nichtlinearer Mehrgrößensysteme. *VDI Fortschrittsberichte Reihe 8. Nr. 664* (1997), 181.

[132] SCHIEHLEN, W., Ed. *Multibody Systems Handbook*. Springer Berlin, 1990.

[133] SCHIEHLEN, W., AND EBERHARD, P. *Technische Dynamik. Modelle für Regelung und Simulation*, 2. ed. B. G. Teubner Stuttgart Leipzig Wiesbaden, 2004.

[134] SCHIEHLEN, W., GUSE, N., AND SEIFRIED, R. Multibody dynamics in computational mechanics and engineering applications. *Computer methods in applied mechanics and engineering 195* (2006), 5509–5522.

[135] SEIFRIED, R. Optimization-Based Design of Feedback Linearizable Underactuated Multibody Systems. In *Proceedings of the ECCOMAS Thematic Conference on Multibody Dynamics, Warsaw, Poland* (2009).

[136] SEIFRIED, R. Optimization-Based Design of Minimum Phase Underactuated Multibody Systems. *Multibody Dynamics: Computational Methods and Applications, Computational Methods in Applied Sciences 23* (2010), 261–282.

[137] SEIFRIED, R. Two Approaches for Designing Minimum Phase Underactuated Multibody Systems. In *Proceedings of the 1st Joint International Conference on Multibody System Dynamics, Lappeenranta, Finland* (2010).

[138] SELLGREN, U. Component Mode Synthesis - A method for efficient dynamic simulation of complex technical systems. Tech. rep., Department of Machine Design, The Royal Institute of Technology (KTH), Stockholm, Sweden, 2003.

[139] SHABANA, A. A. *Dynamics of Multibody Systems*, vol. Third Edition. Cambridge University Press, 2005.

[140] SHABANA, A. A. *Computational Dynamics*, 3 ed. John Wiley & Sons, 2010.

[141] SOHONI, V. N., AND WHITESELL, J. Automatic Linearization of Constrained Dynamical Systems. *ASME Journal of Mechanisms, Transmissions and Automation in Design 108 (8)* (1986), 300–304.

[142] STEINER, W. Multibody Dynamics. Lecture Notes. Upper Austria University of Applied Sciences, WS 2009/10.

[143] STEINER, W., AND REICHL, S. A Contribution to Inverse Dynamical Problems in Multibody Systems. In *Proceedings of the ECCOMAS Thematic Conference on Multibody Dynamics, Brussels, Belgium* (2011).

[144] STEINER, W., AND REICHL, S. The Optimal Control Approach to Dynamical Inverse Problems. *Journal of Dynamic Systems, Measurement, and Control (under review)* (2011).

[145] STEININGER, S. Schwingungsanalyse eines Stahlkonverters. Bachelor Thesis, Upper Austria University of Applied Sciences, Campus Wels, 2009.

[146] SVARICEK, F. Nulldynamik linearer und nichtlinearer Systeme: Definitionen, Eigenschaften und Anwendungen. *Automatisierungstechnik 54* (2006), 310–322.

[147] THOMAS, B. *Konzeption und Simulation eines passiven Kabinenfederungssystems für Traktoren*. PhD thesis, Technische Universität Carolo-Wilhelmina zu Braunschweig, Fakultät für Maschinenbau und Elektrotechnik, 2001.

[148] UHLAR, S. *Energy Consistent Time-Integration of Hybrid Multibody Systems*. PhD thesis, Universität Siegen, Fachbereich Maschinenbau, Institut für Mechanik und Regelungstechnik, Arbeitsgruppe Numerische Mechanik, 2009.

[149] UHLAR, S., AND BETSCH, P. *Conserving Integrators for Parallel Manipulators*. I-Tech Education and Publishing, 2008, ch. 5, pp. 75–108.

[150] UHLAR, S., AND BETSCH, P. A rotationless formulation of multibody dynamics: Modeling of screw joints and incorporation of control constraints. *Multibody System Dynamics 22* (2009), 69–95.

[151] UHLAR, S., AND BETSCH, P. A Unified Modeling Approach for Hybrid Multibody Systems Applying Energy-Momentum Consistent Time Integration. In *Proceedings of the ECCOMAS Thematic Conference on Multibody Dynamics, Warsaw, Poland* (2009).

[152] UHLAR, S., AND BETSCH, P. Energy-consistent integration of multibody systems with friction. *Journal of Mechanical Science and Technology 23* (2009), 901–909.

[153] UNGER, B., AND DANNBAUER, H. Neue Trends in den virtuellen Methoden der Bauteilentwicklung. In *33. Tagung des DVM-Arbeitskreises Betriebsfestigkeit. Betriebsfestigkeit in der virtuellen Produktentwicklung, 11./12. Oktober 2006, Steyr* (2006).

[154] VON DOMBROWSKI, R., AND ULRICH, H. Einbindung von Fahrermodellen in MKS-Fahrzeugmodelle. In *XVII. Deutsch-Polnisches Wissenschaftliches Seminar* (2005).

[155] VON HOLST, C. *Vergleich von Reifenmodellen zur Simulation der Fahrdynamik von Traktoren*. Fortschr.-Ber. VDI Reihe 14 Nr. 102 Düsseldorf, Technische Universität Berlin, Institut für Landmaschinen und Ölhydraulik, 2000.

[156] VON HOLST, C., AND GÖHLICH, H. Simulation von großvolumigen Reifen. *Landtechnik 54 (3)* (1999), 152–153.

[157] VON SCHWERIN, R. *MultiBody System Simulation*. Springer, 1999.

[158] WALLMICHRATH, M., AND JÖCKEL, M. Virtual Test Lab - A simulation based method for the optimisation of test procedures. In *Proceedings of Virtual Product Development in Automotive Engineering* (2005).

[159] WEIGEL, M., LUGNER, P., AND PLÖCHL, M. A Driver Model for a Truck-Semitrailer Combination. In *18. Symposium of the International Association for Vehicle System Dynamics, Kanagawa* (2003).

[160] WEIGEL, M., LUGNER, P., AND PLÖCHL, M. Ein Fahrermodell für einen zweigliedrigen Fahrzeugzug. Grazer Nutzfahrzeug Workshop, 2004.

[161] WEIGEL, N., WEIHE, S., BITSCH, G., AND DRESSLER, K. Einsatz von Simulationswerkzeugen zur Auslegung und Optimierung von Prüfkonzepten. In *33. Tagung des DVM-Arbeitskreises Betriebsfestigkeit. Betriebsfestigkeit in der virtuellen Produktentwicklung* (2006).

[162] WEIGEL, N., WEIHE, S., SING, V., GERLACH, R., FERREIRO DE SOUZA, A., SPECKERT, M., MARQUARDT, A., AND DRESSLER, K. Virtual Iteration for Set Up of Truck Cab Tests. In *Simpack User Meeting, Bonn* (2007).

[163] WITTENBURG, J. *Dynamics of Multibody Systems*. Springer-Verlag Berlin Heidelberg, 2008.

[164] YUASA, T., AND NAKAMARU, T. The Improvement of Accuracy of Road Load Input Prediction for Durability. *VDI-Berichte 1967* (2006), 463–472.

[165] ZIMMERT, N., AND SAWODNY, O. A Trajectory Tracking Control with Disturbance-Observer of a Fire-Rescue Turntable Ladder. In *American Control Conference, Seattle, Washington, USA* (2008).

List of Figures

1.1	Building of virtual product development	11
1.2	Block diagram with input, output and state variables of a MBS	12
1.3	Test rigs in the automotive industry	13
1.4	Test rigs in the agricultural industry (Source: Pöttinger)	13
2.1	Position of a material point with respect to a fixed-body coordinate system	18
2.2	Constraint violation by the numerical solutions $\tilde{\mathbf{q}}(t)$ and $\tilde{\mathbf{v}}(t)$ [20]	35
2.3	Floating reference frame	40
2.4	Master nodes and slave nodes in a FEM of a beam	41
2.5	Normal modes, sketched for a Hermite beam element	44
2.6	Constraint modes, sketched for a Hermite beam element	44
3.1	Generic magnitude plot, computed by noise excitation	50
3.2	Flow chart of the virtual iteration	59
3.3	Algorithm of the virtual iteration	60
4.1	Reactions of passive geometric and control constraints [22]	65
4.2	Subspaces of control constraint realizations [21]	68
5.1	Line search algorithm for a quadratic function	74
5.2	Calculus of variations in mechanics and optimal control theory [115]	86
5.3	Boundary minima in a constrained case $u \in U = [u^-, u^+]$ [77]	87
5.4	Shooting method [14]	90
5.5	Piecewise linear control functions	99
5.6	Optimal control approach which treats the MBS as black box	101
5.7	Performance measure with contour plot and gradient field for different step sizes	102
5.8	Influence of the Tikhonov regularization in the computed control variables	102
5.9	Conjugate gradient method with a line search algorithm	103
5.10	Nassi-Shneiderman diagram of the optimal control approach with a partly discretization	104
5.11	Classification of solution strategies for optimal control problems [115]	106
6.1	Block diagram of the Byrnes-Isidori normal form [83]	110
6.2	Block diagram of the Brunovsky canonical form [83]	113

LIST OF FIGURES

6.3 Two-degree-of-freedom design of a trajectory tracking control [91, 112] 117

7.1 Nonlinear Oscillator 125
7.2 Target signals $y_{1,d}(t)$, $y_{2,d}(t)$ 126
7.3 Contour plot and gradient field within one time interval 128
7.4 Inputs of the nonlinear oscillator 130
7.5 Outputs of the nonlinear oscillator 130
7.6 State variables of the nonlinear oscillator 131
7.7 Costate variables of the nonlinear oscillator 131
7.8 Planar overhead crane 133
7.9 Desired trajectory with time derivatives 139
7.10 Contour plot and gradient field within one time interval 140
7.11 Inputs of the planar overhead crane 144
7.12 Lagrange multiplier and cable tension force of the planar overhead crane 144
7.13 Outputs, redundant coordinates of the planar overhead crane 145
7.14 Inputs of the planar overhead crane computed by the Kelley-Bryson method 146
7.15 Convergence of the performance measure for $\chi = 5$ and $\chi = 0$ 146
7.16 Snapshots of the motion of the planar overhead crane 147
7.17 Rotary crane 148
7.18 Desired trajectory of the load of the 3D rotary crane 149
7.19 Coordinate systems of the rotary crane 150
7.20 Polynomial for the desired trajectory with its time derivatives 156
7.21 Inputs and Lagrange multiplier of the rotary crane 160
7.22 Redundant coordinates and outputs of the rotary crane 161
7.23 Snapshots of the motion of the rotary crane 161

8.1 AOD converter (Source: Siemens VAI) 164
8.2 Main parts of AOD converter 165
8.3 Finite element model of the AOD converter 165
8.4 Positions of accelerometers and strain gauges 166
8.5 Pole zero map of the AOD converter 166
8.6 Eigenfrequencies and mode shapes of the AOD converter 167
8.7 Magnitude plots of the transfer matrix, AOD converter 168
8.8 Phase plots of the transfer matrix, AOD converter 168
8.9 Magnitude plots of the pseudoinverse of the transfer matrix, AOD converter 169
8.10 Phase plots of the pseudoinverse of the transfer matrix, AOD converter 169
8.11 Measured target signals of the AOD converter 170
8.12 Frequency spectra of target signals, AOD converter 170
8.13 Measured unfiltered strains with drift effect 171
8.14 Computed excitations in the AOD converter 171
8.15 Frequency spectra of the excitations in the AOD converter 172
8.16 Comparison between measured targets and simulation outputs, AOD converter 172

8.17 Comparison of frequency spectra between measured targets and simulation outputs, AOD converter 173
8.18 Sensitivity of modal damping 173
8.19 Estimation of modal damping for 1st Eigenfrequency 174
8.20 Forces and torques at the trunnion ring 174
8.21 Forces and torques at the torsion bar 175
8.22 Frequency spectra of forces and torques at the trunnion ring 175
8.23 Frequency spectra of forces and torques at the torsion bar 176
8.24 Movement of the vessel 176
8.25 Trailed cultivator: Synkro 6003T (Source: Pöttinger) 178
8.26 MBS model of the virtual test rig with the trailed cultivator Synkro 6003T 179
8.27 Tire test rig (Source: Pöttinger) 180
8.28 Sensor positions of the Synkro 6003T 181
8.29 Pole zero map of the Synkro 6003T 181
8.30 Measured and simulated Eigenfrequencies of Synkro 6003T 182
8.31 Eigenfrequencies and mode shapes of the Synkro6003T 183
8.32 Magnitude plots of the transfer matrix, Synkro 6003T 184
8.33 Phase plots of the transfer matrix, Synkro 6003T 184
8.34 Magnitude plots of the pseudoinverse transfer matrix, Synkro 6003T . 185
8.35 Phase plots of the pseudoinverse transfer matrix, Synkro 6003T . . . 185
8.36 Measured targets of the Synkro 6003T 186
8.37 Frequency spectra of the target signals of the Synkro 6003T 186
8.38 Targets, simulation results and error in the 1st iteration, Synkro 6003T 187
8.39 Frequency spectra of targets and system outputs in the 1st iteration, Synkro 6003T 188
8.40 Targets, simulation results and error in the 5th iteration, Synkro 6003T 188
8.41 Frequency spectra of targets and system outputs in the 5th iteration, Synkro 6003T 189
8.42 Computed drives signals and drives from the physical test rig, Synkro 6003T 189
8.43 Frequency spectra of the drive signals, Synkro 6003T 190
8.44 Convergence curves of the Synkro 6003T 190
8.45 Plough: Servo 6.50 (Source: Pöttinger) 191
8.46 MBS model of the virtual test rig with the plough Servo 6.50 192
8.47 Sensor positions of the Servo 6.50 193
8.48 Pole zero map of Servo 6.50 193
8.49 Measured and simulated Eigenfrequencies of Servo 6.50 194
8.50 Eigenfrequencies and mode shapes of Servo 6.50 195
8.51 Magnitude plots of the transfer matrix, Servo 6.50 196
8.52 Phase plots of the transfer matrix, Servo 6.50 196
8.53 Magnitude plots of pseudoinverse transfer matrix, Servo 6.50 197
8.54 Phase plots of pseudoinverse transfer matrix, Servo 6.50 197
8.55 Measured target signals of Servo 6.50 198
8.56 Frequency spectra of the measured target signals, Servo 6.50 198
8.57 Targets, simulation results and error in the 1st iteration, Servo 6.50 . 199

8.58 Frequency spectra of targets and system outputs in the 1st iteration, Servo 6.50 199
8.59 Targets, simulation results and error in the 3rd iteration, Servo 6.50 . 200
8.60 Frequency spectra of targets and system outputs in the 3rd iteration, Servo 6.50 200
8.61 Computed drives signals and drives from the physical test rig, Servo 6.50 201
8.62 Frequency spectra of the drive signals, Servo 6.50 201
8.63 Convergence curves of the Servo 6.50 202

Die VDM Verlagsservicegesellschaft sucht für wissenschaftliche Verlage abgeschlossene und herausragende

Dissertationen, Habilitationen, Diplomarbeiten, Master Theses, Magisterarbeiten usw.

für die kostenlose Publikation als Fachbuch.

Sie verfügen über eine Arbeit, die hohen inhaltlichen und formalen Ansprüchen genügt, und haben Interesse an einer honorarvergüteten Publikation?

Dann senden Sie bitte erste Informationen über sich und Ihre Arbeit per Email an *info@vdm-vsg.de*.

Sie erhalten kurzfristig unser Feedback!

VDM Verlagsservicegesellschaft mbH
Dudweiler Landstr. 99
D - 66123 Saarbrücken
www.vdm-vsg.de

Telefon +49 681 3720 174
Fax +49 681 3720 1749

Die VDM Verlagsservicegesellschaft mbH vertritt

Printed by Books on Demand GmbH, Norderstedt / Germany